Numerical Methods for Non-linear Optimization

Numerical Methods for Non-linear Optimization

Conference sponsored by the
Science Research Council
University of Dundee, Scotland, 1971

Edited by

F. A. LOOTSMA

Philips Research Laboratories
Eindhoven, The Netherlands

1972

Academic Press   London · New York

ACADEMIC PRESS INC. (LONDON) LTD.
24/28 Oval Road,
London NW1

United States Edition published by
ACADEMIC PRESS INC.
111 Fifth Avenue
New York, New York 10003

Copyright © 1972 by
ACADEMIC PRESS INC. (LONDON) LTD.

All Rights Reserved

No part of this book may be reproduced in any form by photostat, microfilm, or any other means, without written permission from the publishers

Library of Congress Catalog Card Number: 72-84446
ISBN: 0-12-455650-7

PRINTED IN GREAT BRITAIN BY
WILLIAM CLOWES & SONS LIMITED,
LONDON, COLCHESTER AND BECCLES

Contributors

BEALE, E. M. L., *Scientific Control Systems Ltd., Sanderson House, London, England*

BIGGS, M. C., *Numerical Optimization Centre, The Hatfield Polytechnic, Hatfield, Hertfordshire, England*

BOROWSKI, N., *Department of Electrical Engineering, The University of British Columbia, Vancouver, British Columbia, Canada*

BRANIN, F. H., JR., *Systems Development Division Laboratory, I.B.M. Corporation, Kingston, New York, U.S.A.*

BROYDEN, C. G., *Computing Centre, University of Essex, Colchester, Essex, England*

DAVIES, M., *Department of Mathematics, University of Surrey, Guildford, Surrey, England*

DENNIS, J. E., JR., *Department of Computer Science, Cornell University, Ithaca, New York, U.S.A.*

DIXON, L. C. W., *The Numerical Optimization Centre, The Hatfield Polytechnic, Hatfield, Hertfordshire, England*

ECKHARDT, U., *Zentralinstitut für Angewandte Mathematik, Kernforschungsanlage Jülich, Jülich, Germany*

FLETCHER, R., *Theoretical Physics Division, U.K.A.E.A. Research Group, Atomic Energy Research Establishment, Harwell, England*

GOLDFARB, D., *The City College of the City University of New York, New York, New York, U.S.A.*

GONÇALVES, A. S., *Universidade de Coimbra, Instituto de Matemática, Coimbra, Portugal*

GOULD, F. J., *Department of Statistics and Curriculum in Operations Research and Systems Analysis, University of North Carolina at Chapel Hill, North Carolina, U.S.A.*

HANSEN, P., *Institut d'Economie Scientifique et de Gestion, Lille, France*

HIMMELBLAU, D. M., *Department of Chemical Engineering, The University of Texas at Austin, Austin, Texas, U.S.A.*

HOO, S. K., *Systems Development Division Laboratory, I.B.M. Corporation, Kingston, New York, U.S.A.*

HUTCHINSON, D., *Department of Computational Science, The University of Leeds, Leeds, England*

JOHNSON, M. P., *Department of Mathematics, Mid-Essex Technical College, Chelmsford, Essex, England*

KREUSER, J., *University of Minnesota, Minneapolis, Minnesota, U.S.A.*
LILL, S. A., *Department of Computational and Statistical Science, University of Liverpool, Liverpool, England*
LONEY, S. T., *South of Scotland Electricity Board, Glasgow, Scotland*
LOOTSMA, F. A., *Philips Research Laboratories, Eindhoven, The Netherlands*
MCCORMICK, G. P., *George Washington University, Washington, D.C., U.S.A.*
MIFFLIN, R., *Department of Administrative Sciences, Yale University, New Haven, Connecticut, U.S.A.*
OSBORNE, M. R., *Australian National University, Canberra, Australia*
PARKINSON, J. M., *Department of Computational Science, The University of Leeds, Leeds, England*
POWELL, M. J. D., *Mathematics Branch, Theoretical Physics Division, U.K.A.E.A. Research Group, Atomic Energy Research Establishment, Harwell, England*
ROSEN, J. B., *University of Minnesota, Minneapolis, Minnesota, U.S.A.*
RYAN, D. M., *Australian National University, Canberra, Australia*
SARGENT, R. W. H., *Department of Chemical Engineering, Imperial College, London, England*
SCHRACK, G., *The University of British Columbia, Department of Electrical Engineering, Vancouver, British Columbia, Canada*
SEBASTIAN, D. J., *Department of Chemical Engineering, Imperial College, London, England*
UEING, U., *University of Bonn, Institut für Gesellschafts- und Wirtschaftswissenschaften, Bonn, Germany*
WHITTING, I. J., *Operational Research Department, The Gas Council, London, England*

Preface

Organization

This volume contains a collection of papers presented at the Conference on Numerical Methods for Non-linear Optimization, which was held at the University of Dundee (Scotland), from 28th June to 1st July 1971. The choice of Dundee was not a coincidence. The Science Research Council had generously decided to give considerable financial support to the University of Dundee for the purpose of supporting a symposium on the theory of Numerical Analysis during the period September 1970 to August 1971. Many pure and applied mathematicians, numerical analysts etc. visited Dundee, and two major gatherings were organized: a Research Conference on Numerical Analysis and the above named Conference (the idea originated from the Mathematics Department in Dundee), which fitted very well into the framework of the 'Numerical Analysis Year'.

The Dundee Conference on Numerical Methods for Non-linear Optimization was the successor to the Conference on Optimization (sponsored by the British Computer Society and the Institute of Mathematics and its Applications), held in March 1968 at the University of Keele (England). It was also confined to the area of non-linear, unconstrained and constrained optimization with the emphasis on the numerical aspects; methods for integer, linear programming have been left out of consideration. The papers of the Keele conference have been edited by R. Fletcher, and published by Academic Press, London, 1969, under the title "Optimization".

The Programme Committee for the Dundee Conference consisted of M. J. D. Powell (Atomic Energy Research Establishment, Harwell; Chairman of the Committee), G. A. Watson (University of Dundee; secretary of the Committee), M. R. Osborne (Australian National University, Canberra; visiting professor at the University of Dundee), and the editor (Philips Research Laboratories, Eindhoven, The Netherlands; temporarily at Mullard Research Laboratories, Redhill, Surrey). We have cooperated in a most pleasant atmosphere. An encouraging factor was the rapid and favourable response of the invited speakers: P. Wolfe (IBM, New York), R. W. H. Sargent (Imperial College, London), G. P. McCormick (The George Washington University, Washington), E. M. L. Beale (Scientific Control Systems, London), R. Fletcher (Atomic Energy Research Establishment, Harwell), J. Abadie (Electricité de France, Paris), and J. B. Rosen (University of Minnesota,

Minneapolis); in addition M. R. Osborne and the editor agreed to present a survey on their favourite subject.

The programme was finally composed during a meeting in Dundee, on 5th and 6th May 1971. This completed the task of the Programme Committee. The local organization was further carried out by G. A. Watson assisted by many others too numerous to mention by name, both academic and secretarial staff and students of the Mathematics Department at the University of Dundee. They gave all the time and effort that was needed to ensure a well run conference. We thank all of them for the smooth organisation and for their excellent hospitality.

Outline of the Present Volume

Surveying the contents one will find that the contributions fall into a few, distinct categories.

First, there are the papers dealing with methods for unconstrained optimization. The theoretical aspects are further explored by M. J. D. Powell, J. B. Dennis, C. G. Broyden and M. P. Johnson, and E. M. L. Beale; the authors are mainly concerned with the successful variable-metric and conjugate-gradient methods. Numerical experiments to compare algorithms for unconstrained optimization are presented by R. W. H. Sargent and D. J. Sebastian, D. M. Himmelblau, J. M. Parkinson and D. Hutchinson, G. Schrack and N. Borowski, and L. C. W. Dixon. It is interesting to note that some of the authors devote considerable attention to the non-gradient methods. In the last few years this important class of methods has been somewhat neglected; at least, this is our impression if we inspect the leading journals and compare the number of papers on this subject with the abundant literature on gradient methods.

Problems of non-linear least squares and curve fitting are presented by M. R. Osborne, M. Davies and I. J. Whitting, and S. T. Loney. These problems have a wide range of applications, and the numerical methods for solving them are so closely connected with optimization methods that we gladly received the papers for presentation at the conference.

An appealing direction for future research is the design of methods to find global minima of problems that may have local, non-global minima. This volume contains three papers on this subject presented by G. P. McCormick, U. Ueing, and F. H. Branin and S. K. Hoo. Particularly the lively presentation by Branin should be mentioned here; it was one of the highlights of the conference.

Lastly, the reader will find a collection of papers on constrained optimization. Methods for quadratic programming are discussed by D. Goldfarb and A. S. Gonçalves, and P. Hansen presents a quadratic-programming method for the particular problem where the variables are restricted to the values 0 and 1. Methods tailored to problems with linear constraints are surveyed by R. Fletcher. Surprisingly enough, there is only one other paper, by J. B. Rosen and J. Kreuser, that deals extensively with methods of this kind. The Programme Committee expected more papers on this subject: in the literature these methods are often considered as effective tools if the (majority of the) constraints are linear. A broad treatise on complementary algorithms for programming problems with an infinite number of constraints is given by U. Eckhardt. Finally, keen interest is still being shown in the penalty-function approach for solving constrained minimization problems. This volume contains seven papers on this subject, presented by F. J. Gould, R. Mifflin, R. Fletcher, Shirley Lill, M. R. Osborne and D. M. Ryan, M. C. Biggs, and by the editor.

As in many other fields of scientific research there is an astonishing proliferation of names and terms. During the conference this was alluded to by F. H. Branin, who suddenly interrupted his talk to ask his audience a teasing question; an immediate answer was given by D. J. Wilde.

Branin: By the way, everybody seems to know who Newton is, but who is Raphson?
Wilde: He is Newton's programmer, I guess.
Branin: That is the best answer I have ever heard.

Presentation of Numerical Results

The literature on optimization has shown a regrettable lack of uniformity in the presentation of numerical results. When the number of function evaluations is recorded, it may include the gradient evaluations as well; it is not always obvious whether the derivatives were obtained by numerical differentiation or not; the accuracy required *a priori* is sometimes not mentioned so that one does not know what is understood by a solution to a given test problem; the execution time is often omitted, although this quantity might be of interest if a number of algorithms are compared on the same computer. The Programme Committee therefore suggested some guidelines to the contributors, and it is gratifying that many authors have followed the recommendations. Nevertheless, there are several inherent difficulties which should be mentioned here.

The number of function, gradient, and possibly Hessian-matrix evaluations, sometimes lumped together in the number of equivalent function evaluations, is a machine-independent measure of efficiency. However, it is an unsatisfactory yardstick for comparing optimization algorithms, since it does not accurately inform the reader of the total effort necessary to solve a given test problem. The gradient projection methods, for instance, will entail many time-consuming array manipulations which are often not negligible with respect to the function and gradient evaluations. Execution times and the computer used could be mentioned in order to provide a basis for overall comparison. And this is also a reasonable basis: the ultimate purpose of many investigations in this field is the development of algorithms that are faster than the existing ones. We cannot ignore the following problems, however.

First, computers with multi-programming and time-sharing facilities do not always inform the user of the execution time for the total job or, which is even more desirable, for specified parts of the job. Several contributors (Dixon, Biggs) have expressed their concern about this state of affairs, and we deplore a development that might destroy an important criterion for the comparison of algorithms.

Even if the computer used and the compiler are specified, a comparison of various optimization algorithms on different computers remains cumbersome. To our knowledge, only one such attempt, with the execution time as performance criterion, has been made. The study was carried out by Colville (see A. R. Colville, "A comparative study on non-linear programming codes". IBM NYSC Report No. 320-2949, June 1968), and his report gives an impressive list of specialists in the field of optimization cooperating in this project. We appreciate the attempt, but we have found many deficiencies in the report. The standard timing programme (the basis for the comparison) is lengthy and not very representative; basically, it consists of a number of arithmetic operations and array manipulations (in the proper balance?), but subroutine calls are practically missing. Furthermore, a detailed account of the number of function and derivative evaluations is not given, and the accuracy of the calculated solutions is sometimes rather poor.

Nevertheless, we feel that the comparison of optimization algorithms on the basis of execution times is very important. The papers by Himmelblau and Biggs, for instance, show that the difference in execution times is sometimes not very striking, even if there are considerable variations in the number of function and derivative evaluations. In those circumstances the properties of the computers and compilers cannot be neglected, and it is our conviction that significant progress in optimization methods can only be made if more attention is given to these aspects, as well as to the mathematical approach.

Acknowledgement

It is a pleasure to acknowledge the considerable support I have received from the referees. It is gratifying to record that hardly anyone who was asked to review a contribution refused to do so, notwithstanding their many professional commitments. I am also most grateful to the contributors for their cooperation and their willingness to follow the suggestions of the referees.

I wish to express my gratitude to the Board of Directors and to my colleagues at the Computer Departments of Mullard Research Laboratories (Redhill, Surrey, England) and Philips Research Laboratories (Eindhoven, The Netherlands). The time I spent on the Dundee Conference and the proceedings could not have been available without their acceptance that other tasks would have to be postponed.

I would like to dedicate the book to my wife Ricky for her whole-hearted support and encouragement.

Eindhoven, June 1972   F. A. LOOTSMA

Contents

    Contributors v
    Preface vii
1. Some Properties of the Variable Metric Algorithm
   M. J. D. Powell 1
2. On Some Methods Based on Broyden's Secant Approximation to the Hessian
   J. E. Dennis, Jr. 19
3. A Class of Rank–1 Optimization Algorithms
   C. G. Broyden and M. P. Johnson 35
4. A Derivation of Conjugate Gradients
   E. M. L. Beale 39
5. Numerical Experience with Algorithms for Unconstrained Minimization
   R. W. H. Sargent and D. J. Sebastian 45
6. A Uniform Evaluation of Unconstrained Optimization Techniques
   D. M. Himmelblau. 69
7. A Consideration of Non-gradient Algorithms for the Unconstrained Optimization of Functions of High Dimensionality
   J. M. Parkinson and D. Hutchinson 99
8. An Investigation into the Efficiency of Variants on the Simplex Method
   J. M. Parkinson and D. Hutchinson 115
9. An Experimental Comparison of Three Random Searches
   G. Schrack and N. Borowski 137
10. The Choice of Step Length, a Crucial Factor in the Performance of Variable Metric Algorithms
    L. C. W. Dixon 149
11. Some Aspects of Non-linear Least Squares Calculations
    M. R. Osborne 171
12. A Modified Form of Levenberg's Correction
    M. Davies and I. J. Whitting 191
13. A Dynamic Programming Algorithm for Load Duration Curve Fitting
    S. T. Loney 203

14. Attempts to Calculate Global Solutions of Problems that may have Local Minima
    G. P. McCormick 209
15. A Combinatorial Method to Compute a Global Solution of Certain Non-convex Optimization Problems
    U. Ueing 223
16. A Method for Finding Multiple Extrema of a Function of n Variables
    F. H. Branin, Jr. and S. K. Hoo 231
17. Extensions of Newton's Method and Simplex Methods for Solving Quadratic Programs
    D. Goldfarb 239
18. A Primal-Dual Method for Quadratic Programming with Bounded Variables
    A. S. Gonçalves 255
19. Quadratic Zero-One Programming by Implicit Enumeration
    P. Hansen 265
20. Minimizing General Functions Subject to Linear Constraints
    R. Fletcher 279
21. A Gradient Projection Algorithm for Non-linear Constraints
    J. B. Rosen and J. Kreuser 297
22. Pseudo-Complementary Algorithms for Mathematical Programming
    U. Eckhardt 301
23. A Survey of Methods for Solving Constrained Minimization Problems via Unconstrained Minimization
    F. A. Lootsma 313
24. Non-linear Tolerance Programming
    F. J. Gould 349
25. On the Convergence of the Logarithmic Barrier Function Method
    R. Mifflin 367
26. A Class of Methods for Non-linear Programming. III: Rates of Convergence
    R. Fletcher 371
27. Generalization of an Exact Method for Solving Equality Constrained Problems to deal with Inequality Constraints
    S. A. Lill 383
28. A Hybrid Algorithm for Non-linear Programming
    M. R. Osborne and D. M. Ryan 395
29. Constrained Minimization Using Recursive Equality Quadratic Programming
    M. C. Biggs 411
    Author Index 429
    Subject Index 435

1. Some Properties of the Variable Metric Algorithm

M. J. D. POWELL

Mathematics Branch, Theoretical Physics Division,
U.K.A.E.A. Research Group, Atomic Energy Research Establishment,
Harwell, England

Summary

The variable metric algorithm for calculating the least value of a function $F(\mathbf{x})$ is usually successful in practice, but it has not been analysed theoretically, except when $F(\mathbf{x})$ is a uniformly convex function. Therefore, in this paper some preliminary theorems are given, that require only that a level set $\{\mathbf{x}|F(\mathbf{x}) \leq F(\mathbf{x}^{(1)})\}$ is bounded, and that $F(\mathbf{x})$ has bounded second derivatives. In this paper we are unable to show convergence in general, but an interesting corollary of the theorems is that we can prove convergence if $F(\mathbf{x})$ is convex, whereas the previous theorems depend on *uniform* convexity.

1. Introduction

To solve the problem of calculating on a computer the least value of a function $F(x_1, x_2, \ldots, x_n) = F(\mathbf{x})$, say, the variable metric algorithm (Davidon, 1959; Fletcher and Powell, 1963) is often used, and it is usually successful. This algorithm is described briefly in Section 2, and some of its properties, taken from previously published papers, are given in Section 3. However, there are no theorems that explain the success of the algorithm, except in the special case when $F(\mathbf{x})$ is a uniformly convex function. So this paper describes some preliminary results that depend on much less restrictive conditions on $F(\mathbf{x})$.

The algorithm is iterative, and, given a starting point $\mathbf{x}^{(1)}$, it generates a sequence of points $\mathbf{x}^{(k)}$ ($k = 1, 2, \ldots$), that is intended to converge to the point at which $F(\mathbf{x})$ is least. The vector $\mathbf{g}^{(k)}$ is defined to be the gradient of $F(\mathbf{x})$ at $\mathbf{x}^{(k)}$.

The conditions on $F(\mathbf{x})$ that we impose are that the level set $\{\mathbf{x}|F(\mathbf{x}) \leq F(\mathbf{x}^{(1)})\}$ is bounded, and that $F(\mathbf{x})$ has bounded second derivatives. The notation $G(\mathbf{x})$ stands for the second derivative matrix of $F(\mathbf{x})$ at $\mathbf{x}$, and we let the bound be the inequality

$$\|G(\mathbf{x})\| \leq M. \tag{1}$$

Throughout this paper the vector norms are Euclidean, and the matrix norms are induced by the Euclidean vector norm.

Most of the given results are derived from the doubtful conjecture: 'There exist functions $F(\mathbf{x})$, satisfying the conditions of the last paragraph, for which the sequence of numbers $\|\mathbf{g}^{(k)}\|$ ($k = 1,2,...$) is bounded away from zero'. We would like to show that this conjecture is false, because then it would follow that the limit points of the sequence $\mathbf{x}^{(k)}$ ($k = 1,2,...$) include at least one stationary point of $F(\mathbf{x})$. We have used the term stationary point, instead of local minimum, because Wolfe (1971) has found a function for which the variable metric algorithm converges to a saddle point.

Therefore, in Section 4 we suppose that for some positive constant c the inequality $\|\mathbf{g}^{(k)}\| \geqslant c$ ($k = 1,2,...$) holds, and we deduce a number of consequences of this hypothesis. Some of these deductions are surprising, but unfortunately we have not been able to show that they are contradictory.

However, in Section 5 we note that the deductions are contradictory if we include the extra condition that $F(\mathbf{x})$ is convex. Thus we prove that the variable metric algorithm converges for convex functions, whereas the previous theorems (Powell, 1971) depend on *uniform* convexity.

In Section 6 a different method of analysis is used, and it is shown that if $F(\mathbf{x})$ is a function of only two variables, then the sequence $\mathbf{x}^{(k)}$ ($k = 1,2,...$) cannot converge to a point at which the gradient of $F(\mathbf{x})$ is non-zero.

Finally, in Section 7, there is a discussion of our theorems, and it is pointed out that they are relevant to a number of published algorithms (Dixon, 1971).

2. The Variable Metric Algorithm

The kth iteration of the algorithm calculates the point $\mathbf{x}^{(k+1)}$ from the point $\mathbf{x}^{(k)}$. It depends on a positive definite matrix $H^{(k)}$. To begin the first iteration, $H^{(1)}$ is frequently set to the unit matrix, but any other positive definite matrix may be used instead. As well as calculating $\mathbf{x}^{(k+1)}$, the kth iteration calculates the matrix $H^{(k+1)}$.

If $\mathbf{g}^{(k)} = \mathbf{0}$ the iterative process is terminated. Otherwise the point $\mathbf{x}^{(k+1)}$ is defined by the formula

$$\mathbf{x}^{(k+1)} = \mathbf{x}^{(k)} - \lambda^{(k)} H^{(k)} \mathbf{g}^{(k)}, \tag{2}$$

where $\lambda^{(k)}$ is a multiplier. The value of $\lambda^{(k)}$ is obtained by considering the function of one variable

$$\phi(\lambda) = F(\mathbf{x}^{(k)} - \lambda H^{(k)} \mathbf{g}^{(k)}), \tag{3}$$

and ideally $\lambda^{(k)}$ should be the value of λ that minimizes $\phi(\lambda)$ subject to $\lambda \geqslant 0$. Because $H^{(k)}$ is positive definite, and because the level set $\{\mathbf{x} | F(\mathbf{x}) \leqslant F(\mathbf{x}^{(1)})\}$ is bounded, $\lambda^{(k)}$ is positive and finite. If it happens that more than one value of

λ minimizes $\phi(\lambda)$, we remove the ambiguity by letting $\lambda^{(k)}$ be the least positive value of λ that minimizes $\phi(\lambda)$.

However, in practice it is impossible to calculate $\lambda^{(k)}$ precisely, and it seems to be a good strategy to tolerate rather large errors in the value of $\lambda^{(k)}$. But in this paper we presume the ideal case where $\lambda^{(k)}$ does minimize $\phi(\lambda)$ subject to $\lambda \geqslant 0$.

The matrix $H^{(k+1)}$ is defined by the formula

$$H^{(k+1)} = H^{(k)} - \frac{H^{(k)}\gamma^{(k)}\gamma^{(k)T}H^{(k)}}{(\gamma^{(k)}, H^{(k)}\gamma^{(k)})} + \frac{\delta^{(k)}\delta^{(k)T}}{(\delta^{(k)}, \gamma^{(k)})}, \quad (4)$$

where $\delta^{(k)}$ and $\gamma^{(k)}$ are the vectors

$$\left.\begin{array}{l}\delta^{(k)} = \mathbf{x}^{(k+1)} - \mathbf{x}^{(k)},\\ \gamma^{(k)} = \mathbf{g}^{(k+1)} - \mathbf{g}^{(k)},\end{array}\right\} \quad (5)$$

and where the superscript T distinguishes a row vector from a column vector. When equations (2) and (4) have been applied the $(k+1)$th iteration is begun.

3. Properties of the Variable Metric Algorithm

In this section we summarize the properties of the variable metric algorithm that have been published already, and that are required in later sections.

The numbers $(\delta^{(k)}, \gamma^{(k)})$ are positive, all the matrices $H^{(k)}$ $(k = 1, 2, ...)$ are positive definite, and the definition of $\lambda^{(k)}$ implies the equation

$$(\mathbf{g}^{(k+1)}, \delta^{(k)}) = 0, \quad (k = 1, 2, ...), \quad (6)$$

(Fletcher and Powell, 1963). Therefore, because $\delta^{(k)}$ is a positive multiple of $-H^{(k)}\mathbf{g}^{(k)}$, we have the identity

$$(\delta^{(k)}, \gamma^{(k)}) = \|\delta^{(k)}\|(\mathbf{g}^{(k)}, H^{(k)}\mathbf{g}^{(k)})/\|H^{(k)}\mathbf{g}^{(k)}\|. \quad (7)$$

The determinants of the matrices $H^{(k)}$ satisfy the recurrence relation

$$\det(H^{(k+1)}) = \det(H^{(k)}) \frac{(\delta^{(k)}, \gamma^{(k)})}{(\gamma^{(k)}, H^{(k)}\gamma^{(k)})}, \quad (8)$$

(Pearson, 1969).

The equation

$$\frac{1}{(\mathbf{g}^{(k+1)}, H^{(k+1)}\mathbf{g}^{(k+1)})} = \frac{1}{(\mathbf{g}^{(k)}, H^{(k)}\mathbf{g}^{(k)})} + \frac{1}{(\mathbf{g}^{(k+1)}, H^{(k)}\mathbf{g}^{(k+1)})} \quad (9)$$

holds, and, because $H^{(k)}$ is positive definite, it implies that the sequence $(\mathbf{g}^{(k)}, H^{(k)}\mathbf{g}^{(k)})$ $(k = 1, 2, ...)$ decreases strictly monotonically (Wolfe, 1969, private communication).

Defining $\Gamma^{(k)}$ to be the inverse of $H^{(k)}$, we have the recurrence relation

$$\Gamma^{(k+1)} = \left(I - \frac{\gamma^{(k)}\delta^{(k)T}}{(\delta^{(k)},\gamma^{(k)})}\right)\Gamma^{(k)}\left(I - \frac{\delta^{(k)}\gamma^{(k)T}}{(\delta^{(k)},\gamma^{(k)})}\right) + \frac{\gamma^{(k)}\gamma^{(k)T}}{(\delta^{(k)},\gamma^{(k)})}, \quad (10)$$

(Fletcher, 1970).

The trace of $\Gamma^{(k+1)}$ is related to the trace of $\Gamma^{(k)}$ by the equation

$$\mathrm{Tr}(\Gamma^{(k+1)}) = \mathrm{Tr}(\Gamma^{(k)}) + \frac{\|g^{(k+1)}\|^2 - \|g^{(k)}\|^2}{(g^{(k)}, H^{(k)}g^{(k)})} + \frac{\|\gamma^{(k)}\|^2}{(\delta^{(k)},\gamma^{(k)})}, \quad (11)$$

(Powell, 1971).

If $F(\mathbf{x})$ is uniformly convex, then the sequence $\mathbf{x}^{(k)}$ ($k = 1, 2, \ldots$) converges to the point at which $F(\mathbf{x})$ is least (Powell, 1971).

The condition (1) implies the inequality

$$\|\gamma^{(k)}\| \leq M\|\delta^{(k)}\|, \quad (k = 1, 2, \ldots). \quad (12)$$

This result is not special to the variable metric algorithm, but it follows from the definition of a derivative. It is proved, for instance, in Section 3.2.3 of Ortega and Rheinboldt (1970).

4. Consequences of a Conjecture

Although the variable metric algorithm usually works well, in this section we consider the conjecture that for some functions $F(\mathbf{x})$ there exists a positive constant c such that

$$\|\mathbf{g}^{(k)}\| \geq c, \quad (k = 1, 2, \ldots). \quad (13)$$

We deduce a number of lemmas from inequality (13) that I hoped would deny the conjecture, but the truth of the conjecture is still an open question.

To prove these lemmas we introduce two more definitions. We let Δ be the diameter of the level set $\{\mathbf{x} | F(\mathbf{x}) \leq F(\mathbf{x}^{(1)})\}$, so we have the inequality

$$\|\delta^{(k)}\| \leq \Delta, \quad (k = 1, 2, \ldots), \quad (14)$$

and we let m be the bound

$$m = \sup \|\mathbf{g}^{(k)}\|, \quad (k = 1, 2, \ldots). \quad (15)$$

This bound exists because of condition (1) and because the points $\mathbf{x}^{(k)}$ belong to a bounded set.

Lemma 1

There exist positive constants, c_1 and c_2 say, such that the trace of $\Gamma^{(k+1)}$ is bounded by the inequalities

$$c_1 \sum_{j=1}^{k} \frac{\|\gamma^{(j)}\|^2}{(\delta^{(j)},\gamma^{(j)})} \leq \mathrm{Tr}(\Gamma^{(k+1)}) \leq c_2 \sum_{j=1}^{k} \frac{\|\gamma^{(j)}\|^2}{(\delta^{(j)},\gamma^{(j)})}. \quad (16)$$

1. SOME PROPERTIES OF THE VARIABLE METRIC ALGORITHM

Proof

Equations (9) and (11) imply the identity

$$\mathrm{Tr}(\Gamma^{(k+1)}) = \mathrm{Tr}(\Gamma^{(k)}) + \frac{\|\mathbf{g}^{(k+1)}\|^2 - c^2}{(\mathbf{g}^{(k+1)}, H^{(k+1)}\mathbf{g}^{(k+1)})} - \frac{\|\mathbf{g}^{(k)}\|^2 - c^2}{(\mathbf{g}^{(k)}, H^{(k)}\mathbf{g}^{(k)})}$$

$$- \frac{\|\mathbf{g}^{(k+1)}\|^2 - c^2}{(\mathbf{g}^{(k+1)}, H^{(k)}\mathbf{g}^{(k+1)})} + \frac{\|\mathbf{\gamma}^{(k)}\|^2}{(\mathbf{\delta}^{(k)}, \mathbf{\gamma}^{(k)})}, \tag{17}$$

and we apply this equation k times to express $\mathrm{Tr}(\Gamma^{(k+1)})$ in terms of $\mathrm{Tr}(\Gamma^{(1)})$. Thus we obtain the relation

$$\mathrm{Tr}(\Gamma^{(k+1)}) = \mathrm{Tr}(\Gamma^{(1)}) + \frac{\|\mathbf{g}^{(k+1)}\|^2 - c^2}{(\mathbf{g}^{(k+1)}, H^{(k+1)}\mathbf{g}^{(k+1)})} - \frac{\|\mathbf{g}^{(1)}\|^2 - c^2}{(\mathbf{g}^{(1)}, H^{(1)}\mathbf{g}^{(1)})}$$

$$+ \sum_{j=1}^{k} \left(- \frac{\|\mathbf{g}^{(j+1)}\|^2 - c^2}{(\mathbf{g}^{(j+1)}, H^{(j)}\mathbf{g}^{(j+1)})} + \frac{\|\mathbf{\gamma}^{(j)}\|^2}{(\mathbf{\delta}^{(j)}, \mathbf{\gamma}^{(j)})} \right)$$

$$< \mathrm{Tr}(\Gamma^{(1)}) + \frac{\|\mathbf{g}^{(k+1)}\|^2 - c^2}{(\mathbf{g}^{(k+1)}, H^{(k+1)}\mathbf{g}^{(k+1)})} + \sum_{j=1}^{k} \frac{\|\mathbf{\gamma}^{(j)}\|^2}{(\mathbf{\delta}^{(j)}, \mathbf{\gamma}^{(j)})}, \tag{18}$$

the last line being a consequence of inequality (13). Now because the trace of a matrix is equal to the sum of its eigenvalues, and because $\Gamma^{(k+1)}$ is positive definite, we infer the inequality

$$\frac{\|\mathbf{g}^{(k+1)}\|^2}{(\mathbf{g}^{(k+1)}, H^{(k+1)}\mathbf{g}^{(k+1)})} < \mathrm{Tr}(\Gamma^{(k+1)}), \tag{19}$$

and therefore expression (18) implies the bound

$$\mathrm{Tr}(\Gamma^{(k+1)}) < \mathrm{Tr}(\Gamma^{(1)}) + \frac{\|\mathbf{g}^{(k+1)}\|^2 - c^2}{\|\mathbf{g}^{(k+1)}\|^2} \mathrm{Tr}(\Gamma^{(k+1)}) + \sum_{j=1}^{k} \frac{\|\mathbf{\gamma}^{(j)}\|^2}{(\mathbf{\delta}^{(j)}, \mathbf{\gamma}^{(j)})}, \tag{20}$$

which is equivalent to the condition

$$\mathrm{Tr}(\Gamma^{(k+1)}) < \frac{\|\mathbf{g}^{(k+1)}\|^2}{c^2} \left(\mathrm{Tr}(\Gamma^{(1)}) + \sum_{j=1}^{k} \frac{\|\mathbf{\gamma}^{(j)}\|^2}{(\mathbf{\delta}^{(j)}, \mathbf{\gamma}^{(j)})} \right). \tag{21}$$

Because $\|\mathbf{g}^{(k+1)}\|$ is bounded above, the right-hand inequality of expression (16) is proved.

To prove the left-hand inequality of expression (16), we replace c^2 by m^2 in equation (17), and then instead of inequality (18) we find the expression

$$\operatorname{Tr}(\Gamma^{(k+1)}) = \operatorname{Tr}(\Gamma^{(1)}) + \frac{\|\mathbf{g}^{(k+1)}\|^2 - m^2}{(\mathbf{g}^{(k+1)}, H^{(k+1)}\mathbf{g}^{(k+1)})} - \frac{\|\mathbf{g}^{(1)}\|^2 - m^2}{(\mathbf{g}^{(1)}, H^{(1)}\mathbf{g}^{(1)})}$$

$$+ \sum_{j=1}^{k}\left(-\frac{\|\mathbf{g}^{(j+1)}\|^2 - m^2}{(\mathbf{g}^{(j+1)}, H^{(j)}\mathbf{g}^{(j+1)})} + \frac{\|\mathbf{\gamma}^{(j)}\|^2}{(\mathbf{\delta}^{(j)}, \mathbf{\gamma}^{(j)})}\right)$$

$$> \frac{\|\mathbf{g}^{(k+1)}\|^2 - m^2}{(\mathbf{g}^{(k+1)}, H^{(k+1)}\mathbf{g}^{(k+1)})} + \sum_{j=1}^{k} \frac{\|\mathbf{\gamma}^{(j)}\|^2}{(\mathbf{\delta}^{(j)}, \mathbf{\gamma}^{(j)})}, \tag{22}$$

the last line being a consequence of definition (15). We now use inequality (19) again, and deduce the bound

$$\operatorname{Tr}(\Gamma^{(k+1)}) > \frac{\|\mathbf{g}^{(k+1)}\|^2 - m^2}{\|\mathbf{g}^{(k+1)}\|^2} \operatorname{Tr}(\Gamma^{(k+1)}) + \sum_{j=1}^{k} \frac{\|\mathbf{\gamma}^{(j)}\|^2}{(\mathbf{\delta}^{(j)}, \mathbf{\gamma}^{(j)})}, \tag{23}$$

which gives the condition

$$\operatorname{Tr}(\Gamma^{(k+1)}) > \frac{\|\mathbf{g}^{(k+1)}\|^2}{m^2} \sum_{j=1}^{k} \frac{\|\mathbf{\gamma}^{(j)}\|^2}{(\mathbf{\delta}^{(j)}, \mathbf{\gamma}^{(j)})}. \tag{24}$$

Because of inequality (13), the proof of lemma 1 is complete.

Lemma 2

There exists a constant, c_3 say, such that $\|H^{(k+1)}\mathbf{g}^{(k+1)}\|$ is bounded by the inequality

$$\|H^{(k+1)}\mathbf{g}^{(k+1)}\| \leq c_3 + \sum_{j=1}^{k} \|\mathbf{\delta}^{(j)}\|. \tag{25}$$

Proof

Equations (4) and (7) imply the inequality

$$\operatorname{Tr}(H^{(k+1)}) < \operatorname{Tr}(H^{(k)}) + \frac{\|\mathbf{\delta}^{(k)}\|^2}{(\mathbf{\delta}^{(k)}, \mathbf{\gamma}^{(k)})} = \operatorname{Tr}(H^{(k)}) + \frac{\|\mathbf{\delta}^{(k)}\| \|H^{(k)}\mathbf{g}^{(k)}\|}{(\mathbf{g}^{(k)}, H^{(k)}\mathbf{g}^{(k)})}, \tag{26}$$

and by applying it k times we obtain the bound

$$\operatorname{Tr}(H^{(k+1)}) < \operatorname{Tr}(H^{(1)}) + \sum_{j=1}^{k} \frac{\|\mathbf{\delta}^{(j)}\| \|H^{(j)}\mathbf{g}^{(j)}\|}{(\mathbf{g}^{(j)}, H^{(j)}\mathbf{g}^{(j)})}. \tag{27}$$

1. SOME PROPERTIES OF THE VARIABLE METRIC ALGORITHM

Moreover, by using an argument like the one that gave expression (19) we deduce the condition

$$\frac{\|H^{(k+1)}\mathbf{g}^{(k+1)}\|^2}{(\mathbf{g}^{(k+1)}, H^{(k+1)}\mathbf{g}^{(k+1)})} < \text{Tr}(H^{(k+1)}), \tag{28}$$

so we obtain the expression

$$\frac{\|H^{(k+1)}\mathbf{g}^{(k+1)}\|^2}{(\mathbf{g}^{(k+1)}, H^{(k+1)}\mathbf{g}^{(k+1)})} < \text{Tr}(H^{(1)}) + \sum_{j=1}^{k} \frac{\|\boldsymbol{\delta}^{(j)}\|\,\|H^{(j)}\mathbf{g}^{(j)}\|}{(\mathbf{g}^{(j)}, H^{(j)}\mathbf{g}^{(j)})}. \tag{29}$$

We rearrange this inequality to the form

$$\|H^{(k+1)}\mathbf{g}^{(k+1)}\| < \frac{\text{Tr}(H^{(1)})(\mathbf{g}^{(1)}, H^{(1)}\mathbf{g}^{(1)})}{\|H^{(1)}\mathbf{g}^{(1)}\|} \frac{\|H^{(1)}\mathbf{g}^{(1)}\|}{\|H^{(k+1)}\mathbf{g}^{(k+1)}\|} \frac{(\mathbf{g}^{(k+1)}, H^{(k+1)}\mathbf{g}^{(k+1)})}{(\mathbf{g}^{(1)}, H^{(1)}\mathbf{g}^{(1)})}$$

$$+ \sum_{j=1}^{k} \|\boldsymbol{\delta}^{(j)}\| \frac{\|H^{(j)}\mathbf{g}^{(j)}\|}{\|H^{(k+1)}\mathbf{g}^{(k+1)}\|} \frac{(\mathbf{g}^{(k+1)}, H^{(k+1)}\mathbf{g}^{(k+1)})}{(\mathbf{g}^{(j)}, H^{(j)}\mathbf{g}^{(j)})}. \tag{30}$$

We let c_3 be the constant

$$c_3 = \frac{\text{Tr}(H^{(1)})(\mathbf{g}^{(1)}, H^{(1)}\mathbf{g}^{(1)})}{\|H^{(1)}\mathbf{g}^{(1)}\|}, \tag{31}$$

and therefore, because the numbers $(\mathbf{g}^{(k)}, H^{(k)}\mathbf{g}^{(k)})$ decrease monotonically (see Section 3), we have the condition

$$\|H^{(k+1)}\mathbf{g}^{(k+1)}\| < c_3 \frac{\|H^{(1)}\mathbf{g}^{(1)}\|}{\|H^{(k+1)}\mathbf{g}^{(k+1)}\|} + \sum_{j=1}^{k} \|\boldsymbol{\delta}^{(j)}\| \frac{\|H^{(j)}\mathbf{g}^{(j)}\|}{\|H^{(k+1)}\mathbf{g}^{(k+1)}\|}. \tag{32}$$

Now the definition (31) and expression (28) imply the bound

$$\|H^{(1)}\mathbf{g}^{(1)}\| < c_3, \tag{33}$$

so, by using this inequality if $\|H^{(2)}\mathbf{g}^{(2)}\| \leqslant \|H^{(1)}\mathbf{g}^{(1)}\|$, and by using expression (32) if $\|H^{(2)}\mathbf{g}^{(2)}\| > \|H^{(1)}\mathbf{g}^{(1)}\|$, we deduce that inequality (25) holds for $k = 1$.

For the larger values of k we use an inductive argument, so now we suppose that inequality (25) is satisfied for $k = 1, 2, \ldots, K-1$, and we prove that it is true for $k = K$. If $\|H^{(K+1)}\mathbf{g}^{(K+1)}\| \leqslant \|H^{(1)}\mathbf{g}^{(1)}\|$ then inequality (25) is true for $k = K$ by expression (33), if $\|H^{(K+1)}\mathbf{g}^{(K+1)}\| \leqslant \|H^{(k+1)}\mathbf{g}^{(k+1)}\|$ for at least one value of k in $1 \leqslant k \leqslant K-1$ then inequality (25) is true for $k = K$ by expression (25), and if $\|H^{(K+1)}\mathbf{g}^{(K+1)}\| > \|H^{(j)}\mathbf{g}^{(j)}\|$ for every j in $1 \leqslant j \leqslant K$ then inequality (25) is true for $k = K$ by expression (32). The induction, and therefore the proof of lemma 2, is now complete.

Lemma 3

There exists a constant, c_4 say, such that the trace of $\Gamma^{(k+1)}$ is bounded by the inequalities

$$\frac{\|\mathbf{g}^{(k+1)}\|^2}{(\mathbf{g}^{(k+1)}, H^{(k+1)}\mathbf{g}^{(k+1)})} < \mathrm{Tr}\,(\Gamma^{(k+1)}) \leq \frac{c_4 k^2}{(\mathbf{g}^{(k)}, H^{(k)}\mathbf{g}^{(k)})}. \qquad (34)$$

Proof

The left-hand inequality is just expression (19). To prove the right-hand inequality we first substitute expression (7) in inequality (16) and obtain the bound

$$\mathrm{Tr}\,(\Gamma^{(k+1)}) \leq c_2 \sum_{j=1}^{k} \frac{\|\mathbf{\gamma}^{(j)}\|^2 \|H^{(j)}\mathbf{g}^{(j)}\|}{\|\mathbf{\delta}^{(j)}\|(\mathbf{g}^{(j)}, H^{(j)}\mathbf{g}^{(j)})}. \qquad (35)$$

Now the numbers $(\mathbf{g}^{(j)}, H^{(j)}\mathbf{g}^{(j)})$ decrease monotonically, so from expressions (12), (25) and (35) we deduce the inequality

$$\mathrm{Tr}\,(\Gamma^{(k+1)}) \leq \frac{c_2 M^2}{(\mathbf{g}^{(k)}, H^{(k)}\mathbf{g}^{(k)})} \sum_{j=1}^{k} \|\mathbf{\delta}^{(j)}\| \left(c_3 + \sum_{t=1}^{j-1} \|\mathbf{\delta}^{(t)}\| \right). \qquad (36)$$

The truth of lemma 3 now follows from the bound (14).

Lemma 4

There exists a constant, c_5 say, such that the inequality

$$k^{3/2}(\mathbf{g}^{(k)}, H^{(k)}\mathbf{g}^{(k)}) < c_5 \qquad (37)$$

holds.

Proof

The key to the proof is the recurrence relation (8). By substituting expression (7) for $(\mathbf{\delta}^{(k)}, \mathbf{\gamma}^{(k)})$ and expression (5) for $\mathbf{\gamma}^{(k)}$ in equation (8), and by using equation (6), we obtain the identity

$$\det(H^{(k+1)}) = \det(H^{(k)}) \frac{\|\mathbf{\delta}^{(k)}\|}{\|H^{(k)}\mathbf{g}^{(k)}\|} \frac{(\mathbf{g}^{(k)}, H^{(k)}\mathbf{g}^{(k)})}{(\mathbf{g}^{(k)}, H^{(k)}\mathbf{g}^{(k)}) + (\mathbf{g}^{(k+1)}, H^{(k)}\mathbf{g}^{(k+1)})}$$

$$= \det(H^{(k)}) \frac{\|\mathbf{\delta}^{(k)}\|}{\|H^{(k)}\mathbf{g}^{(k)}\|} \left(1 - \frac{(\mathbf{g}^{(k+1)}, H^{(k+1)}\mathbf{g}^{(k+1)})}{(\mathbf{g}^{(k)}, H^{(k)}\mathbf{g}^{(k)})} \right), \qquad (38)$$

1. SOME PROPERTIES OF THE VARIABLE METRIC ALGORITHM

where to deduce the last line we have used equation (9). This recurrence relation is applied k times to give the identity

$$\det(H^{(k+1)}) = \det(H^{(1)}) \frac{(\mathbf{g}^{(k+1)}, H^{(k+1)}\mathbf{g}^{(k+1)})}{(\mathbf{g}^{(1)}, H^{(1)}\mathbf{g}^{(1)})} \left\{ \prod_{j=1}^{k} \|\boldsymbol{\delta}^{(j)}\| \right\} \left\{ \prod_{j=1}^{k} \frac{(\mathbf{g}^{(j)}, H^{(j)}\mathbf{g}^{(j)})}{\|H^{(j)}\mathbf{g}^{(j)}\|} \right\}$$

$$\times \left\{ \prod_{j=1}^{k} \left(\frac{1}{(\mathbf{g}^{(j+1)}, H^{(j+1)}\mathbf{g}^{(j+1)})} - \frac{1}{(\mathbf{g}^{(j)}, H^{(j)}\mathbf{g}^{(j)})} \right) \right\}. \tag{39}$$

We have divided the product into three terms, because we consider each term separately.

For the first product term in expression (39), we use inequality (14) to obtain the bound

$$\prod_{j=1}^{k} \|\boldsymbol{\delta}^{(j)}\| \leq \Delta^k. \tag{40}$$

To bound the second product term we make use of a result due to Wolfe (1971), that holds generally for minimization methods with linear searches, provided that $F(\mathbf{x})$ has bounded second derivatives. Specifically, Wolfe's Theorem 5 and inequality (13) imply that the sum

$$\sum_{k=1}^{\infty} \left(\frac{(\mathbf{g}^{(k)}, H^{(k)}\mathbf{g}^{(k)})}{\|\mathbf{g}^{(k)}\| \|H^{(k)}\mathbf{g}^{(k)}\|} \right)^2$$

is finite. Because $\|\mathbf{g}^{(k)}\|$ is bounded we deduce that there exists a constant ρ such that the inequality

$$\sum_{j=1}^{k} \left(\frac{(\mathbf{g}^{(j)}, H^{(j)}\mathbf{g}^{(j)})}{\|H^{(j)}\mathbf{g}^{(j)}\|} \right)^2 < \rho, \quad (k=1,2,\ldots), \tag{41}$$

holds, and therefore, by some straightforward algebra, we obtain the bound

$$\prod_{j=1}^{k} \frac{(\mathbf{g}^{(j)}, H^{(j)}\mathbf{g}^{(j)})}{\|H^{(j)}\mathbf{g}^{(j)}\|} < \left(\frac{\rho}{k}\right)^{k/2}. \tag{42}$$

And to bound the last product term of expression (39) we ask the question: given that $(\mathbf{g}^{(1)}, H^{(1)}\mathbf{g}^{(1)})$ and $(\mathbf{g}^{(k+1)}, H^{(k+1)}\mathbf{g}^{(k+1)})$ are fixed, but that $(\mathbf{g}^{(j)}, H^{(j)}\mathbf{g}^{(j)})$ $(j=2,3,\ldots,k)$ can take any values subject to the condition that $(\mathbf{g}^{(j)}, H^{(j)}\mathbf{g}^{(j)})$ $(j=1,2,\ldots,k+1)$ is a monotonically decreasing sequence, what is the largest possible value of the product term? Thus we obtain the bound

$$\prod_{j=1}^{k} \left(\frac{1}{(g^{(j+1)}, H^{(j+1)} g^{(j+1)})} - \frac{1}{(g^{(j)}, H^{(j)} g^{(j)})} \right) \leq$$

$$\left\{ \frac{1}{k} \left(\frac{1}{(g^{(k+1)}, H^{(k+1)} g^{(k+1)})} - \frac{1}{(g^{(1)}, H^{(1)} g^{(1)})} \right) \right\}^k < \left\{ \frac{1}{k(g^{(k+1)}, H^{(k+1)} g^{(k+1)})} \right\}^k. \quad (43)$$

It follows from expressions (39)–(43) that the determinant of $H^{(k+1)}$ is bounded by the inequality

$$\det(H^{(k+1)}) < \det(H^{(1)}) \frac{(g^{(k+1)}, H^{(k+1)} g^{(k+1)})}{(g^{(1)}, H^{(1)} g^{(1)})} \left\{ \frac{\Delta \rho^{1/2}}{k^{3/2} (g^{(k+1)}, H^{(k+1)} g^{(k+1)})} \right\}^k. \quad (44)$$

Now every eigenvalue of $H^{(k+1)}$ is greater than the inverse of the trace of $\Gamma^{(k+1)}$, so from inequality (34) we obtain the bound

$$\det(H^{(k+1)}) > [1/\mathrm{Tr}(\Gamma^{(k+1)})]^n$$
$$\geq [(g^{(k)}, H^{(k)} g^{(k)})/c_4 k^2]^n$$
$$> [(g^{(k+1)}, H^{(k+1)} g^{(k+1)})/c_4 k^2]^n. \quad (45)$$

Therefore, using inequality (44) also, we deduce the expression

$$(g^{(k+1)}, H^{(k+1)} g^{(k+1)})^{n+k-1} < \frac{\det(H^{(1)})(c_4 k^2)^n}{(g^{(1)}, H^{(1)} g^{(1)})} \left\{ \frac{\Delta \rho^{1/2}}{k^{3/2}} \right\}^k, \quad (46)$$

from which we infer the inequality

$$(k+1)^{3/2} (g^{(k+1)}, H^{(k+1)} g^{(k+1)})$$

$$< \left[\frac{\det(H^{(1)})(c_4 k^2)^n}{(g^{(1)}, H^{(1)} g^{(1)})} \left\{ \frac{k^{3/2}}{\Delta \rho^{1/2}} \right\}^{n-1} \right]^{1/(n+k-1)} \left[\frac{(k+1)^{3/2}}{k} \right]^{3/2} \Delta \rho^{1/2}. \quad (47)$$

As k tends to infinity, the logarithm of the first term on the right hand side tends to $\frac{1}{2}(7n-3)\log k/(n+k-1)$, and the second term in square brackets tends to one, so the right-hand side of expression (47) is bounded. Lemma 4 is proved.

Lemma 5

$$\sum_k \|\boldsymbol{\delta}^{(k)}\| \text{ diverges.}$$

Proof

We suppose that the sum of the step lengths is bounded, and deduce a contradiction. If $\sum \|\boldsymbol{\delta}^{(k)}\|$ converges, then lemma 2 implies that the quantities $\|H^{(j)}\mathbf{g}^{(j)}\|$ ($j = 1, 2, \ldots$) are bounded, and therefore from inequalities (12) and (35) there exists a constant, c_6 say, such that the inequality

$$\operatorname{Tr}(\Gamma^{(k+1)}) \leqslant c_6 \sum_{j=1}^{k} \frac{\|\boldsymbol{\delta}^{(j)}\|}{(\mathbf{g}^{(j)}, H^{(j)}\mathbf{g}^{(j)})} \tag{48}$$

holds. Further, the assumption that $\sum \|\boldsymbol{\delta}^{(k)}\|$ converges implies that we can find an integer, K say, satisfying the condition

$$\sum_{j=K+1}^{\infty} \|\boldsymbol{\delta}^{(j)}\| < \tfrac{1}{2} c^2 / c_6, \tag{49}$$

where c is the bound (13). It is convenient to define c_7 to be the constant

$$c_7 = c_6 \sum_{j=1}^{K} \frac{\|\boldsymbol{\delta}^{(j)}\|}{(\mathbf{g}^{(j)}, H^{(j)}\mathbf{g}^{(j)})}, \tag{50}$$

and then it follows from the monotonicity of $(\mathbf{g}^{(j)}, H^{(j)}\mathbf{g}^{(j)})$ ($j = 1, 2, \ldots$), and from inequalities (48) and (49), that for all $k > K$ we have the bound

$$\operatorname{Tr}(\Gamma^{(k+1)}) \leqslant c_7 + c_6 \sum_{j=K+1}^{k} \frac{\|\boldsymbol{\delta}^{(j)}\|}{(\mathbf{g}^{(j)}, H^{(j)}\mathbf{g}^{(j)})}$$

$$< c_7 + \frac{c_6}{(\mathbf{g}^{(k+1)}, H^{(k+1)}\mathbf{g}^{(k+1)})} \sum_{j=K+1}^{k} \|\boldsymbol{\delta}^{(j)}\|$$

$$< c_7 + \tfrac{1}{2} c^2 / (\mathbf{g}^{(k+1)}, H^{(k+1)}\mathbf{g}^{(k+1)}). \tag{51}$$

Now lemma 4 shows that $(\mathbf{g}^{(k)}, H^{(k)}\mathbf{g}^{(k)})$ tends to zero, and therefore for sufficiently large k the inequality

$$(\mathbf{g}^{(k+1)}, H^{(k+1)}\mathbf{g}^{(k+1)}) \operatorname{Tr}(\Gamma^{(k+1)}) < c^2 \tag{52}$$

is satisfied. But expressions (13) and (19) imply the bound

$$(\mathbf{g}^{(k+1)}, H^{(k+1)}\mathbf{g}^{(k+1)}) \operatorname{Tr}(\Gamma^{(k+1)}) > c^2, \tag{53}$$

which is a contradiction. Lemma 5 is proved.

5. The Case when $F(\mathbf{x})$ is Convex

If $F(\mathbf{x})$ is a convex function, and satisfies the conditions stated in Section 1, then by using the lemmas of Section 4 it can be shown that inequality (13) leads to a contradiction. Thus we prove in this section that the variable metric algorithm converges. This result is an advance on the convergence theorem of Powell (1971), because he requires $F(\mathbf{x})$ to be uniformly convex.

We make use of the following lemma, which is proved by a method due to Dr. D. Goldfarb (private communication, 1971).

Lemma 6

If the function $F(\mathbf{x})$ is convex, then the inequality

$$(\boldsymbol{\delta}^{(k)}, \boldsymbol{\gamma}^{(k)}) \geq \|\boldsymbol{\gamma}^{(k)}\|^2/M, \qquad (k = 1, 2, \ldots), \tag{54}$$

holds, where M is the bound (1).

Proof

The definition of a derivative gives the identity

$$\begin{aligned}
\boldsymbol{\gamma}^{(k)} &= \int_{\theta=0}^{1} G(\mathbf{x}^{(k)} + \theta \boldsymbol{\delta}^{(k)}) \boldsymbol{\delta}^{(k)} \, d\theta \\
&= \left\{ \int_{\theta=0}^{1} G(\mathbf{x}^{(k)} + \theta \boldsymbol{\delta}^{(k)}) \, d\theta \right\} \boldsymbol{\delta}^{(k)} \\
&= \bar{G}^{(k)} \boldsymbol{\delta}^{(k)},
\end{aligned} \tag{55}$$

say. Because $F(\mathbf{x})$ is convex, the matrix $\bar{G}^{(k)}$ is positive definite or positive semi-definite, and therefore it has a square root. We let $\mathbf{z}^{(k)}$ be the vector

$$\mathbf{z}^{(k)} = [\bar{G}^{(k)}]^{1/2} \boldsymbol{\delta}^{(k)}. \tag{56}$$

Moreover expression (1) and the definition of $\bar{G}^{(k)}$ give the bound

$$\|\bar{G}^{(k)}\| \leq M, \tag{57}$$

which implies the inequality

$$(\mathbf{z}^{(k)}, \bar{G}^{(k)} \mathbf{z}^{(k)}) \leq M \|\mathbf{z}^{(k)}\|^2. \tag{58}$$

By substituting the definition of $\mathbf{z}^{(k)}$ in this expression, and by using equation (55), we obtain inequality (54). Lemma 6 is proved.

Theorem 1

If $F(\mathbf{x})$ is a convex function, having continuous second derivatives bounded by inequality (1), and if the level set $\{\mathbf{x} | F(\mathbf{x}) \leq F(\mathbf{x}^{(1)})\}$ is bounded, then, if the

1. SOME PROPERTIES OF THE VARIABLE METRIC ALGORITHM

variable metric algorithm is applied to $F(\mathbf{x})$, the sequence of function values $F(\mathbf{x}^{(k)})$ ($k = 1, 2, \ldots$) terminates at, or converges to, the least value of $F(\mathbf{x})$.

Proof

First we prove that inequality (13) gives a contradiction. It it were true then lemmas 1 and 6 would imply the inequality

$$\text{Tr}(\Gamma^{(k+1)}) \leqslant c_2 kM \tag{59}$$

and therefore from inequalities (13) and (19) we could deduce the bound

$$(\mathbf{g}^{(k+1)}, H^{(k+1)} \mathbf{g}^{(k+1)}) > c^2/\{c_2 kM\}. \tag{60}$$

However, lemma 4 implies the bound

$$(\mathbf{g}^{(k+1)}, H^{(k+1)} \mathbf{g}^{(k+1)}) < c_5/(k+1)^{3/2}, \tag{61}$$

which contradicts expression (60) when k becomes large. It follows that, if $F(\mathbf{x})$ is a convex function, then either the iterations of the variable metric algorithm terminate because some $\mathbf{g}^{(k)}$ is zero, or we find the limit

$$\liminf_{k \to \infty} \|\mathbf{g}^{(k)}\| = 0. \tag{62}$$

Because the sequence $\mathbf{x}^{(k)}$ ($k = 1, 2, \ldots$) is in a compact set, namely $\{\mathbf{x} | F(\mathbf{x}) \leqslant F(\mathbf{x}^{(1)})\}$, and because the first derivatives of $F(\mathbf{x})$ are continuous, we deduce from expression (62) that there is a limit point, $\mathbf{x}^*$ say, of the sequence $\mathbf{x}^{(k)}$ ($k = 1, 2, \ldots$), at which the gradient of $F(\mathbf{x})$ is zero. And if the iterations of the algorithm terminate it is convenient to use the notation $\mathbf{x}^*$ for the point $\mathbf{x}^{(k)}$ at which $\mathbf{g}^{(k)} = 0$. Moreover, the algorithm ensures that the function values $F(\mathbf{x}^{(k)})$ ($k = 1, 2, \ldots$) decrease monotonically, so we have the limit

$$\lim F(\mathbf{x}^{(k)}) = F(\mathbf{x}^*). \tag{63}$$

Now, it is well known that if $F(\mathbf{x})$ is convex, and if the first derivative vector of $F(\mathbf{x})$ is zero at $\mathbf{x} = \mathbf{x}^*$, then $F(\mathbf{x}^*)$ is the least value of $F(\mathbf{x})$ (for instance see Section 4.2.8 of Ortega and Rheinboldt (1970)), so statement (63) completes the proof of Theorem 1.

Note that if $F(\mathbf{x})$ is least at only one point, $\mathbf{x}^*$ say (which is the case if $F(\mathbf{x})$ is strictly convex), then Theorem 1 implies that the sequence $\mathbf{x}^{(k)}$ ($k = 1, 2, \ldots$) converges to $\mathbf{x}^*$. However if it happens that $F(\mathbf{x})$ is least for a set of points, X^* say, then Theorem 1 implies that every limit point of the sequence $\mathbf{x}^{(k)}$ ($k = 1, 2, \ldots$) is in X^*.

6. Functions of two Variables

For the purpose of this section we say that the direction $\mathbf{s}^{(k)}$ at the point $\mathbf{x}^{(k)}$ is a descent direction if the inequality

$$(\mathbf{g}^{(k)}, \mathbf{s}^{(k)}) < 0 \tag{64}$$

holds, and we say that an iterative method for minimizing a function $F(\mathbf{x})$ is a 'descent algorithm' if, for $k = 1, 2, \ldots$, the point $\mathbf{x}^{(k+1)}$ is defined by the formula

$$\mathbf{x}^{(k+1)} = \mathbf{x}^{(k)} + \lambda^{(k)} \mathbf{s}^{(k)}, \tag{65}$$

where $\mathbf{s}^{(k)}$ is a descent direction, and where $\lambda^{(k)}$ minimizes the function of one variable

$$\phi(\lambda) = F(\mathbf{x}^{(k)} + \lambda \mathbf{s}^{(k)}), \quad (\lambda \geqslant 0). \tag{66}$$

Therefore, corresponding to condition (6), the equation

$$(\mathbf{g}^{(k+1)}, \mathbf{s}^{(k)}) = 0, \quad (k = 1, 2, \ldots), \tag{67}$$

holds. Further we say that an algorithm is a 'conjugate descent algorithm' if it is a descent algorithm, and if the directions $\mathbf{s}^{(k)}$ satisfy the equation

$$(\mathbf{g}^{(k)} - \mathbf{g}^{(k-1)}, \mathbf{s}^{(k)}) = 0, \quad (k = 2, 3, \ldots) \tag{68}$$

Because equations (4) and (6) give the identity

$$(\mathbf{\gamma}^{(k)}, H^{(k+1)} \mathbf{g}^{(k+1)}) = 0, \tag{69}$$

it is straightforward to verify that the variable metric method is a conjugate descent algorithm. Now if we are minimizing a function of only two variables, the direction of $\mathbf{s}^{(k)}$ is determined unambiguously by expressions (64) and (68), so, except for the procedure for starting the iterative process, all conjugate descent algorithms are equivalent when applied to functions of two variables. We make this point to emphasize that Theorem 2 is relevant not only to the variable metric method, but also to some other minimization algorithms, including the Polak-Ribière conjugate gradient algorithm (Polak, 1971).

Theorem 2

If a conjugate descent algorithm is applied to a continuously differentiable function $F(\mathbf{x})$ of two variables, and if the calculated sequence of points $\mathbf{x}^{(k)}$ $(k = 1, 2, \ldots)$ converges to a limit, $\mathbf{x}^*$ say, then $\mathbf{x}^*$ is a stationary point of $F(\mathbf{x})$.

Proof

We suppose that the theorem is false, and deduce a contradiction. If it is false then the sequence of gradients $\mathbf{g}^{(k)}$ $(k = 1, 2, \ldots)$ converges to a non-zero limit

$$\mathbf{g}^* = \begin{pmatrix} \lambda^* \\ \mu^* \end{pmatrix} \tag{70}$$

say. We suppose that $\lambda^* \neq 0$, and this does not lose any generality, because if $\lambda^* = 0$ then we can reorder the co-ordinate directions so that the two components of every vector are interchanged. Also we use the notation

$$\mathbf{g}^{(k)} = \begin{pmatrix} \lambda^{(k)} \\ \mu^{(k)} \end{pmatrix}, \quad (k = 1, 2, \ldots). \tag{71}$$

1. SOME PROPERTIES OF THE VARIABLE METRIC ALGORITHM

Equations (67) and (68) show that the vector $\mathbf{g}^{(k+1)}$ is parallel to $(\mathbf{g}^{(k)} - \mathbf{g}^{(k-1)})$, so we have the recurrence relation

$$\lambda^{(k+1)}\{\mu^{(k)} - \mu^{(k-1)}\} = \mu^{(k+1)}\{\lambda^{(k)} - \lambda^{(k-1)}\}, \quad (k \geq 2). \tag{72}$$

We prove the theorem by showing that it is not possible to satisfy this recurrence relation, and for $\lambda^{(k)}$ ($k = 1, 2, \ldots$) to converge to a non-zero limit.

If $\lambda^* \neq 0$, then the ratio $\{\lambda^{(k)} - \lambda^{(k-1)}\}/\lambda^{(k+1)}$ tends to zero, and we let K be an integer such that the inequality

$$|\lambda^{(k+1)}| > 4|\lambda^{(k)} - \lambda^{(k-1)}| \tag{73}$$

holds for all $k \geq K$. If it happens that $\mu^{(K)} = 0$, then we make the definition

$$\sigma^{(k)} = \mu^{(k)}, \tag{74}$$

but in the more usual case when $\mu^{(K)} \neq 0$, we make the definition

$$\sigma^{(k)} = \lambda^{(k)} - \mu^{(k)}\{\lambda^{(K)}/\mu^{(K)}\}, \tag{75}$$

so in both cases $\sigma^{(K)}$ has the value

$$\sigma^{(K)} = 0. \tag{76}$$

In the first case $\sigma^{(k)}$ ($k = 1, 2, \ldots$) should converge to μ^*, and in the second case this sequence should converge to the number

$$\sigma^* = \lambda^* - \mu^*\{\lambda^{(K)}/\mu^{(K)}\}, \tag{77}$$

but now we show that the sequence $\sigma^{(k)}$ ($k = 1, 2, \ldots$) diverges, which is the contradiction that proves the theorem.

Because of the identity (76), we have the inequality

$$|\sigma^{(K+1)}| > 2|\sigma^{(K)}|, \tag{78}$$

unless $\sigma^{(K+1)} = 0$. However, if $\sigma^{(K+1)} = 0$, then equation (74) or (75) implies that $\mathbf{g}^{(K+1)}$ is parallel to $\mathbf{g}^{(K)}$, which contradicts expression (64) or (67) in the case $k = K$. Therefore, condition (78) is satisfied.

Now the definition of $\sigma^{(k)}$ and equation (72) give the identity

$$\lambda^{(k+1)}\{\sigma^{(k)} - \sigma^{(k-1)}\} = \sigma^{(k+1)}\{\lambda^{(k)} - \lambda^{(k-1)}\}, \quad (k \geq 2), \tag{79}$$

and therefore from inequality (73) we deduce the bound

$$|\sigma^{(k+1)}| \geq 4|\sigma^{(k)} - \sigma^{(k-1)}|, \quad (k \geq K). \tag{80}$$

In particular in the case $k = K + 1$, we use expressions (78) and (80) to obtain the inequality

$$|\sigma^{(K+2)}| \geq 4|\sigma^{(K+1)}| - 4|\sigma^{(K)}| > 2|\sigma^{(K+1)}|. \tag{81}$$

Similarly, by applying inequality (80) in the case $k = K+2$, and by using expression (81), we find the bound

$$|\sigma^{(K+3)}| > 2|\sigma^{(K+2)}|. \tag{82}$$

This argument is continued in an inductive way, and thus we find that the condition

$$|\sigma^{(k+1)}| > 2|\sigma^{(k)}| \tag{83}$$

is satisfied for all $k \geq K$. It follows that the sequence $\sigma^{(k)}$ ($k = 1, 2, \ldots$) diverges, which proves Theorem 2.

7. Discussion

The lemmas and theorems of this paper are the result of about eighteen months of intermittent work to try to explain and understand the behaviour of the variable metric algorithm when $F(\mathbf{x})$ is not convex. They make only a small contribution to this problem, because all the lemmas of Section 4 depend on condition (13), which may never be satisfied. Indeed, I believe that the lemmas may be useful only in showing that inequality (13) cannot hold, if $F(\mathbf{x})$ satisfies the conditions of Section 1. For example these lemmas do lead to a contradiction in Section 5, where we give a theorem that is stronger than previously published theorems on the convergence of the variable metric algorithm.

I would like to know if the following conjecture is true: 'There exist differentiable functions $F(\mathbf{x})$, such that the sequence $\mathbf{x}^{(k)}$ ($k = 1, 2, \ldots$) converges to a point that is not a stationary point of $F(\mathbf{x})$'. I have tried to prove that it is false, and to find examples to show that it is true, but, as described in Section 6, I have proved only that the conjecture is false in two dimensions.

However, recently Dixon (1971) made a remarkable discovery about the variable metric algorithm, which will almost certainly be of great value in trying to solve these questions of convergence. It is that, given exact linear searches, there are many iterative algorithms that calculate the sequence of points $\mathbf{x}^{(k)}$ ($k = 1, 2, \ldots$) obtained by the variable metric algorithm.

Specifically if we make the following two changes to the algorithm described in Section 2, then the sequence of points $\mathbf{x}^{(k)}$ ($k = 1, 2, \ldots$) is unaltered. The first change is that instead of minimizing the function (3) subject to the condition $\lambda \geq 0$, we minimize $\phi(\lambda)$ subject to the condition that the sign of λ is the same as the sign of $(\mathbf{g}^{(k)}, H^{(k)}\mathbf{g}^{(k)})$. And the second change is that instead of defining $H^{(k+1)}$ by formula (4), we use the formula

$$H^{(k+1)} = H^{(k)} - \frac{H^{(k)}\boldsymbol{\gamma}^{(k)}\boldsymbol{\gamma}^{(k)T}H^{(k)}}{(\boldsymbol{\gamma}^{(k)}, H^{(k)}\boldsymbol{\gamma}^{(k)})} + \frac{\boldsymbol{\delta}^{(k)}\boldsymbol{\delta}^{(k)T}}{(\boldsymbol{\delta}^{(k)}, \boldsymbol{\gamma}^{(k)})} + \theta^{(k)}\mathbf{z}^{(k)}\mathbf{z}^{(k)T}, \tag{84}$$

where $\mathbf{z}^{(k)}$ is the vector

$$\mathbf{z}^{(k)} = (\boldsymbol{\gamma}^{(k)}, H^{(k)}\boldsymbol{\gamma}^{(k)})\boldsymbol{\delta}^{(k)} - (\boldsymbol{\delta}^{(k)}, \boldsymbol{\gamma}^{(k)})H^{(k)}\boldsymbol{\gamma}^{(k)}, \tag{85}$$

and where, for $k = 1, 2, \ldots$, the value of $\theta^{(k)}$ is *any* number except for the quantity

$$\frac{-(\mathbf{g}^{(k)}, H^{(k)} \mathbf{g}^{(k)})}{(\boldsymbol{\delta}^{(k)}, \boldsymbol{\gamma}^{(k)})^2 (\boldsymbol{\gamma}^{(k)}, H^{(k)} \boldsymbol{\gamma}^{(k)}) (\mathbf{g}^{(k+1)}, H^{(k)} \mathbf{g}^{(k+1)})}, \qquad (86)$$

the purpose of this last condition being to ensure that the matrix $H^{(k+1)}$ is not singular.

Different choices of $\theta^{(k)}$ in equation (84) give the different members of Broyden's (1967) linear family of updating formulae, so it follows that our theorems apply to many algorithms, including, for example, Broyden's (1970) recent method for minimizing a function of several variables.

Because the choice of $\theta^{(k)}$ in equation (84) does not affect the sequence of points $\mathbf{x}^{(k)}$ ($k = 1, 2, \ldots$), it would be worthwhile to look for a choice that makes the matrices $H^{(k)}$ ($k = 1, 2, \ldots$) easy to handle from the point of view of proving convergence theorems. Thus we may find that Dixon's (1971) theorem leads to a much better understanding of the convergence properties of the variable metric algorithm.

References

Broyden, C. G. (1967). Quasi-Newton methods and their application to function minimization. *Math. Comput.* **21**, 368–381.

Broyden, C. G. (1970). The convergence of a class of double-rank minimization algorithms 2. The new algorithm. *J. Inst. Maths Applics* **6**, 222–231.

Davidon, W. C. (1959). 'Variable Metric Method for Minimization'. A. E. C. Research and Development Report ANL-5990 (Rev.).

Dixon, L. C. W. (1971). 'Variable Metric Algorithms: Necessary and Sufficient Conditions for Identical Behaviour on Non-quadratic Functions'. N.O.C. (Hatfield Polytechnic) Report No. 26.

Fletcher, R. (1970). A new approach to variable metric algorithms. *Comput. J.* **13**, 317–322.

Fletcher, R., and Powell, M. J. D. (1963). A rapidly convergent descent method for minimization. *Comput. J.* **6**, 163–168.

Ortega, J. M., and Rheinboldt, W. C. (1970). 'Iterative Solution of Non-linear Equations in Several Variables. Academic Press, New York.

Pearson, J. D. (1969). Variable metric methods of minimization. *Comput. J.* **12**, 171–178.

Polak, E. (1971). 'Computational Methods in Optimization: a Unified Approach'. Academic Press, New York.

Powell, M. J. D. (1971). On the convergence of the variable metric algorithm. *J. Inst. Maths Applics* **7**, 21–36.

Wolfe, P. (1971). Convergence conditions for ascent methods II: some corrections. *SIAM Review* **13**, 185–188.

2. On Some Methods Based on Broyden's Secant Approximation to the Hessian

J. E. DENNIS JR.
*Department of Computer Science, Cornell University,
Ithaca, New York, U.S.A.*

1. Introduction

The secant method is well known to be an excellent computational tool for the solution of non-linear equations in a single variable. It is therefore quite understandable that much effort has gone into finding multivariate analogues. Historically this work has been directed toward generalizations which preserve the order of convergence of the scalar method. Most of these methods tend to be computationally inconvenient if not actually unstable, although for fairly wide classes of problems they are quite useful. (See Collatz, 1965; Hofmann, 1970.) The secant method is not really a high-order scalar method, instead its main value lies in being fast relative to the small amount of computing it requires. (See Ralston, 1965.) For this reason, the so-called update methods seem to be the real spiritual heirs of the secant method. They require very little computation and they certainly are eligible for the name, since they can be written

$$x_{n+1} = x_n - A_n^{-1} F(x_n), \qquad (1)$$

where $F = (f_1,\ldots,f_N)^T$ is the system whose zero we desire and A_n is chosen to satisfy the divided difference relation

$$A_{n+1}(x_{n+1} - x_n) = F(x_{n+1}) - F(x_n). \qquad (2)$$

This is sometimes called the 'quasi-Newton equation', though of course, Newton's method, i.e. $A_{n+1} = F'(x_{n+1})$, does not satisfy it. We will call (2) the 'secant equation'.

Quite often the update methods are implemented not in the form (1) but rather as

$$x_{n+1} = x_n - t_n A_n^{-1} F(x_n), \qquad (3)$$

where t_n is chosen by some descent criterion. Our interest here is in performance near a root, when $t_n = 1$. It is certainly granted that to get near enough, t_n or indeed A_n, may need to be chosen differently.

2. A General Theorem

In this section we present the general theorem and notation which will be useful in the sequel.

Sometimes in an implementation of (1) we wish to leave A_n constant for some number of iterations, and re-evaluate it only at irregular intervals in the computation. This can be quite easily incorporated into the theory and so we make the following definition (Dennis, 1969a, 1971).

2.1 Definition

A recalculation sequence $\{\alpha_n\}$ is a non-decreasing sequence of non-negative integers such that $\alpha_0 = 0$ and $\alpha_n = \alpha_{n-1}$ or $\alpha_n = n$. Formula (1) with recalculation sequence $\{\alpha_n\}$ is

$$x_{n+1} = x_n - A_{\alpha_n}^{-1} F(x_n). \tag{4}$$

Let $\|\cdot\|$ denote a vector norm and the subordinate matrix norm. The set D will be a convex set.

2.2 Definition

A method of bounded deterioration is any method (1) such that as long as x_n and $F'(x_n)$ are defined,

$$\|A_n - F'(x_n)\| \leq \delta_0 + \gamma \sum_{j=1}^{n} \|x_{j-1} - x_j\|. \tag{4}$$

2.3 Definition

$F' \in \mathrm{Lip}_K(D)$ for some set D if and only if there exists some $K \geq 0$ such that for every pair $x, y \in D$,

$$\|F'(x) - F'(y)\| \leq K\|x - y\|. \tag{5}$$

The following theorem shows that a method of bounded deterioration is generally locally convergent.

2.4 Theorem (Dennis, 1971)

Let $F' \in \mathrm{Lip}_K D$, $x_0 \in D$, and let A_0 be a non-singular $N \times N$ matrix with

$$\|A_0^{-1} F(x_0)\| \leq \eta, \|A_0^{-1}\| \leq \beta.$$

Assume that there are non-negative real numbers δ and γ such that for every n for which $x_0, \ldots, x_n$ as defined by (1), are in D,

$$\|A_n - F'(x_n)\| \leq \delta_n + \gamma \sum_{j=1}^{n} \|x_j - x_{j-1}\|, \quad \delta_n \leq \delta \text{ for } n > 0. \tag{6}$$

2. METHODS BASED ON BROYDEN'S SECANT APPROXIMATION

If, in addition

$$1 > \beta\delta_0 + 2\beta\delta \tag{7}$$

$$\tfrac{1}{2} \geq h \equiv \frac{(2\gamma + K)\beta\eta}{(1 - 2\beta\delta - \beta\delta_0)^2} \tag{8}$$

and

$$N(x_0, r_0) \subset D,$$

where

$$r_0 = \frac{1 - (1 - 2h)^{1/2}}{\beta(2\gamma + K)}(1 - 2\beta\delta - \beta\delta_0),$$

then $\{x_n\}$ generated by (1) exists in $N(x_0, r_0)$ and converges to x^*. The point x^* is the unique root of F in

$$\bar{N}\left(x_0, \frac{1 - (1 - 2h')^{1/2}}{\beta K}(1 - \beta\delta_0)\right),$$

where

$$h' \equiv \frac{\beta K \eta}{(1 - \beta\delta)^2}$$

and if $h < \tfrac{1}{2}$, then x^* is unique in

$$D \cap N\left(x_0, \frac{1 + (1 - 2h')^{1/2}}{\beta K}(1 - \beta\delta_0)\right).$$

If one applies the previous theorem to the special case of Newton's method then $\delta_n = 0 = \gamma$ and the result is the Kantorovich theorem. Note that the existence of a solution is asserted not assumed.

3. The Single Rank Methods

Broyden (1965) described a class of secant methods based on the following solutions to (2), the secant equation.

$$A_{n+1} = A_n - F(x_{n+1})\frac{d_n^T A_n}{d_n^T F(x_n)}$$

or

$$A_{n+1}^{-1} = A_n^{-1} - A_n^{-1} F(x_{n+1})\frac{d_n^T}{d_n^T [F(x_{n+1}) - F(x_n)]}.$$

The vector d_n must be chosen so that neither denominator is zero. Broyden (1965) suggested $d_n = F(x_{n+1}) - F(x_n)$ and $d_n = (A_n^{-1})^T A_n^{-1} F(x_n)$. These approximations to the Jacobian and its inverse are usually denoted by B_n and H_n respectively. We conform to this convention and also generalize the formulae to give a Jacobian approximation at x' given an approximation B and $H = B^{-1}$ at x.

$$B' = B + [F(x') - F(x) - B(x' - x)] \frac{d^T B}{d^T B(x' - x)}. \qquad (9)$$

$$H' = H - \{H[F(x') - F(x)] - (x' - x)\} \frac{d^T}{d^T[F(x') - F(x)]}. \qquad (10)$$

In Dennis (1969b) we proved the following theorem. Let $\|\cdot\| = \|\cdot\|_2$.

3.1 Theorem

Let $F \in \text{Lip}_K D$ and let $x, x' \in D$ with B a non-singular matrix. If B' is defined by (9),

$$\|B' - F'(x)\| \leq q\|B - F'(x)\| + K(1 + q/2)\|x' - x\|,$$

where

$$q = \frac{\|x' - x\| \cdot \|d^T B\|}{|d^T B(x' - x)|}.$$

This allows a straightforward induction proof of the following.

3.2 Theorem (Dennis, 1969b)

Let $F' \in \text{Lip}_K D$ and let $x_0, \ldots, x_{n+1}, B_0, \ldots, B_{n+1}$ be generated by any single rank method. If $\{x_i : i = 0, \ldots, n+1\} \subset D$, then

$$\|B_{n+1} - F'(x_{n+1})\| \leq \left(\prod_{i=0}^n q_i\right) \|B_0 - F'(x_0)\|$$

$$+ K \sum_{j=0}^n \left(\prod_{i=j+1}^n q_i\right) (1 + q_j/2) \|x_{j+1} - x_j\|, \qquad (11)$$

where q_j is defined as

$$\frac{\|x_{j+1} - x_j\| \cdot \|d_j^T B_j\|}{|d_j^T B_j(x_{j+1} - x_j)|},$$

for $j \geq 0$ and

$$\prod_{i=r+1}^r q_i = 1, \quad r = 0, 1, \ldots.$$

Now in order for the single rank method to be of bounded deterioration (according to this analysis) we need

$$\prod_{i=0}^{n} q_i$$

to be uniformly bounded. There is no problem in the case of Broyden's method, $d_j = H_i^T(x_{j+1} - x_j)$, since then $q_i = 1$, its minimum value. Otherwise, it is necessary and sufficient that

$$\sum_{j=0}^{\infty} (q_j - 1)$$

converge (Knopp, 1947). It is an open problem to characterize d_j in terms of the convergence of

$$\sum_{j=0}^{\infty} (q_j - 1).$$

Some special cases of interest will be mentioned in the next section. We finish this section with a theorem giving conditions for convergence of Broyden's method, $d_n = H_n^T(x_{n+1} - x_n)$, although it should be clear how to give a theorem for the case

$$\sum_{j=0}^{\infty} (q_j - 1) < \infty.$$

This theorem strengthens one in Dennis (1969b).

3.3 *Theorem*

Let $F' \in \text{Lip}_K D$, $x_0 \in D$, and H_0 be a non-singular $N \times N$ matrix bounded in norm by β. Assume also that $\eta \geq \|H_0 F(x_0)\|$, and $\|B_0 - F'(x_0)\| \leq \delta$. If $1 > 3\beta\delta$,

$$\frac{1}{8} \geq h \equiv \frac{\beta K \eta}{(1 - 3\beta\delta)^2}$$

and $N(x_0, r_0) \subset D$ for

$$r_0 = \frac{1 - (1 - 2h)^{1/2}}{4\beta K} (1 - 3\beta\delta)$$

then Broyden's method with arbitrary recalculation sequence converges to x^*, a zero of F in $\bar{N}(x_0, r_0)$.

Proof

The proof consists simply of noticing that by Theorem 3.2, Broyden's method satisfies the hypotheses of Theorem 2.4 with $\delta_n = \delta$ and $\gamma = 3K/2$.

In Dennis (1971), we give theoretical justification as well as numerical examples in support of the error bounds

$$\tfrac{1}{2}\|x_{n+1} - x_n\| \leq \|x^* - x_n\| \leq 4\frac{\|H_0\|}{\|H_{\alpha_n}\|}\|x_{n+1} - x_n\|$$

for the case when δ is very small and the hypotheses of the previous theorem are satisfied.

4. A Class of Double Rank Formulae

Broyden's method works very well for the solution of general non-linear vector equations, but in the unconstrained minimization problem it suffers from the fact that even though $F'(x) = (\nabla f)'(x) = f''(x)$ is symmetric and even if H_n is, H_{n+1} need not be symmetric. In fact, a glance at (10) shows that the method with $d = \eta$ is the only single rank method with this property.

Powell (1970c) gave the following procedure. Given G, a symmetric non-singular matrix, update $G = G^{(0)}$ using the Broyden single rank formula to obtain $G^{(1/2)}$. This Hessian approximation satisfies the secant equation (2), but it is not symmetric. Set $G^{(1)} = (G^{(1/2)} + G^{(1/2)T})/2$. The matrix $G^{(1)}$ does not satisfy (2), so repeat the process. Powell gives the limit of $G^{(k)}$ as

$$G^* = G + \frac{\mu\delta^T + \delta\mu^T}{\|\delta\|^2} - \frac{\delta^T\mu\delta\delta^T}{\|\delta\|^4}, \tag{12}$$

where $\delta = x' - x$, $\gamma = \nabla f(x') - \nabla f(x)$, and $\mu = \gamma - G\delta$.

D. Goldfarb in a private communication was the first to notice that the analysis of the previous section could be applied to (12). Before making this analysis, we give the results obtained by applying Powell's procedure to (9) and (10).

$$G^* = G + \frac{\mu d^T G + G d \mu^T}{d^T G \delta} - \frac{\delta^T \mu G d d^T G}{(d^T G \delta)^2}, \tag{13}$$

$$H^* = H - \frac{\eta d^T + d\eta^T}{d^T \gamma} + \frac{\gamma^T \eta d d^T}{(d^T \gamma)^2}, \tag{14}$$

where $\eta = H\mu$. Notice that even if $GH = I$, unless we consider the special symmetric single rank case, $d = \eta$, $G^*H^* \neq I$ in general. Formula (14) intersects the class of double rank formulae defined in Broyden (1967), only if $d = c_1 H\gamma + c_2 \delta$. We could refer to (13) and (14) as 'dual' update formulae.

First consider the $H\gamma = d$ row in Table 1. One can easily find that $(G^*)^{-1}$ is the Davidon-Fletcher-Powell update. (See Powell, 1971.) Its dual is an

2. METHODS BASED ON BROYDEN'S SECANT APPROXIMATION

update method discovered by Greenstadt and given favourable reports in a recent work (Greenstadt, 1970). The matrix H^* of the $d = \delta$ row is a method found independently by Broyden (1969), Goldfarb (1970), Shanno (1970) and Fletcher (1970). It is interesting to note that the underlying single rank method is due to Pearson (1969). The bottom right-hand element was found by Greenstadt (1970) and is the symmetrization of Broyden's (1965) poor method. We return to Table 1 later, in Section 6.

TABLE 1

Some examples of dual formulae

d	G^*	H^*
$H\delta$	$G + \dfrac{\mu\delta^T\delta\mu^T}{\|\delta\|^2} - \dfrac{\delta^T\mu\delta\delta^T}{\|\delta\|^4}$	$H - \dfrac{\eta\delta^T H + H\delta\eta^T}{\delta^T H\gamma} + \dfrac{\gamma^T\eta H\delta\delta^T H}{(\delta^T H\gamma)^2}$
$H\gamma$	$G + \dfrac{\mu\gamma^T + \gamma\mu^T}{\gamma^T\delta} - \dfrac{\delta^T\mu\gamma\gamma^T}{(\gamma^T\delta)^2}$	$H - \dfrac{\eta\gamma^T H + H\gamma\eta^T}{\gamma^T H\gamma} + \dfrac{\gamma^T\eta H\gamma\gamma^T H}{(\gamma^T H\gamma)^2}$
δ	$G + \dfrac{\mu\delta^T G + G\delta\mu^T}{\delta^T G\delta} - \dfrac{\delta^T\mu G\delta\delta^T G}{(\delta^T G\delta)^2}$	$H - \dfrac{\eta\delta^T + \delta\eta^T}{\delta^T\gamma} + \dfrac{\gamma^T\eta\delta\delta^T}{(\delta^T\gamma)^2}$
γ	$G + \dfrac{\mu\gamma^T G + G\gamma\mu^T}{\gamma^T G\delta} - \dfrac{\delta^T\mu G\gamma\gamma^T G}{\gamma^T G\delta}$	$H - \dfrac{\eta\gamma^T + \gamma\eta^T}{\|\gamma\|^2} + \dfrac{\gamma^T\eta\gamma\gamma^T}{\|\gamma\|^4}$

Let us now proceed to analyse Newton-like methods based on (13). Powell (1970c) mentions that for his update G^* for $d = H\delta$, if f is a quadratic with Hessian $\bar{G}$, G^* satisfies

$$G^* - \bar{G} = \left(I - \frac{\delta\delta^T}{\|\delta\|^2}\right)(G - \bar{G})\left(I - \frac{\delta\delta^T}{\|\delta\|^2}\right).$$

The importance of this identity is clearly that G^* is at least as good an approximation as G in the quadratic case. One can easily show that (13) satisfies the identity

$$G^* - \bar{G} = \left(I - \frac{Gd\delta^T}{d^T G\delta}\right)(G - \bar{G})\left(I - \frac{\delta d^T G}{d^T G\delta}\right).$$

Note that the l_2 norm of the projectors is $q \geqslant 1$. The following double rank analogue of Theorem 3.1 generalizes this identity to the non-linear case.

4.1 Theorem

Let $f'' \in \text{Lip}_K D$ and let G^* be determined by (13) for $x, x' \in D$, then
$$\|G^* - f''(x')\| \leq q^2 \cdot \|G - f''(x)\| + (K/2)(2 + q + q^2)\|x' - x\|$$
where
$$q = \frac{\|Gd\| \|x' - x\|}{|d^T G(x' - x)|}.$$

Proof

Straightforward algebra gives
$$G^* - f''(x') = G - f''(x) + f''(x) - f''(x')$$
$$+ \frac{[\gamma - f''(x)\delta]d^T G + Gd[\gamma - f''(x)\delta]^T}{d^T G \delta}$$
$$- \frac{\delta^T[\gamma - f''(x)\delta]Gdd^T G}{(d^T G \delta)^2}$$
$$- \frac{[G - f''(x)]\delta d^T G + Gd\delta^T[G - f''(x)]}{d^T G \delta}$$
$$+ \frac{\delta^T[G - f''(x)]\delta Gdd^T G}{(d^T G \delta)^2}.$$

Hence
$$G^* - f''(x') = \left(I - \frac{Gd\delta^T}{d^T G \delta}\right)[G - f''(x)]\left(I - \frac{\delta d^T G}{d^T G \delta}\right) + \frac{[\gamma - f''(x)\delta]d^T G}{d^T G \delta}$$
$$+ \frac{Gd}{d^T G \delta}[\gamma - f''(x)\delta]^T\left(I - \frac{\delta d^T G}{d^T G \delta}\right) + f''(x) - f''(x').$$

Now
$$\left\|I - \frac{Gd\delta^T}{d^T G \delta}\right\| = \left\|I - \frac{\delta d^T G}{d^T G \delta}\right\| = q,$$
(Broyden, 1970), and $\|\gamma - f''(x)\delta\| \leq (K/2)\|\delta\|^2$, so
$$\|G^* - f''(x')\| \leq q^2 \|G - f''(x)\| + (K/2)\frac{\|\delta\|^2 \|d^T G\|}{|d^T G \delta|} + \frac{\|Gd\|}{|d^T G \delta|}(K/2)\|\delta\|^2 q + K\|\delta\|,$$
and the result follows.

This time, in analogy with Theorem 3.2 we need

$$\prod_{i=0}^{n} q_i^2$$

to be uniformly bounded. Certainly, for Powell's (1970) update, $q_i \equiv 1$, it is.

4.2 Theorem

Let $f'' \in \text{Lip}_K D$, $x_0 \in D$ and H_0 be a symmetric non-singular $N \times N$ matrix bounded in norm by β. Assume also that $\eta \geq \|H_0 \nabla f(x_0)\|$ and $\|G_0 - f''(x_0)\| \leq \delta$. If $1 > 3\beta\delta$,

$$0{\cdot}1 \geq h \equiv \frac{\beta K \eta}{(1 - 3\beta\delta)^2}$$

and $N(x_0, r_0) \subset D$ for

$$r_0 = \frac{1 - (1 - 2h)^{1/2}}{\delta \beta K}(1 - 3\beta\delta),$$

then Powell's (1970c) update procedure with arbitrary recalculation sequence converges to x^*, a stationary point of f in $\bar{N}(x_0, r_0)$.

Proof

Since the previous theorem for the case of Powell's (1970) update reduces to $\|G^* - f''(x')\| \leq \|G' - f''(x)\| + 2K\|x' - x\|$, it needs only an easy induction argument to show that the analogue of (11) is

$$\|G_{n+1} - f''(x_{n+1})\| \leq \|G_0 - f''(x_0)\| + 2K \sum_{j=0}^{n} \|x_{j+1} - x_j\|.$$

Now apply Theorem 2.4, with $\delta = \delta$, $\gamma = 2K$.

The same error estimates noted at the end of Section 3 for Broyden's method should be of value here. They have not been tested.

5. A Class of Methods for Non-linear Least Squares

Suppose we wish to minimize

$$f(x) = \sum_{i=1}^{M} [\phi_i(x)]^2$$

where ϕ_i is a scalar function defined on a subset of E^N, $M \geq N$. Let us assume that we are willing to calculate $\nabla f(x) = 2\Phi'(x)\Phi(x)$, where $\Phi = (\phi_1, \ldots, \phi_M)^T$. One way to apply the update methods is the rather obvious way of applying a correction to an approximation to the Hessian of f. A better, though less obvious, way was given in Brown and Dennis (1970a). We sketch it below.

$$f''(x) = \sum_{i=1}^{M} \phi_i(x)\phi_i''(x) + \Phi'(x)^T \Phi'(x).$$

Notice that we have assumed $\Phi'(x)$ can be computed, and so its ith row, $\nabla \phi_i(x)$, is available. Hence an update method can be used to approximate $\phi_i''(x_{n+1})$, given $\Phi'(x_{n+1})$ and $\Phi'(x_n)$. If one of the formulae of Section 4 is used, then, since the $\phi_i''(x_{n+1})$ approximations will be symmetric, the algorithm requires the same storage as Newton's method applied to ∇f but none of the $(NM^2/2) + NM$ cross partials of that method. In the reference examples are given which show the method to out-perform Levenberg-Marquardt when f is large at the solution, i.e. $M \gg N$, which justifies the additional arithmetic in that case. A theorem based on the analysis of Section 2 is also given which shows the method to be second order when the Powell update is used and f is zero at the minimum. This last will usually correspond to the case $M = N$ and one then has another update method due to Powell (1970b) available which does not require Φ'. One also has the finite difference analogue of Levenberg-Marquardt given in Brown and Dennis (1970b) available, which seems to be very good when the minimum of f is small.

6. Numerical Results and Discussion

When the results of Section 3 were first given it did not really seem very important whether or not Broyden's (1965) method was the only single rank method of bounded deterioration. Generally the other work on single rank methods was more concerned with the minimization problem than the solution of vector equations. The relationship between the single and double rank methods shown in Section 4 makes this question of much more interest. One certainly cannot dismiss the methods given in Table 1.

Whether or not q for say the Davidon-Fletcher-Powell method satisfies

$$Q \geqslant \prod_{i=0}^{n} q_i, \qquad (n = 0, 1, \ldots),$$

it would be interesting to know if the update generates a locally convergent Newton-like method; that is, can one add to Powell's (1971) theorem that t_n in (3) can eventually be taken as 1? This becomes even more interesting because of L. C. W. Dixon's results presented later in this volume (p. 149). (See the computational results in this context.) If a counter-example can be found, this will certainly mean that one can not be completely half-hearted in his choice of t_n.

2. METHODS BASED ON BROYDEN'S SECANT APPROXIMATION

It is also possible to be irreverent and ask if possibly one of these methods could be better than the Davidon-Fletcher-Powell, even when used in connection with iteration (3).

The examples we report are of course subject to error, and even if they are correct, they prove nothing. All the computations were done on the Cornell University IBM 360/65. The programs were written in FORTRAN, and the WATFIV compiler was used. All gradients were obtained analytically.

Problem I refers to Rosenbrock's function:

$$f(x) = 100(x_2 - x_1)^2 + (1 - x_1)^2.$$

$$x_0 = (-1{\cdot}2, 1{\cdot}0), \qquad f_0 = 24{\cdot}2,$$

$$x^* = (1, 1), \qquad \min f = 0.$$

Problem II is:

$$f(x) = x_1^2 + 2x_2^2 + 3x_3^2 + 4x_4^2 + (x_1 + x_2 + x_3 + x_4)^4.$$

$$x_0 = (1, -1, -1, 1), \qquad f_0 = 10,$$

$$x^* = (0, 0, 0, 0), \qquad \min f = 0.$$

Problem III is the Wood-Colville problem:

$$f(x) = 100(x_2 - x_1^2)^2 + (1 - x_1)^2 + 90(x_4 - x_3^2)^2 + (1 - x_3)^2$$
$$+ 10{\cdot}1\{(x_2 - 1)^2 + (x_4 - 1)^2\} + 19{\cdot}8(x_2 - 1)(x_4 - 1).$$

$$x_0 = (-3, -1, -3, -1), \qquad f_0 = 10135{\cdot}9,$$

$$x^* = (1, 1, 1, 1), \qquad \min f = 0.$$

Problem IV refers to Powell's four variable function:

$$f(x) = (x_1 + 10x_2)^2 + 5(x_3 - x_4)^2 + (x_2 - 2x_3)^4 + 10(x_1 - x_4)^4.$$

$$x_0 = (3, -1, 0, 1), \qquad f_0 = 215,$$

$$x^* = (0, 0, 0, 0), \qquad \min f = 0.$$

Algorithm I

Algorithm I is based on Powell's MINFA routine (1970d); Powell's (1970c) update (13) with $d = H\delta$ is used to approximate the Hessian and (14) with $d = \gamma$ furnishes the inverse Hessian approximation. The program used did not have a singularity monitor.

Algorithm II

This algorithm used a modification of Powell's MINFA with Powell's (1970c) Hessian approximation and the dual approximation to the inverse Hessian, i.e. the first row of Table 2.

Algorithm III

This algorithm is MINFA using the bottom right hand element of Table 1 as the inverse Hessian approximation and its inverse matrix as the Hessian approximation.

Algorithm IV

This algorithm was MINFA using row two of Table 1, i.e. the DFP and Greenstadt's methods.

Algorithm V

This algorithm was Powell's MINFA (1970d).

These five algorithms are based on the assumption that the underlying method is locally convergent. Algorithms II and V, both based on Broyden's (1965) method, seem, on the strength of admittedly thin evidence, to perform best in this context.

We summarize these results in Table 2. Perhaps it is in order to mention that a comparison of execution times would have been of greater interest if we had stopped the iteration when, say, $f < 10^{-5}$. This would not have told us as much about the local behaviour of the methods as we wished to know, however (see Table 3 in this context).

TABLE 2
Performance of modified MINFA routines

	PROBLEM I		
Algorithm	Final f value	Number of f values[a]	Execution time (sec)
I	0.24×10^{-10}	203	7·79
II	0.2×10^{-9}	50	1·94
III	3·6	95[b]	0·81
IV	0·7	299[b]	5·00
V	0·0	49	0·88

2. METHODS BASED ON BROYDEN'S SECANT APPROXIMATION

TABLE 2—*continued*

Algorithm	Final f value	Number of f values[a]	Execution time (sec)
	PROBLEM II		
I	0.14×10^{-21}	21	0.99
II	0.8×10^{-24}	19	0.86
III	1.0×10^{-19}	142	5.91
IV	2.3×10^{-22}	22	0.88
V	0.95×10^{-21}	19	0.83
	PROBLEM III		
II	0.39×10^{-12} [c]	60	3.0
III	0.18×10^{-1}	180[b]	8.17
V	0.59×10^{-12} [c]	105	5.0
	PROBLEM IV		
II	0.33×10^{-16}	115	5.61
III	0.82×10^{-2}	145[b]	7.21
V	0.19×10^{-13}	300	13.69

[a] This is also the number of iterations and gradient values.
[b] Termination because H_n became singular.
[c] The times given here are estimations of the times at which the given value of f was reached. Neither method could reduce f any further.

Algorithms I and IV are sufficiently bad that we have dropped some of their results to conserve space in the table. Both Algorithms II and V seemed ultimately superlinear, except for Problem IV. They could not really be expected to be superlinear in that case because the Hessian is singular at the minimum point. In Problem III the convergence seemed superlinear until the sequences respectively converged to points just away from the minimum.

The next question we investigate is how these methods compare in a standard descent algorithm. The updates were all used in iteration (3), where t_n was chosen by subroutine VD02A from the A.E.R.E Harwell subroutine library. The results are summarized in Table 3.

Tests were also run to judge the utility of averaging an update with its dual. The resulting formula always mimicked the behaviour of its worse parent.

As a final synthesis of computational results we estimate the performance of some of the apparently better algorithms from Tables 2 and 3 in a standard way in Table 4.

Thus the new update,

$$H^* = H - \frac{\eta \delta^T H + H \delta \eta^T}{\delta^T H \gamma} + \frac{\gamma^T \eta H \delta \delta^T H}{(\delta^T H \gamma)^2}$$

shows promise of being a very versatile formula worthy of further investigation. The first project probably should be to fit this update more snugly into the MINFA framework, since the higher run times for Algorithm II are probably due to the author's sloppy coding.

TABLE 3
Performance of descent routines based on various updates

Update	Final f value	Number of ITNS/f values[a]	Execution time (sec)
		PROBLEM I	
I_{11}[b]	0	22/120	1·28
I_{12}	0	21/121	1·26
I_{21}	0	18/ 96	0·98
I_{22}	0	21/121	1·51
		PROBLEM II	
I_{11}	$0·37 \times 10^{-32}$	12/41	0·79
I_{12}	$0·28 \times 10^{-34}$	12/41	0·74
I_{21}	$0·52 \times 10^{-34}$	10/34	0·59
I_{22}	$0·52 \times 10^{-34}$	10/34	0·59
		PROBLEM III	
I_{11}	1·8	30/143	1·97
I_{12}	1·8	30/141	1·82
I_{21}	0·4	30/249	2·67
I_{22}	$0·26 \times 10^{21}$	30/158	2·06
		PROBLEM IV	
I_{11}	$0·18 \times 10^{-5}$	30/128	2·11
I_{12}	$0·89 \times 10^{-9}$	30/159	2·24
I_{21}	$0·24 \times 10^{-14}$	30/171	2·19
I_{22}	$0·10 \times 10^{-13}$	30/159	1·99

[a] This is also the number of gradient evaluations.
[b] I_{ij} refers to the i,j entry of Table 1, for example, update I_{32} is
$$H_{n+1} = H_n - \frac{\eta_n \delta_n^T + \delta_n \eta_n^T}{\delta_n^T \gamma_n} + \frac{\gamma_n^T \eta_n \delta_n \delta_n^T}{(\delta_n^T \gamma_n)^2}.$$

TABLE 4
Function evaluations/run time to obtain $|f| \leq 10^{-5}$

Method	Problem I	Problem II	Problem III	Problem IV
II[a]	48/1·86[d]	13/0·75	53/2·6	28/1·36
V[a]	46/0·81	12/0·55	100/4·51	30/1·37
I_{12}[b]	100/1·04	24/0·43	Fail[c]	85/1·2
I_{21}[b]	89/0·88	23/0·4	Fail[c]	79/1·01
I_{22}[b]	113/1·32	23/0·4	Fail[c]	73/0·94

[a] Methods from Table 2.
[b] Methods from Table 3.
[c] See Table 3.
[d] In seconds.

Acknowledgements

This work was supported by NSF Grant GJ-27528. I wish to thank J. J. Moré for useful suggestions and R. Schumann and A. Chien for programming help. I would also like to thank M. J. D. Powell for making a listing of VD02A available.

References

Brown, K. M., and Dennis, J. E. (1970a). 'New Computational Algorithms for Minimizing a Sum of Squares of Nonlinear Functions'. Yale University technical report (submitted for publication).
Brown, K. M., and Dennis, J. E. (1970b). 'Derivative Free Analogues of the Levenberg-Marquardt and Gauss Algorithms for Nonlinear Least Squares Approximation'. IBM Phila. Scientific Center technical report (in *Numer. Math.* (1972). **18**, 289.
Broyden, C. G. (1965). A class of methods for solving nonlinear simultaneous equations. *Math. Comput.* **19**, 577–593.
Broyden, C. G. (1967). Quasi-Newton methods and their application to function minimization. *Math. Comput.* **21**, 368–381.
Broyden, C. G. (1969). A new double-rank minimization algorithm. *AMS Notices* **16**, 670.
Broyden, C. G. (1970). The convergence of a class of double-rank minimization algorithms, Parts I and II. *J. Inst. Maths Applics.* **6**, 66–90; 222–231.
Collatz, L. (1966). 'Functional Analysis and Numerical Mathematics'. Academic Press, New York.
Davidon, W. C. (1959). 'Variable Metric Methods for Minimization'. Argonne Nat. Labs. report ANL-5990 Rev.
Dennis, J. E. (1969a). On the Kantorovich hypothesis for Newton's method. *SIAM J. Numer. Anal.* **6**, 493–507.

Dennis, J. E. (1969b). 'On the Convergence of Broyden's Method for Nonlinear Systems òf Equations'. Cornell Comp. Sci. report 69–48 (also in *Math. Comput.* (1971). **25**, 559).

Dennis, J. E. (1971). Toward a unified convergence theory for Newton-like methods. *In* 'Nonlinear Functional Analysis and Applications' (L. B. Rall, ed.). Academic Press, New York.

Fletcher, R. (1970). A new approach to variable metric algorithms. *Comput. J.* **13**, 317–322.

Goldfarb, D. (1970). A family of variable-metric methods derived by variational means. *Math. Comput.* **24**, 23–26.

Greenstadt, J. (1970). Variations on variable-metric methods. *Math. Comput.* **24**, 1–18.

Hofmann, W. (1970). 'Regula-falsi Verfahren in Banachraumen'. Dissertation, Hamburg.

Knopp, K. (1947). 'Theory of Functions, Part II' (translated by F. Bagemihl). Dover Publications, New York.

Pearson, J. D. (1969). Variable metric methods of minimization. *Comput. J.* **12**, 171–178.

Powell, M. J. D. (1970a). A hybrid method for nonlinear equations. *In* 'Numerical Methods for Nonlinear Algebraic Equations' (P. Rabinowitz, ed.). Gordon and Breach, London.

Powell, M. J. D. (1970b). A Fortran subroutine for solving systems of nonlinear algebraic equations. *In*: 'Numerical Methods for Nonlinear Algebraic Equations' (P. Rabinowitz, ed.). Gordon and Breach, London.

Powell, M. J. D. (1970c). A new algorithm for unconstrained optimization. *In* 'Nonlinear Programming' (J. B. Rosen, O. L. Mangasarian, and K. Ritter, eds.). Academic Press, New York.

Powell, M. J. D. (1970d). 'A Fortran Subroutine for Unconstrained Minimization, Requiring First Derivatives of the Objective Functions'. A.E.R.E. Harwell report R64-69.

Powell, M. J. D. (1971). On the convergence of the variable metric algorithm. *J. Inst. Maths Applics* **7**, 21–36.

Ralston, A. (1965). 'A First Course in Numerical Analysis'. McGraw-Hill, NewYork.

Shanno, D. F. (1970). Conditioning of quasi-Newton methods for function minimization. *Math. Comput.* **24**, 647–656.

3. A Class of Rank–1 Optimization Algorithms

C. G. BROYDEN AND M. P. JOHNSON
*Computing Centre, University of Essex,
Colchester, Essex, England and Department of Mathematics,
Mid-Essex Technical College, Chelmsford, Essex, England*

We consider some of the convergence properties of a class of single-rank formulae of quasi-Newton type when applied to the problem of minimizing the positive definite quadratic function

$$\phi(x) \equiv \tfrac{1}{2} x^T A x - b^T x + c. \tag{1}$$

Let

$$g(x) \equiv Ax - b, \tag{2}$$

and let the subscript i refer to quantities appropriate to the ith iteration. Then if x_i is the ith approximation to the solution and H_i is the ith iteration matrix the algorithm is substantially defined by

$$x_{i+1} = x_i + s_i \tag{3}$$

where

$$s_i = -H_i^T g_i \alpha_i \tag{4}$$

and α_i is chosen so that

$$g_{i+1}^T s_i = 0, \tag{5}$$

i.e. exact line search is assumed, and

$$H_{i+1} = H_i - (H_i y_i - s_i) \left[\frac{\beta_i y_i^T H_i}{y_i^T H_i y_i} + \frac{(1-\beta_i) s_i^T}{s_i^T y_i} \right] \tag{6}$$

where

$$y_i = g_{i+1} - g_i \tag{7}$$

and β_i is an arbitrary parameter. Note that choosing $\beta_i = 0$ or $\beta_i = 1$ gives two algorithms given by Pearson (1969). Note also that since the class of algorithms defined by equation (6) is a subclass of the class defined by Huang (1970) the algorithm has the property of quadratic termination for all β_i despite the fact that H_i is in general not symmetric. In order to show that equation (6) defines a sub-class of the Huang class we note that the latter is given by

$$H_{i+1} = H_i + s_i \rho_i u_i^T - H_i y_i v_i^T$$

where
$$u_i = s_i \gamma_i + H_i^T y_i \delta_i$$
and
$$v_i = s_i \lambda_i + H_i^T y_i \mu_i.$$

The scalar ρ_i is chosen arbitrarily and the scalars γ_i, δ_i, λ_i and μ_i are also arbitrary but subject to the requirement that $u_i^T y_i = v_i^T y_i = 1$. Putting

$$\rho_i = 1,$$
$$\gamma_i = \lambda_i = (1 - \beta_i)/s_i^T y_i,$$
$$\delta_i = \mu_i = \beta_i / y_i^T H_i y_i$$

gives equation (6), establishing our result.

To try to determine a good value of β we let

$$A = B^2 \tag{8}$$

(possible since by hypothesis A is positive definite) and define z_i and K_i by

$$z_i = B s_i \tag{9}$$

and

$$K_i = B H_i B. \tag{10}$$

Since, from equations (2), (3) and (7),

$$y_i = A s_i, \tag{11}$$

it follows from equations (6) and (10) that

$$K_{i+1} = K_i - (K_i - I) z_i \left[\frac{\beta_i z_i^T K_i}{z_i^T K_i z_i} + (1 - \beta_i) \frac{z_i^T}{z_i^T z_i} \right]. \tag{12}$$

Now, the philosophy underlying the quasi-Newton methods requires H to resemble the local inverse Jacobian as closely as possible. For the quadratic function being considered this resemblance is exact if $H_i = A^{-1}$, and this is equivalent, by equations (8) and (10), to $K_i = I$. We may thus define an error E_i by

$$E_i = K_i - I \tag{13}$$

and regard the norm of this error as a quantity that it would be desirable to reduce. Equations (12) and (13) then give

$$E_{i+1} = E_i(I - z_i q_i^T) \tag{14a}$$

where

$$q_i = \frac{K_i^T z_i \beta_i}{z_i^T K_i z_i} + \frac{z_i(1 - \beta_i)}{z_i^T z_i}. \tag{14b}$$

Now if $\|\cdot\|$ denotes the spectral norm, it follows from a lemma due to Broyden (1970a) that, since $q_i^T z_i = 1$,

$$\|E_{i+1}\| \leq \|q_i\| \|z_i\| \|E_i\| \tag{15}$$

where

$$\|q_i\| \|z_i\| \geq 1. \tag{16}$$

Since inequality (16) becomes an equality if and only if q_i is some multiple of z_i (Cauchy's Inequality) it would appear that a good choice of β_i would be zero, since then the inequality $\|E_{i+1}\| \leq \|E_i\|$ can be guaranteed. Thus one would expect the update

$$H_{i+1} = H_i - (H_i y_i - s_i) \frac{s_i^T}{s_i^T y_i} \tag{17}$$

to be at least near-optimal, and to possess very good stability properties in the neighbourhood of the solution.

We now consider one or two properties of this update. We first record an observation of Dennis (1971) that if the update is symmetrized by using the technique used by Powell (1970), the symmetric update due to Broyden (1970b), Fletcher (1970), Goldfarb (1970) and Shanno (1970) is obtained. This is of interest, as not only does it relate two theoretically good algorithms, but it suggests the possibility that every symmetric rank-2 algorithm may be obtained by a similar symmetrization. Our next and final result concerns the behaviour of update (17) when $\alpha_i = 1$, i.e. when no linear search is carried out. We show that if the initial error norm $\|E_0\|$ is less than unity then the vector error norms $\|x_i - A^{-1}b\|$ are bounded above by a superlinearly convergent sequence.

Theorem

Let update (17) be applied in the manner described above to the quadratic function defined by equation (1), except that $\alpha_i = 1$ in equation (4) for all i. Then, if $\|E_0\| < 1$ and e_r is defined by $e_r = x_r - A^{-1}b$,

$$\|e_r\| \leq \|A^{1/2}\| \|A^{-1/2}\| (k/r^{1/2})^r \|e_0\|$$

where

$$k = \frac{(1 + \|E_0\|)}{1 - \|E_0\|} \|E_0\|_{\text{Euc}}.$$

Proof

This follows closely the proof of a similar theorem relating to Broyden's update (Broyden, 1970a) and is omitted.

We thus see that update (17) has some attractive theoretical properties. We suggest that it is a method of 'bounded deterioration' (Dennis, 1971) and that a local convergence proof for a more general function than quadratic might be possible. The method was not, however, outstanding in Pearson's (1969) tests, and we attribute this to lack of stability when far from the solution.

References

Broyden, C. G. (1970a). The convergence of single-rank quasi-Newton methods. *Math. Comput.* **24**, 365–382.

Broyden, C. G. (1970b). The convergence of a class of double-rank minimization algorithms. *J. Inst. Maths Applics* **6**, 222–231.

Dennis, J. E. (1971). 'On Some Methods Based on Broyden's Secant Approximation to the Hessian'. Conference on Numerical Methods for Nonlinear Optimization, Dundee, June/July 1971.

Fletcher, R. (1970). A new approach to variable metric algorithms. *Comput. J.* **13**, 317–322.

Goldfarb, D. (1970). A family of variable-metric methods derived by variational means. *Math. Comput.* **24**, 23–26.

Huang, H. (1970). A unified approach to quadratically convergent algorithms for function minimisation. *J. Optim. Theory Applns* **5**, 405–423.

Pearson, J. D. (1969). On variable metric methods of minimization. *Comput. J.* **12**, 171–178.

Powell, M. J. D. (1970). 'A new algorithm for unconstrained optimization'. *In*: 'Nonlinear Programming' (J. B. Rosen, O. L. Mangasarian, and K. Ritter, eds.). Academic Press, New York.

Shanno, D. F. (1970). Conditioning of quasi-Newton methods for function minimization. *Math. Comput.* **24**, 647–656.

4. A Derivation of Conjugate Gradients

E. M. L. BEALE

*Scientific Control Systems Ltd., Sanderson House,
London, England*

Summary

This paper consists of an elementary account of the method of conjugate gradients, due to Hestenes and Stiefel (1952) and first used for general optimization problems by Fletcher and Reeves (1964).

The approach is similar to that of Section 11 of Hestenes and Stiefel (1952), and of Sections 6 and 7 of Hestenes (1969). But it is hopefully easier to follow as a minimization algorithm derived from first principles, rather than as an adaptation of a method for solving linear equations.

The only new feature is the derivation of formulae that apply when the initial search direction is not the initial gradient direction. These formulae could be used to devise a sophisticated restart procedure when using conjugate gradients to minimize a non-quadratic objective function. But no specific proposals are developed here.

1. Introduction

This paper consists of an elementary account of the method of conjugate gradients, due to Hestenes and Stiefel (1952) and first used for general optimization problems by Fletcher and Reeves (1964). In Section 2 the concept of conjugate directions is reviewed. Then in Section 3 formulae are derived for producing a set of mutually conjugate directions from an arbitrary set of linearly independent directions, given the Hessian matrix of the objective function. These are in effect the standard Gram-Schmidt formulae. Then in Section 4 we see that these formulae can be applied without any explicit evaluation of second derivatives if we observe the way the gradient vector changes. We also see how the formulae simplify to yield the standard formulae for conjugate gradients. The full simplification is seen to depend crucially on using the gradient direction at the first trial solution as the initial search direction. A two-term correction formula instead of the standard one-term correction formula is derived for use with a more general initial search direction.

This may be useful with the periodic restarts required when the objective function is not strictly quadratic.

2. The Concept of Conjugate Directions

Modern methods for unconstrained optimization in n dimensions are designed to produce the answer in a finite number of steps when the objective function is quadratic, this being a necessary (though not sufficient) condition for a method to be efficient when the objective function has slowly varying second derivatives. We therefore concentrate on quadratic functions, and write the objective function $f(\mathbf{x})$ as

$$f(\mathbf{x}) = \mathbf{c}^T \mathbf{x} + \tfrac{1}{2} \mathbf{x}^T \mathbf{D} \mathbf{x} \tag{1}$$

where $\mathbf{x}$ is a column vector, $\mathbf{c}$ is a column vector representing the gradient of f at the origin, and $\mathbf{D}$ is a symmetric square matrix representing the Hessian of the objective function.

A set of search direction $\boldsymbol{\xi}_i$ ($i = 1, 2, \ldots$) are said to be mutually conjugate with respect to $\mathbf{D}$ if

$$\boldsymbol{\xi}_i^T \mathbf{D} \boldsymbol{\xi}_j = 0 \tag{2}$$

whenever $i \neq j$. It is well known that if these directions are non-null and are taken as successive search directions along which f is minimized, then

  (a) After any number r of steps, we have the minimum value of f in the hyperplane through our current trial solution spanned by the search directions used.
  (b) After at most n steps we have the unconstrained optimum.

[These results are intuitively obvious when $\mathbf{D}$ is a unit matrix and the conjugate directions become orthogonal directions. The general result can be proved geometrically by orthogonal projection or algebraically, see for example Fletcher and Reeves (1964)].

3. Constructing Conjugate Directions

If we knew the Hessian matrix $\mathbf{D}$, and knew that it was positive definite, we could construct a set of mutually conjugate directions $\boldsymbol{\xi}_1, \ldots, \boldsymbol{\xi}_n$ from an arbitrary set of linearly independent directions $\boldsymbol{\eta}_1, \ldots, \boldsymbol{\eta}_n$ by a Gram-Schmidt process, as follows. Put

$$\boldsymbol{\xi}_1 = \boldsymbol{\eta}_1.$$

Then for $i = 2, 3, \ldots, n$ successively put

$$\boldsymbol{\xi}_i = \boldsymbol{\eta}_i + \sum_{j=1}^{i-1} c_{ij} \boldsymbol{\xi}_j, \tag{3}$$

4. DERIVATION OF CONJUGATE GRADIENTS

where the coefficients c_{ij} are chosen to make $\xi_i^T \mathbf{D} \xi_k = 0$ for all $k = 1, \ldots, i-1$. This means that the c_{ij} must satisfy the equations

$$\eta_i^T \mathbf{D} \xi_k + \sum_{j=1}^{i-1} c_{ij} \xi_j^T \mathbf{D} \xi_k = 0. \tag{4}$$

In principle (4) defines $i-1$ simultaneous equations for the $i-1$ unknowns c_{ij}. But if we use the fact that $\xi_1, \ldots, \xi_{i-1}$ are mutually conjugate we find that they reduce to the $i-1$ simple equations

$$c_{ij} = -\eta_i^T \mathbf{D} \xi_j / \xi_j^T \mathbf{D} \xi_j \qquad (j = 1, \ldots, i-1). \tag{5}$$

The denominator in (5) cannot vanish if $\mathbf{D}$ is positive definite, since the η_j are assumed to be linearly independent. Hence the ξ_j cannot vanish.

4. Constructing Conjugate Directions from Gradient Vectors

Suppose now that we wish to minimize f without calculating the Hessian matrix $\mathbf{D}$, but that we are prepared to calculate the gradient vector $\mathbf{g}_i$ at each point $\mathbf{x}_i$ reached at the end of each linear search, and also at the initial point $\mathbf{x}_0$.

There is nothing magical about the fact that, if the second derivatives are constant, we can find out about them by observing the way the first derivatives change. But we shall see that the process is gratifyingly convenient. We have the general relationship

$$\mathbf{g} = \mathbf{c} + \mathbf{D}\mathbf{x}, \tag{6}$$

so that, if $\mathbf{x}_i = \mathbf{x}_{i-1} + \lambda_i \xi_i$, we see that

$$\mathbf{g}_i - \mathbf{g}_{i-1} = \lambda_i \mathbf{D} \xi_i. \tag{7}$$

Let us now take an arbitrary first search direction ξ_1, subject only to the condition that $\xi_1^T \mathbf{g}_0 < 0$, i.e. that the objective function decreases as we move in this direction. If we cannot find such a direction then $\mathbf{x}_0$ is optimum.

If we now use (3) to define ξ_i for $i > 1$, with $\eta_i = -\mathbf{g}_{i-1}$, we see, using (7), that the c_{ij} defined by (5) can be computed from

$$c_{ij} = \mathbf{g}_{i-1}^T (\mathbf{g}_j - \mathbf{g}_{j-1}) / \xi_j^T (\mathbf{g}_j - \mathbf{g}_{j-1}), \tag{8}$$

without any explicit evaluation of the Hessian. We can be sure that the denominator of (8) does not vanish, since $\xi_j^T \mathbf{g}_j = 0$ if $\mathbf{x}_j$ minimizes f along the line $\mathbf{x}_{j-1} + \lambda \xi_j$, and $\xi_j^T \mathbf{g}_{j-1} < 0$ if ξ_j is an improving direction from the point $\mathbf{x}_{j-1}$.

But the formulae can be further simplified if we use the fact that after j linear searches along conjugate directions we have optimized f in a hyperplane spanned by these directions. It follows that $\mathbf{g}_j$ must be orthogonal to this hyperplane and hence to $\xi_1, \ldots, \xi_j$; and since these directions were constructed from $g_1, \ldots, g_{j-1}$ we see that $g_i^T g_j = 0$ for $0 < i < j$. Hence $c_{ij} = 0$ for $j = 2, 3, \ldots, i-2$. So (3) reduces to

$$\xi_2 = -g_1 + c_{21} \xi_1, \tag{9}$$

$$\xi_i = -g_{i-1} + c_{i1} \xi_1 + c_{i,i-1} \xi_{i-1} \qquad (i > 2). \tag{10}$$

In the standard conjugate gradient method, the initial direction ξ_1 is not arbitrary. It is taken as $-g_0$, and the coefficient c_{i1} in (10) then vanishes. We therefore obtain the very simple formula

$$\xi_i = -g_{i-1} + c_{i,i-1} \xi_{i-1} \tag{11}$$

for all $i > 1$. But the analysis shows that we should not use (11) unless $\xi_1 = g_0$, and we can reasonably assume that all subsequent search directions are approximately conjugate to each other. There is therefore no mathematical justification for using (11) for more than n iterations. In fact, it is well known that when minimizing an objective function that is not truly quadratic one should restart the process of constructing conjugate directions after about n iterations. The process then consists of several 'major iterations' each containing about n 'minor iterations'. It might then be sensible to use the last search direction of one major iteration as the first search direction of the next major iteration. The analysis shows that if we then use (10) rather than (11), we can save one minor iteration on each major iteration after the first. But the purpose of the paper is to point out that we can use the concepts of conjugate gradients with any initial search direction, at a significant but not disastrous cost in terms of extra computation, rather than to recommend any particular algorithm.

In view of the relationship between the search directions and the gradient directions, there are alternative formulae equivalent to (8) for strictly quadratic objective functions. But (8) is recommended by Sorenson (1969) as the most satisfactory version for practical work.

5. Acknowledgement

I am grateful to Dr. J. W. Daniel for his helpful comments on the first draft of this paper.

References

Fletcher, R., and Reeves, C. M. (1964). Function minimization by conjugate gradients. *Comput. J.* 7, 147–154.

Hestenes, M. R. (1969). Multiplier and gradient methods. *In*: 'Computing Methods in Optimization Problems—2' (L. A. Zadeh, L. W. Neustadt and A. V. Balakrishnan, eds.), pp. 143–164. Academic Press, New York and London.

Hestenes, M. R. and Stiefel, E. (1952). Methods of conjugate gradients for solving linear systems. *J. Res. Natl Bureau Standards* **49**, 409–436.

Sorenson, H. W. (1969). Comparison of some conjugate direction procedures for function minimization. *J. Franklin Inst.* **288**, 421–441.

5. Numerical Experience with Algorithms for Unconstrained Minimization

R. W. H. SARGENT AND D. J. SEBASTIAN

*Department of Chemical Engineering,
Imperial College, London, England*

Summary

The paper surveys briefly the development of algorithms for unconstrained minimization which use function and gradient evaluations. Theoretical guidance is usually incomplete and practical algorithms often contain parameters whose values must be fixed from numerical experience. The paper therefore reviews numerical work designed to test the effect of such parameters or to provide insight into the working of the algorithms for several well known and recent algorithms. Consideration is given to such questions as the accuracy of one-variable minimizations, the replacement of one-variable minimization by a stability test and comparison of step-reduction rules, the effect of restart frequency in the Fletcher-Reeves algorithm, the importance of positive-definiteness in quasi-Newton methods, and the effect of accuracy of gradient evaluations, especially when difference approximations are used.

1. Introduction

There are broadly two kinds of users of minimization methods. By far the largest category are the occasional users who rely on the availability of a package program in the computer library. The overriding requirement for such a program is reliability; the user will be little concerned about the number of function evaluations required so long as he obtains an answer, and the *main* criterion of performance of the program is the number of queries to the advisory service to which it gives rise. Of course, this is not to say that efficiency is unimportant, but it is decidedly a secondary consideration. There is a smaller group of users who have significant problems, in the sense that these require large amounts of computing time or that the standard packages frequently fail to produce an answer; such users usually have some knowledge of the basis of the techniques used and are prepared to experiment with different techniques and different values of the adjustable parameters present in most algorithms. Theoretical guidance is usually incomplete and experimentation of this sort must also be guided by past numerical experience. It is with numerical experience of this sort, designed to provide insight into the operation of the algorithms, that this paper will be principally concerned.

The problem we are concerned with is the minimization of a real function $f(\mathbf{x})$ of the n real variables $x^1, x^2, \ldots, x^n$ (written as the n-vector $\mathbf{x}$), based on the evaluation of $f(\mathbf{x})$, and possibly also its gradient vector $\mathbf{g}(\mathbf{x})$, at a sequence of points $\mathbf{x}_0, \mathbf{x}_1, \mathbf{x}_2, \ldots$. Normally no other properties of $f(\mathbf{x})$ are explicitly stated, but most algorithms make implicit assumptions, and useful convergence conditions can be derived if we suppose that $f(\mathbf{x})$ is defined on $U \subset E^n$ and satisfies the conditions:

(i) $f(\mathbf{x})$ is continuous on $\Omega = \{\mathbf{x} | f(\mathbf{x}) \leqslant c\} \in U$ for some c, and Ω is closed and bounded.

(ii) $f(\mathbf{x})$ has continuous second derivatives on $\Omega' = \{\mathbf{x} | f(\mathbf{x}) < c\}$ and there is a finite Λ such that the norm of the Hessian matrix

$$\|\mathbf{H}(\mathbf{x})\| \leqslant \Lambda \text{ for all } \mathbf{x} \in \Omega'.$$

It is fair to say that the real development of a theoretical foundation for unconstrained minimization methods began with Davidon's (1959) report on his variable-metric algorithm. Before then, apart from work on the classical steepest descent and Newton methods, the methods which appeared were largely based on intuitive ideas arising from visualization of the problem as hill climbing in ordinary three-dimensional space. Many of these ideas were, nevertheless, successful and are discussed in other papers of this work, notably by Parkinson and Hutchinson (1971), and by Borowski and Schrack (1971); they are also included in Himmelblau's comparative evaluations (1971). This paper is concerned mainly with the development of ideas for algorithms which use gradients, leaving aside consideration of algorithms using function evaluations only.

2. The DFP Algorithm

Davidon's algorithm introduced a variable metric into the steepest descent formula by means of the equations

$$\mathbf{x}_{k+1} - \mathbf{x}_k = \mathbf{p}_{k+1} = -\alpha_k \mathbf{S}_k \mathbf{g}_k \tag{1}$$

where $\mathbf{S}_k$ is an $n \times n$ matrix defined by the recursion formula

$$\mathbf{S}_k = \mathbf{S}_{k-1} - \frac{\mathbf{S}_{k-1} \mathbf{q}_k \mathbf{q}_k^T \mathbf{S}_{k-1}}{\mathbf{q}_k^T \mathbf{S}_{k-1} \mathbf{q}_k} + \frac{\mathbf{p}_k \mathbf{p}_k^T}{\mathbf{p}_k^T \mathbf{q}_k} \tag{2}$$

where $\mathbf{q}_k = \mathbf{g}_k - \mathbf{g}_{k-1}$, and α_k is chosen to minimize the one-variable function

$$\phi(\alpha) = f(\mathbf{x}_k - \alpha \mathbf{S}_k \mathbf{g}_k). \tag{3}$$

5. NUMERICAL EXPERIENCE WITH ALGORITHMS

The matrix S_0 can be chosen arbitrarily, but was chosen to be the unit matrix I.

Fletcher and Powell (1963) proved a number of interesting properties of this algorithm and greatly clarified its theoretical basis. When $f(x)$ is a quadric with a positive-definite Hessian matrix H, they showed that the algorithm is one of a class for which $S_n = H^{-1}$, that the successive directions p_k are mutually conjugate with respect to the quadric and hence linearly independent, and that the minimum of the function is attained at the nth step. They also showed that for general functions S_k is positive definite for all k provided that S_0 is positive-definite, and deduced from this that the denominators in (2) are always non-zero and that the algorithm is stable in the sense that a function decrease is obtained at every step.

This last assertion requires qualification. The positive-definiteness of S_k ensures that the search direction is never orthogonal to the gradient and hence the step is non-zero (unless the gradient is zero), but if $\phi(\alpha)$ is not unimodal and the procedure used for minimizing it finds only a local minimum this can easily have a higher function value than the starting point. Examples of such functions are those of curving valley type, such as the test problems of Rosenbrock and Wood (cf. Appendix), although in our own tests of the algorithm on these particular problems this situation did not in fact occur.

A further possible difficulty is that although S_k remains positive-definite, one or more of its eigenvalues can become arbitrarily small and in practical computation it is then effectively singular. Thus the denominators in (2) can become arbitrarily small, causing overflow in forming S_k or a very large step. Also $s_k = -S_k g_k$ can become small independently of g_k; it is thus not sufficient to terminate the algorithm when $\|s_k\|$ becomes small, and an independent test on $\|g_k\|$ or the function decrease is necessary. Furthermore, if this situation arises, either a different search direction must be chosen or the algorithm restarted with a new S_0; in either case any sequence of conjugate directions in a quadratic region is broken.

In practice these theoretical difficulties very rarely arise and the algorithm has proved to be extremely effective on a wide variety of problems. However, Broyden (1967) and Bard (1970) both found evidence that S_k does not always remain positive-definite, and attributed this to rounding error. Pearson (1969) and McCormick and Pearson (1969) report some surprising results, showing that regular restarting of the algorithm with $S_0 = I$ can greatly improve the rate of convergence on Wood's function and objective functions containing logarithmic barrier functions, especially in the early stages, and Broyden (1970) points out that this behaviour is probably also attributable to the same difficulty. In discussing these difficulties Powell (1970) shows that $g_k^T S_k g_k$ is monotonically decreasing, so that if g_k is small at any stage, for example by passing close to a stationary saddle-point, $S_k g_k$ is forced to be small thereafter; this is just the situation which occurs with Wood's function.

It is evident that the one-variable minimization is an important factor in the algorithm, for most of the theoretical properties depend on attaining the minimum, and in any case an inefficient method can consume a large number of function evaluations. Davidon (1959) and Fletcher and Powell (1963) both used cubic fitting to values of the function and projected gradients at two points along the line, and Johnson and Myers (1967) propose a combination of the golden section rule and cubic fitting. Murtagh and Sargent (1969, 1970) compared quadratic fitting with cubic fitting and concluded that cubic fitting was more efficient. More recently, Gentry (1970) has proposed a new subdivision technique which he shows to be more efficient than the golden section rule, and Biggs (1970) has proposed a promising new method based on fitting a function of the form

$$\phi(\alpha) = A|\alpha - a|^p + b. \qquad (4)$$

In fitting such a function or a polynomial there is also a choice in the data available. The function value and projected gradient are available at the starting point, and either or both can be used; similarly, either or both can be evaluated and used at subsequent points. Our own experience is that it is an inefficient use of a gradient evaluation to use it simply to compute a projected gradient, and the number of steps along the line is not significantly affected if function evaluations only are used.

Ever since the Davidon-Fletcher-Powell (DFP) algorithm first appeared there have been varied opinions about the need for accuracy in terminating the one-variable minimization. In Table 1 we show our own results on a quadratic function and four well-known test functions; details of these, with their starting points, are given in the Appendix. The method used was to fit a quadratic to the first three points and then a cubic, using function values only. The line-search termination criterion was $|\alpha_i - \alpha_{i-1}| < \epsilon |\alpha_{i-1}|$ unless a specified number of iterations was attained, in which case iterations were continued only until an overall function decrease was obtained. The final termination criterion was stronger than is usually imposed, for it was required that both $\|g_k\| \leq 10^{-4}$ and $\|s_k\| \leq 10^{-4}$; this of course implies a condition on the first-order estimate of the function decrease: $|g_k^T s_k| \leq 10^{-8}$.

It should be noted that in all cases the number of function evaluations includes an extra $2n$ evaluations used to check that a minimum had been attained; the check consisted of confirming that the function increased for a positive and negative perturbation along each coordinate direction. The results in this and other tables given in the paper were obtained on a CDC 6400 using double-precision arithmetic.

The poor behaviour on Wood's function has already been discussed. Otherwise, performance uniformly improves as the terminal accuracy is reduced, and even for $\epsilon = 0.1$ on the linear search the overall minimum of the

5. NUMERICAL EXPERIENCE WITH ALGORITHMS

TABLE 1

DFP formula. Effect of terminal accuracy (ϵ) and specified number of iterations in one-variable minimization (cubic without gradients)

Terminal accuracy (ϵ)	Specified number of iterations	Quadratic function			Rosenbrock's valley			Powell's quartic			Wood's function			Helical valley		
		a	b	c	a	b	c	a	b	c	a	b	c	a	b	c
10^{-1}	10	4	21	41	19	96	136	21	105	193	76	410	718	22	98	167
	15	4	21	41	19	101	141	21	105	193	88	533	889	22	103	172
	20	4	21	41	19	106	146	21	105	193	88	538	894	22	108	177
10^{-2}	10	4	21	41	21	118	162	20	118	202	72	482	774	21	116	182
	15	4	21	41	20	156	198	20	145	229	69	574	854	21	126	192
	20	4	21	41	20	153	195	20	155	239	68	590	866	21	136	202
10^{-3}	10	4	21	41	20	134	176	22	124	216	79	545	855	21	132	198
	15	4	21	41	22	162	208	21	128	216	50	483	687	21	142	208
	20	4	21	41	22	172	218	21	133	221	50	498	702	21	152	218
10^{-4}	10	4	21	41	20	140	182	24	184	284	63	473	729	21	153	219
	15	4	21	41	22	168	214	21	171	259	43	416	592	20	174	237
	20	4	21	41	22	178	224	22	190	282	43	502	678	20	199	262
10^{-6}	10	4	21	41	21	182	226	23	218	314	80	721	1045	22	186	255
	15	4	21	41	22	260	306	15	180	244	47	571	763	20	216	279
	20	4	21	41	22	323	369	17	261	333	45	656	840	20	278	341

[a] Number of iterations.
[b] Number of function evaluations.
[c] Sum of number of function and gradient-component evaluations.

quadratic function is attained, in accordance with theory, at the nth iteration. The pattern is not so clear for the specification on the number of iterations along the line, but the lowest number tried (10) seems the most favourable, and in many cases even this limit is not attained. In spite of these results it should be noted that positive-definiteness, or even semi-definiteness, is no longer guaranteed if the minimum is not accurately attained; although on these examples Wood's function gave the only evidence of difficulties from this cause, others—for example Greenstadt (1970) and Kelley et al. (1970)—have also encountered them when accuracy is relaxed. With low accuracy it is even possible to obtain an overall function increase at the apparent minimum; this latter possibility is more likely at a strongly asymmetric minimum, so that objective functions incorporating barrier functions are likely to be particularly susceptible.

3. Conjugate-Gradient Algorithms

The DFP algorithm requires storage of the $n \times n$ matrix, $\mathbf{S}_k$, so for large problems the conjugate-gradient algorithm of Fletcher and Reeves (1964) is very attractive. This requires storage of only the current gradient and search direction, but generates conjugate directions for a quadratic function by the formula

$$\mathbf{s}_{k+1} = -\mathbf{g}_k + \frac{\mathbf{g}_k^T \mathbf{g}_k}{\mathbf{g}_{k-1}^T \mathbf{g}_{k-1}} \cdot \mathbf{s}_k \qquad (5)$$

with $\mathbf{s}_0 = -\mathbf{g}_0$. Quite apart from its simplicity and low storage requirement, this algorithm enjoys most of the theoretical advantages of the DFP formula. The denominator is clearly non-zero if $\mathbf{g}_{k-1}$ is non-zero, and with accurate minimization along the line the new direction is always a descent direction; it thus shares with the DFP algorithm the 'ridge-aligning' property derived from including a component of the previous search direction in each new direction. For a quadratic function the generation of conjugate directions ensures convergence on the nth iteration, but to achieve quadratic terminal convergence on a general function it is necessary to restart the algorithm periodically with the steepest descent direction.

In spite of its similarities to the DFP algorithm, it seems to be generally agreed that the Fletcher-Reeves algorithm cannot compete in practice. This is generally put down to the need for periodic restarts and a greater sensitivity to the accuracy of the line minimization. Results to test these assertions are given in Tables 2 and 3. Comparison of Tables 1 and 2 immediately shows that the DFP algorithm is indeed much more efficient, except on Wood's function, and there is some evidence that the optimum accuracy for the Fletcher-Reeves algorithm occurs near $\epsilon = 0\cdot 01$. Low accuracy again means that a descent

5. NUMERICAL EXPERIENCE WITH ALGORITHMS

TABLE 2

Fletcher-Reeves algorithm. Effect of terminal accuracy (ϵ) and specified number of iterations in one-variable minimization. (Cubic without gradients)

Terminal accuracy (ϵ)	Specified number of iterations	Quadratic function			Rosenbrock's valley			Powell's quartic			Wood's function			Helical valley		
		a	b	c	a	b	c	a	b	c	a	b	c	a	b	c
10^{-1}	10	4	21	41	36	159	233[d]	78	278	594	127	671	1183[d,e]	38	152	269
	15	4	21	41	36	159	233[d]	76	267	575	130	668	1192[d,e]	38	152	269
	20	4	21	41	36	159	233[d]	76	267	575	146	796	1384[d,e]	38	152	269
10^{-2}	10	4	21	41	27	155	211	61	263	511	55	284	508[e]	31	178	274
	15	4	21	41	27	155	211	64	305	565	57	298	530[e]	30	173	266
	20	4	21	41	27	155	211	61	272	520	57	298	530[e]	36	175	268
10^{-3}	10	4	21	41	24	181	230[e]	61	346	594	34	243	383	35	262	370
	15	4	21	41	26	183	237[e]	51	276	484	55	540	764	34	272	377
	20	4	21	41	24	170	220[e]	51	289	497	87	822	1174	34	272	377
10^{-4}	10	4	21	41	25	222	274	71	491	789	28	242	358	71	660	876
	15	4	21	41	30	346	408	49	388	588	37	390	542	37	451	565
	20	4	21	41	30	374	436	51	453	661	37	421	573	33	386	489
10^{-6}	10	4	21	41	23	246	304[d]	59	613	853	57	624	856	62	672	862
	15	4	21	41	26	387	441	63	880	1134	57	866	1098	30	462	555
	20	4	21	41	29	577	637	61	979	1227	40	742	906	33	641	744

[a] Number of iterations.
[b] Number of function evaluations.
[c] Sum of number of function and gradient-component evaluations.
[d] Descent test failure.
[e] Function increase.

TABLE 3

Fletcher-Reeves algorithm. Effect of restart frequency (cubic without gradients; terminal accuracy = 0·01, specified number of iterations = 10)

Restart frequency	Quadratic function			Rosenbrock's valley			Powell's quartic			Wood's function			Helical valley		
	a	b	c	a	b	c	a	b	c	a	b	c	a	b	c
$\tfrac{1}{2}(n+1)$	9	36	76	>200			>200			>200			>200		
$(n+1)$	4	21	41	27	155	211	61	263	511	55	284	508[d]	31	178	274
$2(n+1)$	4	21	41	27	165	221[d]	40	203	367	55	370	594	33	198	300
$3(n+1)$	4	21	41	39	278	358	57	245	477	74	516	816	24	128	203

[a] Number of iterations.
[b] Number of function evaluations.
[c] Sum of number of function and gradient-component evaluations.
[d] Function increase.

direction is no longer guaranteed and that function increases are possible, and the Fletcher-Reeves algorithm does seem to be more sensitive to such failures, although the theoretical performance was again achieved for the quadratic function in all cases. Table 3 shows the effect of restart frequency, and as expected restarting more frequently than every $(n + 1)$ iterations destroys the effectiveness of the algorithm. There is no clear evidence on less frequent restarts, and every $(n + 1)$ iterations seems the best choice overall.

Kelley and Myers (1971) compare three conjugate gradient algorithms, including that of Fletcher and Reeves, with the DFP algorithm, and their results corroborate the above conclusions. Of the two low-storage algorithms the Fletcher-Reeves algorithm seems slightly better and stores one less vector. It is interesting that their comparison of single-precision and double-precision working indicates that the DFP algorithm is more susceptible to lower accuracy, and they conclude that accuracy is important in the matrix update and in the one-dimensional search. This latter conclusion is of course in direct contrast with the results in Tables 1 and 2.

It is also commonly asserted that the Fletcher-Reeves algorithm loses its effectiveness as the number of variables increases, but contrary evidence is beginning to appear. For example, Pollard and Sargent (1970) applied the algorithm to two optimal control problems; the first, with 225 variables, converged in 20 iterations (208 function evaluations) and the second, with 405 variables, also converged in 20 iterations (143 function evaluations). These results are the more surprizing in that variables were frequently set back to their bounds, thus preventing attainment of the minimum in the line search; the active bounds also changed, so there was no question of finding a minimum in a linear subspace. Perhaps there is something special about the structure of optimal control problems, but these results do indicate that further research on the Fletcher-Reeves algorithm could well be worth while.

4. Other Quasi-Newton Formulae

Broyden (1967) was the first to generalize the theoretical results of the DFP algorithm to a class of variable-metric algorithms, and has more recently (Broyden, 1970) given a detailed study of one of them:

$$\mathbf{S}_k = \mathbf{S}_{k-1} + \frac{1}{\mathbf{p}_k^T \mathbf{q}_k} \left[\left(1 + \frac{\mathbf{q}_k^T \mathbf{S}_{k-1} \mathbf{q}_k}{\mathbf{p}_k^T \mathbf{q}_k}\right) \mathbf{p}_k \mathbf{p}_k^T - \mathbf{p}_k \mathbf{q}_k^T \mathbf{S}_{k-1} - \mathbf{S}_{k-1} \mathbf{q}_k \mathbf{p}_k^T \right]. \quad (6)$$

A whole variety of different formulae now exist, but Huang (1970) has shown that most of them can be generated from the formula

$$\mathbf{S}_k = \mathbf{S}_{k-1} - \frac{\mathbf{S}_{k-1} \mathbf{q}_k \mathbf{u}_k^T}{\mathbf{u}_k^T \mathbf{q}_k} + \rho \frac{\mathbf{p}_k \mathbf{v}_k^T}{\mathbf{v}_k^T \mathbf{q}_k} \quad (7)$$

where

$$\mathbf{u}_k = k_1 \mathbf{p}_k + k_2 \mathbf{S}_{k-1} \mathbf{q}_k; \qquad \mathbf{v}_k = c_1 \mathbf{p}_k + c_2 \mathbf{S}_{k-1} \mathbf{q}_k.$$

Here the $\mathbf{S}_k$ can be asymmetric and there are three independent parameters; ρ and the ratios $c_1:c_2$, $k_1:k_2$. Broyden's class, for example, is obtained by setting $\rho = 1$ and restricting the other constants to keep $\mathbf{S}_k$ symmetric, which leaves one free parameter.

Huang has proved a number of properties of the whole class, on the assumption that steps are made according to:

$$\mathbf{p}_{k+1} = -\alpha_k \mathbf{S}_k^T \mathbf{g}_k; \qquad \mathbf{p}_{k+1}^T \mathbf{g}_{k+1} = 0. \tag{8}$$

Different authors have made rival claims for various members of the family, but Dixon (1971) has recently shown that all members with the same value of ρ generate the same sequence of points on any differentiable function, starting with the same $\mathbf{x}_0$ and $\mathbf{S}_0$ and generating steps according to (8). Proper precautions must be taken to identify the same local minimum if more than one occur, and the proof also assumes that no zero denominators occur. Thus, with these provisos, any differences between formulae must be attributable solely to numerical accuracy.

5. Convergence and the Avoidance of One-Variable Minimization

Quite obviously accurate minimization along each search direction can be expensive in function evaluations. The results quoted earlier show that high accuracy is not required in most cases, but that failure to obtain a function decrease can occur if the accuracy is too low. There is also the possibility of finding a local minimum with an increased function value if the function is not unimodal along the search direction, or even of encountering directions along which there is no minimum. There is therefore every incentive to relax the requirement of minimization and replace this with a stability requirement: that the function must decrease at every step.

Now the special properties of the conjugate-gradient and quasi-Newton algorithms are related to their use on quadratic functions, and for such functions a single step resulting from a quadratic or cubic fit attains the minimum along the line. Thus the quadratic terminal convergence will not be impaired if only a single such fit is made at each iteration, and since the special properties do not apply in a non-quadratic region nothing is lost by not continuing to the minimum in this case. It therefore seems worth exploring the strategy of using just enough quadratic or cubic steps to secure a function decrease. Greenstadt (1970) gives some numerical comparisons on the DFP algorithm with accurate minimization and with a 'weak search', consisting of continuing the iterations until a minimum is 'bracketed'. His results do not enable clear conclusions to be drawn on efficiency as the relative performance varies from problem to problem, but the weak search is clearly less reliable in that it frequently fails to terminate in a reasonable number of iterations.

In fact mere stability is not sufficient to ensure convergence, and Murtagh and Sargent (1969, 1970) gave the following sufficient conditions for convergence, for functions satisfying conditions (i) and (ii) of Section 1:

If the sequence $\{x_k\}$ is generated according to

$$x_{k+1} - x_k = p_{k+1} = \alpha_k s_k \tag{9}$$

where $x_0 \in \Omega'$ and s_k satisfies the conditions:

$$\rho \|g_k\| \leq \|s_k\| \leq \sigma \|g_k\| \tag{10}$$

and

$$|g_k^T s_k| \geq \delta \|g_k\| \cdot \|s_k\| \tag{11}$$

with ρ, σ, δ fixed positive constants, then there exists a $\gamma > 0$ such that a sufficiently small α_k can always be chosen, with $|\alpha_k| \geq \gamma$, to satisfy

$$f(x_k) - f(x_{k+1}) \geq \epsilon |g_k^T p_{k+1}| \tag{12}$$

for any ϵ in the range $0 < \epsilon < 1$. With all α_k so chosen, the sequence $\{x_k\}$ remains in Ω' and converges to $\Omega^* = \{x | x \in \Omega'; g(x) = 0\}$. Convergence to a regular minimum in Ω' is weakly linear, and if $H(x)$ is strictly positive definite everywhere on Ω' convergence is at least linear.

Goldstein and Price (1967) give a theorem which shows that if $H(x)$ is Lipschitzian and strictly positive definite, and if $s_k = -S_k g_k$ where

$$\|S_k - H(x_k)^{-1}\| \to 0 \text{ as } k \to \infty,$$

then convergence is superlinear.

These stronger conditions are not readily verifiable for a given problem, but equations (10), (11) and (12) are easily tested and can form the basis of a computational algorithm. With $s_k = -S_k g_k$, condition (10) can always be satisfied by scaling S_k so long as S_k is finite and does not have g_k as a null eigenvector; finiteness is assured if the denominators of the updating formula are non-zero, and the second situation cannot arise if the 'angle test' (11) is satisfied. In fact, scaling of S_k is not really necessary, since α_k fulfils this purpose, but it may be worth incorporating limits to avoid machine overflow or underflow. If a zero denominator occurs, the simplest course of action is not to use the update formula and set $S_k = S_{k-1}$. If the angle test fails, one can always make a step along the steepest descent direction ($s_k = -g_k$), which automatically satisfies conditions (10) and (11). It should be noted however that, using $\delta = 10^{-30}$ and the same tolerance on the magnitude of denominators, neither of these failures has ever occurred in the many tests we have carried out on a great variety of problems and updating formulae; nor has it been necessary to scale the S_k to remain within machine limits.

The sign of α_k in (9) is determined by the sign of $\mathbf{g}_k^T \mathbf{s}_k$ and we have $\alpha_k > 0$ if

$$\mathbf{g}_k^T \mathbf{s}_k < 0. \qquad (13)$$

This 'descent test' is automatically satisfied if $\mathbf{S}_k$ is positive-definite—although this does not also guarantee satisfying the angle test (11)—and it can be argued that if $\mathbf{S}_k$ is not positive-definite, it is an unsatisfactory approximation to the inverse Hessian at the solution; on the other hand, Powell (1969) points out that it may be better to allow $\mathbf{S}_k$ to approximate the current inverse Hessian and impose no restriction of positive-definiteness.

These contrary hypotheses were tested on the DFP formula (2) and Broyden's formula (6), incorporating the conditions (10) and (11) as above and choosing α_k to satisfy (12). A necessary and sufficient condition for positive-definiteness for both formulae is $\mathbf{p}_k^T \mathbf{q}_k > 0$, and when this condition failed the matrix was not updated.

Various methods of choosing α_k to satisfy the stability test (12) were compared. The initial value can be taken as unity, or the rule proposed by Fletcher and Powell (1963) can be adopted. If $f_k > f_{\text{L.B.}}$:

$$\alpha_k = \min(1, 2(f_{\text{L.B.}} - f_k)/\mathbf{g}_k^T \mathbf{s}_k) \qquad (14)$$

where $f_{\text{L.B.}}$ is a user-supplied estimate of a lower bound on the minimum function value. Quadratic or cubic fitting may be used to reduce α_k, but of course the minimum may be attained before (12) is satisfied, and to deal with this two alternatives were tried. In the first case reduction was continued by successive halving

(a) $\alpha_k = \hat{\alpha}_k$ until

$$|\hat{\alpha}_k - \hat{\alpha}_{k-1}| < 0.05|\hat{\alpha}_{k-1}|.$$

(b) Then

$$\alpha_k = 0.5\alpha_{k-1}, \qquad (15)$$

where $\hat{\alpha}_k$ is the prediction from the polynomial fit.

In the second case a significant decrease was ensured at each step by the rule

$$\alpha_k = \min(\hat{\alpha}_k, 0.5\alpha_{k-1}). \qquad (16)$$

A third reduction rule using systematic scaling ($\alpha_k = 0.1\alpha_{k-1}$) was also tested to compare with some results of Fletcher (1970).

The results are given in Tables 4 and 5. The first point to be noted is that in general the stability test is satisfied for the initial step, whichever rule is used for this. Because of this the quadratic and cubic formulae do not give the theoretical convergence for the quadratic function, but it is evidently not generally advantageous to insist on one fit at each iteration as this would almost double the number of function evaluations required. The relative performance varies from problem to problem with respect to the rules for both the initial choice of α_k and its reduction, and no combination emerges as clearly superior.

5. NUMERICAL EXPERIENCE WITH ALGORITHMS 57

TABLE 4
DFP formula with stability test

Initial α	Step reduction	Positive-definiteness test	Quadratic function a b c	Rosenbrock's valley a b c	Powell's quartic a b c	Wood's function a b c	Helical valley a b c
FP formula (14)	Quadratic fit	None	8 19 55	>1000[d]	74 89 389	>1000[d]	57 66 240
		$\mathbf{p}^T\mathbf{q} > 10^{-30}$	8 19 55	>1000[e]	74 89 389	>1000[e]	57 66 240
	Cubic fit (without gradients)	None	8 19 55	>1000	42 53 225	473 521 2417	>1000
		$\mathbf{p}^T\mathbf{q} > 10^{-30}$	8 19 55	>1000	42 53 225	>1000	>1000
	Cubic fit (with gradients)	None	8 19 63	35 47 133	191 213 1033	>1000[d]	80 90 342
		$\mathbf{p}^T\mathbf{q} > 10^{-30}$	8 19 63	35 47 133	191 213 1033	>1000[e]	80 90 342
	Constant factor $\alpha_k = 0.1\alpha_{k-1}$	None	9 19 59	>1000[d]	113 126 582	>1000[d]	98 110 407[d]
		$\mathbf{p}^T\mathbf{q} > 10^{-30}$	9 19 59	119 131 371[e]	113 126 582	>1000[e]	87 100 364
$\alpha_0 = 1$	Quadratic fit	None	5 17 41	>1000[d]	98 113 509	>1000[d]	146 160 601[d]
		$\mathbf{p}^T\mathbf{q} > 10^{-30}$	5 17 41	>1000[e]	98 113 509	>1000[e]	311 326 1262[e]
	Cubic fit (without gradients)	None	10 20 64	>1000[d]	92 124 496	>1000[d]	>1000[d]
		$\mathbf{p}^T\mathbf{q} > 10^{-30}$	10 20 64	>1000[e]	92 124 496	>1000[e]	>1000[e]
	Cubic fit (with gradients)	None	5 17 53	53 70 202[d]	159 183 883	100 129 613	504 520 2062[d]
		$\mathbf{p}^T\mathbf{q} > 10^{-30}$	5 17 53	46 64 184[e]	159 183 883	100 129 613	259 273 1074[e]
	Constant factor $\alpha_k = 0.1\alpha_{k-1}$	None	10 20 64	>1000[d]	133 159 695	>1000[d]	>1000[d]
		$\mathbf{p}^T\mathbf{q} > 10^{-30}$	10 20 64	246 264 758[e]	133 159 695	>1000[e]	>1000[e]

[a] Number of iterations.
[b] Number of function evaluations.
[c] Sum of number of function and gradient-component evaluations.
[d] Descent test failure.
[e] Positive-definiteness test failure.

TABLE 5
Broyden formula with stability test

Initial α	Step reduction	Positive-definiteness test	Quadratic function a b c	Rosenbrock's valley a b c	Powell's quartic a b c	Wood's function a b c	Helical valley a b c
FP formula (14)	Quadratic fit	None	8 19 55	44 63 153	41 51 219	>1000	31 46 142
		$\mathbf{p}^T\mathbf{q} > 10^{-30}$	8 19 55	44 63 153	41 51 219	>1000	31 46 142
	Cubic fit (without gradients)	None	9 20 60	35 51 123	38 49 205	76 110 418	26 37 118
		$\mathbf{p}^T\mathbf{q} > 10^{-30}$	9 20 60	35 51 123	38 49 205	76 110 418	26 37 118
	Cubic fit (with gradients)	None	8 19 63	36 48 136	44 55 243	74 98 458	24 33 114
		$\mathbf{p}^T\mathbf{q} > 10^{-30}$	8 19 63	36 48 136	44 55 243	74 98 458	24 33 114
	Constant factor $\alpha_k = 0\cdot1\alpha_{k-1}$	None	10 20 64	59 88 208	31 41 169	108 161 597[d]	24 33 108
		$\mathbf{p}^T\mathbf{q} > 10^{-30}$	10 20 64	59 88 208	31 41 169	126 204 712[e]	24 33 108
$\alpha_0 = 1$	Quadratic fit	None	5 17 41	42 57 143	42 56 228	>1000	27 42 126
		$\mathbf{p}^T\mathbf{q} > 10^{-30}$	5 17 41	42 57 143	42 56 228	>1000	27 42 126
	Cubic fit (without gradients)	None	11 22 70	33 58 126	37 67 219	45 92 276	29 47 137
		$\mathbf{p}^T\mathbf{q} > 10^{-30}$	11 22 70	33 58 126	37 67 219	45 92 276	29 47 137
	Cubic fit (with gradients)	None	5 17 53	33 50 142	33 54 238	31 65 293	26 39 138
		$\mathbf{p}^T\mathbf{q} > 10^{-30}$	5 17 53	33 50 142	33 54 238	31 65 293	26 39 138
	Constant factor $\alpha_k = 0\cdot1\alpha_{k-1}$	None	11 22 70	39 53 133	43 61 237	133 218 754	32 52 151
		$\mathbf{p}^T\mathbf{q} > 10^{-30}$	11 22 70	39 53 133	43 61 237	118 193 669	32 52 151

[a] Number of iterations.
[b] Number of function evaluations.
[c] Sum of number of function and gradient-component evaluations.
[d] Descent test failure.
[e] Positive-definiteness test failure.

5. NUMERICAL EXPERIENCE WITH ALGORITHMS

However, the rules (15) and (16) gave virtually identical results in all cases, so only the results for rule (15) are given in the tables.

As noted above, if positive-definiteness is not ensured it may be necessary to use negative values of α_k to satisfy (12), and such cases of 'descent test failure' are indicated in the tables. It can be seen that in general the additional positive-definiteness test makes little difference to the results. For the DFP formula there is some evidence that its use is beneficial, but this is marginal and the single failure with the Broyden formula gives an equally marginal reverse indication. It is clear however that Broyden's formula is both more efficient and more reliable than the DFP formula.

Fletcher (1970) has also studied the DFP and Broyden formulae with the stability test (12), using an initial $\alpha_k = 1$ and reduction by scaling with $\alpha_k = 0{\cdot}1\alpha_{k-1}$. He proves some theorems about relative magnitudes of eigenvalues of S_k for the two formulae, and infers from these a rule for choosing between the two formulae at each iteration which is intended to maximize the probability of keeping S_k bounded and non-singular. His analysis shows that the Broyden formula should be less susceptible to singularity, and since S_k shows no tendency to increase without bound it is difficult to see why one should ever switch to the DFP formula. The results of Tables 4 and 5 certainly support this view and indicate that systematic use of the Broyden formula is likely to be preferable. It should be noted incidentally that these results cannot be compared with those given by Fletcher, since he used the termination criterion $\|p_k\| < 5 \times 10^{-5}$; this criterion is dangerous, since even if $S_k g_k$ is not small the scaling rule can easily produce a very small α_k, and hence cause premature termination.

More detailed examination of our results confirms that the poor performance of the DFP formula is largely due to near orthogonality of s_k and g_k. This could be overcome by using a larger value of δ in the angle test (11) and resetting $S_k = I$ if successive failures occur. However, this proposal has not been tested, as it seems preferable to use a formula, such as the Broyden formula, which is less susceptible to this type of failure.

Kelley *et al.* (1970) point out the interesting fact that their 'batch-processing' version of the DFP formula generates conjugate directions for a quadric, with an exact H^{-1} after n steps, without minimization along the line. They use the formula:

$$S_k = S_{k-1} - \frac{S_{k-1} q_k q_k^T S_{k-1}}{q_k^T S_{k-1} q_k} \tag{17}$$

and since this becomes null at the nth step it is then replaced by:

$$S_n = \sum_{k=1}^{n} p_k p_k^T / p_k^T q_k. \tag{18}$$

The steps must be generated by equation (1), but α_k can be arbitrarily chosen. Kelley et al.(1970) do not exploit this in any special way, but give two examples comparing performance with the DFP algorithm with steadily reduced terminal accuracy for the one-variable minimization. The results show that the DFP algorithm is better with accurate minimization but at low accuracies the batch-processing version shows a clear superiority. The method requires the storage of the two $n \times n$ matrices, but for cases where this is not an inconvenience the freedom in choosing α_k can be used for ensuring convergence without vitiating the quadratic termination property.

Murtagh and Sargent (1969) put forward three rank-one updating formulae which generate the exact $\mathbf{H}^{-1}$ for a quadric in at most n linearly independent arbitrary steps, and showed that steps according to equation (1) with arbitrary α_k have the requisite linear independence. Two of these formulae require the storage of two $n \times n$ matrices, but the third requires only $\mathbf{S}_k$ and in any case seems to be the most efficient. It is the symmetric rank-one formula, first mentioned by Davidon (1959) in the appendix to his report:

$$\mathbf{S}_k = \mathbf{S}_{k-1} + \mathbf{z}_k \mathbf{z}_k^T / c_k$$

where (19)

$$\mathbf{z}_k = \mathbf{p}_k - \mathbf{S}_{k-1} \mathbf{q}_k, \qquad c_k = \mathbf{z}_k^T \mathbf{q}_k.$$

As with the DFP and Broyden formulae without minimization, we have no guarantee of positive-definiteness of $\mathbf{S}_k$ nor of non-zero denominator c_k.

Again, we can adopt the point of view that failure of the descent test (13) or lack of positive-definiteness indicates an unsatisfactory $\mathbf{S}_k$. Murtagh and Sargent (1970) showed that there is a simple test for positive-definiteness of $\mathbf{S}_k$ which requires that:

$$\mathbf{g}_{k-1}^T \mathbf{z}_k / c_k < 0. \tag{20}$$

They also reported that in case of failure of (20) there were fewer subsequent failures if $\mathbf{S}_{k-1}$ was updated to $\mathbf{S}_k$ by (19) using $c_k = \mathbf{z}_k^T \mathbf{z}_k$, rather than setting $\mathbf{S}_k = \mathbf{S}_{k-1}$. Results for these two alternatives are given in Table 6 where they are compared with the strategy of making a steepest descent step in case of failure of the descent test (13), and also with the basic algorithm, for which no tests were made other than for zero denominator and angle test (11).

No failure of these latter two tests occurred in any of the problems for any of the four strategies, and there is no indication of difficulties with persistent near orthogonality, requiring resetting of $\mathbf{S}_k$, as occurs with the DFP formula. There is little to choose between the alternative treatments of $\mathbf{S}_{k-1}$ in case of failure of the positive-definiteness test (20), but the results confirm the earlier conclusion that setting $c_k = \mathbf{z}_k^T \mathbf{z}_k$ gives a more reliable algorithm and both are better than taking a steepest descent step on failure of the descent test (13). However, the results clearly show that any of these extra precautions cause a deterioration of performance of the simple basic algorithm.

5. NUMERICAL EXPERIENCE WITH ALGORITHMS

TABLE 6a
Rank-one formula with stability test

Initial α	Step reduction	Extra tests	Quadratic function			Rosenbrock's valley			Powell's quartic			Wood's function			Helical valley		
			a	b	c	a	b	c	a	b	c	a	b	c	a	b	c
FP formula (14)	Quadratic fit	Basic	5	15	39	47	64	160	23	33	129		>1000		29	39	129
		Descent: $p = -g$	5	15	39	74	129	279	37	48	200		>1000		34	50	155
		P.D.: $c_k = z_k^T z_k$	5	15	39	56	82	196	24	34	134		>1000		34	50	155
		P.D.: $S_k = S_{k-1}$	5	15	39	70	86	228	21	31	119		>1000		35	47	155
	Cubic fit (without gradients)	Basic	5	15	39	42	68	154	41	53	221	80	121	445	31	43	139
		Descent: $p = -g$	5	15	39	43	79	167	41	55	223	242	858	1830	35	74	182
		P.D.: $c_k = z_k^T z_k$	5	15	39	43	80	168	41	53	221	88	176	532	29	47	137
		P.D.: $S_k = S_{k-1}$	5	15	39	43	63	151	41	53	221		>1000		32	44	143
	Cubic fit (with gradients)	Basic	5	15	43	43	55	157	44	56	248	81	120	568	25	36	126
		Descent: $p = -g$	5	15	43	47	65	187	45	59	263	310	728	3608	27	43	154
		P.D.: $c_k = z_k^T z_k$	5	15	43	43	62	178	51	63	283	84	128	608	25	39	138
		P.D.: $S_k = {}_kS_{-1}$	5	15	43	43	57	163	51	63	283		>1000		29	39	138
	Constant factor $\alpha_k = 0.1\alpha_{k-1}$	Basic	5	15	39	50	68	170	39	49	209	114	161	621	33	45	147
		Descent: $p = -g$	5	15	39	58	87	205	39	49	209		>1000		35	51	159
		P.D.: $c_k = z_k^T z_k$	5	15	39	56	78	192	39	49	209	117	178	650	34	51	156
		P.D.: $S_k = S_{k-1}$	5	15	39	79	93	253	39	49	209		>1000		39	53	173

[a] Number of iterations.
[b] Number of function evaluations.
[c] Sum of number of function and gradient-component evaluations.

TABLE 6b
Rank-one formula with stability test

Initial α	Step reduction	Extra tests	Quadratic function			Rosenbrock's valley			Powell's quartic			Wood's function			Helical valley		
			a	b	c	a	b	c	a	b	c	a	b	c	a	b	c
$\alpha_0 = 1$	Quadratic fit	Basic	5	17	41	62	95	221	38	50	206	>1000			35	54	162
		Descent: $p = -g$	5	17	41	>1000			41	55	223	>1000			37	62	176
		P.D.: $c_k = z_k^T z_k$	5	17	41	76	118	272	41	53	221	>1000			36	53	164
		P.D.: $S_k = S_{k-1}$	5	17	41	78	110	268	41	53	221	>1000			42	58	187
	Cubic fit (without gradients)	Basic	5	15	39	39	69	149	39	69	229	58	109	345	29	53	143
		Descent: $p = -g$	5	15	39	42	84	170	40	72	236	59	178	418	35	80	188
		P.D.: $c_k = z_k^T z_k$	5	15	39	40	90	172	32	62	194	50	112	316	29	60	150
		P.D.: $S_k = S_{k-1}$	5	15	39	45	72	164	45	75	259	76	127	435	36	57	168
	Cubic fit (with gradients)	Basic	5	17	53	33	64	184	35	60	268	34	72	328	29	45	162
		Descent: $p = -g$	5	17	53	20	40	112	38	61	273	37	87	403	31	50	182
		P.D.: $c_k = z_k^T z_k$	5	17	53	37	63	181	36	58	258	36	80	368	30	47	170
		P.D.: $S_k = S_{k-1}$	5	17	53	29	50	142	35	57	253	172	216	1048	33	48	174
	Constant factor $\alpha_k = 0.1\alpha_{k-1}$	Basic	5	15	39	49	73	173	43	60	236	109	152	592	29	47	137
		Descent: $p = -g$	5	15	39	61	112	236	45	64	248	>1000			32	59	158
		P.D.: $c_k = z_k^T z_k$	5	15	39	60	92	214	40	57	221	114	169	629	42	77	206
		P.D.: $S_k = S_{k-1}$	5	15	39	65	86	218	42	59	231	484	519	2459	36	58	169

[a] Number of iterations.
[b] Number of function evaluations.
[c] Sum of number of function and gradient-component evaluations.

5. NUMERICAL EXPERIENCE WITH ALGORITHMS

TABLE 7
Comparison of minimization with stability test

Update formula	Line search	Quadratic function			Rosenbrock's valley			Powell's quartic			Wood's function			Helical valley		
		a	b	c	a	b	c	a	b	c	a	b	c	a	b	c
DFP	Minimization	4	21	41	19	96	136	21	105	193	76	410	718	22	98	167
	Stability	8	19	55		>1000		42	53	225		>1000			>1000	
Broyden	Minimization	4	21	41	22	90	136	23	94	190	52	201	413	21	87	153
	Stability	9	20	60	35	51	123	38	49	205	76	110	418	26	37	118
Rank-one	Minimization	4	21	41	20	95	137	23	87	183	52	229	441	21	89	155
	Stability	5	15	39	42	68	154	41	53	221	80	121	445	31	43	139

[a] Number of iterations.
[b] Number of function evaluations.
[c] Sum of number of function and gradient-component evaluations.

Again the initial step usually satisfies the stability test, but for the rank-one formula the value given by (14) seems a little better than $\alpha_k = 1$. It is to be noted that the theoretical convergence in $(n + 1)$ iterations is obtained for the quadratic function in all cases. The relative performances of the different reduction formulae are again variable and no clear pattern emerges. As with the DFP and Broyden formulae, rules (15) and (16) yield virtually identical results and only those for rule (15) are given. We find also, as expected, that use of projected gradients in addition to function evaluations does in general increase the computational effort (third columns); however, it is interesting that this does not occur systematically, and for the DFP formula the results are quite erratic.

The results for minimization and use of the stability test are compared in Table 7 for the three matrix-update formulae. In all cases the initial α_k was determined by equation (14), and step reduction was by cubic fitting using function values only. For minimization the termination accuracy was $\epsilon = 0\cdot 1$ with a specified limit of ten iterations; for the stability test $\epsilon = 10^{-30}$ and rule (15) was used. The DFP formula with the stability test runs into difficulties in three of the five problems and must be regarded as unsuccessful. For the other two update formulae use of the stability test has the expected result of reducing the total number of steps made (second column), but since the number of iterations increases more gradient evaluations are required and the difference in total computational effort (third column) is quite marginal. Except for the quadratic function, the Broyden formula is marginally, but consistently, superior.

Although one might have hoped for clearer support of the theoretical considerations from the numerical evidence, it nevertheless seems preferable to use the stability test rather than minimization in order to ensure convergence. Both the theoretical and the numerical evidence condemn the DFP formula, but it is difficult to decide the balance between the numerical evidence which slightly favours the Broyden formula, and the theoretical advantage of finite quadratic termination for the simpler rank-one formula.

Of course, although the example problems are chosen to include particular kinds of difficulty, they represent a very small sample and the numerical evidence must therefore be treated with caution.

6. Numerical Accuracy and Difference Approximations for Derivatives

We have seen that accuracy in the one-variable minimization is not necessary for good performance, but accuracy in updating the matrices in the quasi-Newton methods or in generating the directions in the conjugate-gradient methods is a different question.

Kelley and Myers (1971), as previously noted, came to the conclusion that the DFP algorithm is more sensitive to low accuracy than the conjugate gradient methods, and that the matrix operations should be carried out in double precision. However, they also conclude that high accuracy is necessary in the minimization as well, which casts some doubt on the basis of their conclusions.

Certainly one would expect to require high accuracy, since one is essentially obtaining information on second derivatives from differences of first derivatives, and the steps are small near the solution. However it is not easy to see whether any of the matrix updating formulae are numerically unstable, nor to make any judgement between the different formulae in this respect. Indeed, in view of the observed regular convergence of the matrices as the solution is approached one is inclined to conclude that numerical instability is not a serious problem.

When finite-difference approximations to the derivatives are used instead of evaluation from analytical formulae, the truncation error is involved as well as rounding errors, and one would expect numerical stability to become a more critical factor. Stewart (1967) first suggested the use of such approximations in the DFP algorithm, and his basic technique uses forward differences, with a variable perturbation chosen to balance truncation error against rounding error; if the resulting error estimate is unsatisfactory, he recalculates the derivative approximation from the central difference formula. Stewart's own numerical tests and those of a number of others show that his algorithm gives good results, with performance comparable with that obtained on the same problems using analytical derivatives.

Dennis (1971) studied various methods involving difference approximations, including the steepest descent and DFP algorithms. He shows that serious difficulties can arise unless the variable perturbations are chosen with care, and quotes some analytical results which suggest that the perturbation must be reduced as the permitted error determining the position of the minimum is reduced. His results on error bounds for the DFP algorithm lead to an estimate for the variable perturbation similar to that of Stewart. These conclusions are further supported (Brown and Dennis, 1970), by a similar analysis of finite-difference versions of the Levenberg-Marquardt and Gauss algorithms, with numerical results confirming the analysis, which again shows that the variable perturbation must be decreased as the error decreases.

Gill and Murray (1971), on the other hand, point out that Stewart's estimate may not be reliable, as it involves the diagonal elements of the Hessian matrix which may not be well estimated, especially in the early stages. They recommend the simpler rule of using a constant perturbation throughout (affecting the binary digit $2^{-t/2}$ in a floating point mantissa of t binary digits) and say that this has given encouraging results. They also give a rule for changing from the forward difference to the central difference formula based on a lower limit on the step-size to achieve a function decrease.

Our own experience with these techniques shows that performance can be very erratic, not because of accumulated error in the S_k matrices, but due to false indications from the angle and descent tests and the final termination criterion. Frequently, truncation errors in the computed gradient cause the algorithms to fail to terminate even when they have located the minimum quite precisely. Since these tests are only ancillary to the basic algorithms the difficulties can undoubtedly be overcome, and the balance of evidence is that use of difference approximations in a quasi-Newton algorithm is the best approach to use when analytical differences are not available.

References

Bard, Y. (1970). Comparison of gradient methods for the solution of non-linear parameter estimation problems. *SIAM J. Numer. Anal.* **7**(1), 157.

Biggs, M. C. (1970). 'A New Method of Linear Minimization'. N.O.C. Technical Report No. 8, Hatfield Polytechnic.

Brown, K. M., and Dennis, J. E. Jr. (1970). 'Derivative-free Analogues of the Levenberg-Marquardt and Gauss Algorithms for Non-linear Least-squares Approximation'. IBM Philadelphia Sci. Center. Technical Report No. 320-2994.

Broyden, C. G. (1967). Quasi-Newton methods and their application to function minimization. *Math. Comput.* **21**, 368.

Broyden, C. G. (1970). The convergence of a class of double-rank minimization algorithms, Part I. *J. Inst. Maths Applics* **6**(1), 76; Part II, *ibid.* **6**(3), 222.

Davidon, W. C. (1959). 'Variable Metric Method for Minimization'. A.E.C. Research and Development Report, ANL-5990.

Dennis, J. E. Jr. (1971). "Algorithms for Non-linear Problems which Use Discrete Approximations to Derivatives'. Cornell University, Technical Report 71-98.

Dixon, L. C. W. (1971). 'Variable Metric Algorithms: Necessary and Sufficient Conditions for Identical Behaviour on Non-quadratic Functions'. N.O.C. Technical Report No. 26, Hatfield Polytechnic.

Fletcher, R. (1970). A new approach to variable metric algorithms. *Comput. J.* **13**(3), 317.

Fletcher, R., and Powell, M. J. D. (1963). A rapidly convergent descent method for minimization. *Comput. J.* **6**, 163.

Fletcher, R., and Reeves, C. M. (1964). Function minimization by conjugate gradients. *Comput. J.* **7**, 149.

Gentry, J. W. (1970). A new one-dimensional search technique. *Chem. Eng. Sci.* **25**, 425.

Gill, P. E., and Murray, W. (1971). 'Quasi-Newton Methods for Unconstrained Optimization'. NPL Report, Maths 97.

Goldstein, A. A., and Price, J. F. (1967). An effective algorithm for minimization. *Numer. Math.* **10**, 184.

Greenstadt, J. (1970). Variations on variable-metric methods. *Math. Comput.* **24**(109), 1.

Himmelblau, D. M. (1971). A uniform evaluation of unconstrained optimization techniques. This volume, pp. 69–98.

Huang, H. Y. (1970). A unified approach to quadratically convergent algorithms for function minimization. *J. Optim. Theory Applns* **6**(3), 269.

Johnson, I. L., and Myers, G. E. (1967). 'One-dimensional Minimization Using Search by Golden Section and Cubic Fit Methods'. NASA Manned Spacecraft Center, Internal Note No. 67-FM-172.

Kelley, H. J., and Myers, G. E. (1971). Conjugate direction methods for parameter optimization. *Astronautica Acta*, **16**, 45.

Kelley, H. J., Myers, G. E., and Johnson, I. L. (1970). An improved conjugate direction minimization procedure. *AIAA Journal* **8**(11), 2091.

McCormick, G. P., and Pearson, J. D. (1969). Variable metric methods and unconstrained optimization. *In*: 'Optimization' (R. Fletcher, ed.), p. 307. Academic Press, London.

Murtagh, B. A., and Sargent, R. W. H. (1969). A constrained minimization method with quadratic convergence. *In*: 'Optimization' (R. Fletcher, ed.), p. 215. Academic Press, London.

Murtagh, B. A., and Sargent, R. W. H. (1970). Computational experience with quadratically convergent minimization methods. *Comput. J.* **13**(2), 185.

Parkinson, J. M., and Hutchinson, D. (1971). An investigation into the efficiency of variants of the simplex method. This volume, pp. 115–136.

Pearson, J. D. (1969). Variable metric methods of minimization. *Comput J.* **12**, 171.

Pollard, G. P., and Sargent, R. W. H. (1970). Off-line computation of optimum controls for a plate distillation column. *Automatica* **6**, 59; Dynamic optimization of a distillation column. Conference on 'Optimization Techniques in Circuit and Control Applications' (June, 1970). IEE Conference Publication No. 66, p. 24.

Powell, M. J. D. (1964). An efficient method of finding the minimum of a function of several variables without calculating derivatives. *Comput. J.* **7**, 155.

Powell, M. J. D. (1969). 'Rank One Methods for Unconstrained Optimization'. A.E.R.E. Report T.P. 372.

Powell, M. J. D. (1970). 'Recent Advances in Unconstrained Optimization'. A.E.R.E. Report TPE 430.

Schrack, G. F., and Borowski, N. (1971). A comparison of some random searches. This volume, pp. 137–148.

Stewart, G. W. (1967). A modification of Davidon's minimization method to accept difference approximations of derivatives. *SIAM J.* **14**, 72.

APPENDIX: Test Problems

1. *Quadratic function*, 4 variables.

$$f(\mathbf{x}) = x_1(2x_1 - x_3 - 1) + x_2(x_2 - 3) + x_3(2x_3 + x_4 + 1) + x_4(x_4 - 1).$$

Starting point: $x_1 = 20$, $x_2 = 20$, $x_3 = 20$, $x_4 = 20$.

2. *Rosenbrock's valley*, 2 variables.

$$f(\mathbf{x}) = 100(x_1^2 - x_2)^2 + (1 - x_1)^2.$$

Starting point: $x_1 = -1.2$, $x_2 = 1.0$.

3. *Powell's quartic*, 4 variables.

$$f(\mathbf{x}) = (x_1 + 10x_2)^2 + 5(x_3 - x_4)^2 + (x_2 - 2x_3)^4 + 10(x_1 - x_4)^4.$$

Starting point: $x_1 = 3$, $x_2 = -1$, $x_3 = 0$, $x_4 = 1$.

4. *Wood's function*, 4 variables.
$$f(\mathbf{x}) = 100(x_1^2 - x_2)^2 + (1 - x_1)^2 + 90(x_3^2 - x_4)^2 + (1 - x_3)^2 + 10 \cdot 1(x_2 - 1)^2 \\ + 10 \cdot 1(x_4 - 1)^2 + 19 \cdot 8(x_2 - 1)(x_4 - 1).$$

Starting point: $x_1 = -3$, $x_2 = -1$, $x_3 = -3$, $x_4 = -1$.

5. *Helical valley*, 3 variables.
$$f(\mathbf{x}) = 100((x_3 - 10\theta)^2 + (r - 1)^2) + x_3^2,$$
where
$$2\pi\theta = \tan^{-1}(x_2/x_1),$$
$$r = (x_1^2 + x_2^2)^{1/2}.$$

Starting point: $x_1 = -1$, $x_2 = x_3 = 0$.

6. A Uniform Evaluation of Unconstrained Optimization Techniques

D. M. HIMMELBLAU

Department of Chemical Engineering, The University of Texas at Austin, Austin, Texas, U.S.A.

Summary

To compare the performance of 15 different unconstrained non-linear programming algorithms, 15 minimization test problems were executed on a CDC 6600 computer. In each test the same degree of precision was sought in the relative value of the objective function, the vector of variables, and the gradient of the objective function. The criteria used in the evaluation were robustness and relative overall ranking for time of execution. It was concluded that Fletcher's algorithm was superior to the others.

1. Introduction

The evaluation of the performance of unconstrained non-linear programming algorithms on an equitable basis requires a suitable combination of appropriate test problems, programming skill, and reasonable criteria of comparison. Algorithms can be examined from a theoretical viewpoint as well as being tested by experimentation. The former can only be applied to a rather restricted class of problems; hence we will be concerned here with the evaluation of the effectiveness of algorithms by experimentation, i.e., by solving test problems. Algorithms can be tested on problems with both a small and large number of variables, on problems with varying degrees of non-linearity, and on problems evolving from practical applications, such as least squares, solution of sets of non-linear equations, and the like. By examining the effectiveness of an algorithm in treating a variety of problems, one can hope to predict the general effectiveness of an algorithm in solving other problems of a like and also of a different nature.

For a test problem to be useful it is best that the problem have a single extremum or at least a restricted number of extrema. We cannot yet expect at the current stage of development that an algorithm will pick out the global minimum if a problem has more than one minimum, but it should at least reach a local minimum to be considered successful. Nevertheless, if one considers the type of problem in which one local minimum exists at some x, the vector of the variables, and in addition the objective function, $f(x)$, has a global minimum at $-\infty$, if a particular minimization technique proceeds towards the global minimum, should it be deemed a success or a failure?

Any experimental comparison of algorithms depends to a considerable extent on how the algorithms are programmed for the computer. Small details of the programming can exert a considerable influence on the effectiveness of an algorithm. Slight changes in the termination criteria, the unidimensional search technique, test of matrices for singularity, matrix inversion procedures, reset, and the like make a big difference in the performance of an algorithm. Just a simple change in the initial step in the unidimensional search can be shown to exert quite an impact on the search trajectory in minimizing Rosenbrock's function. Many of these factors have been ignored by authors in reporting their test results for an algorithm because the factors were deemed to be unrelated to the basic algorithm, yet their contribution to the workability of an algorithm should never go unrecognized. Many algorithms to be successful also require that heuristic logic be introduced into the algorithm in addition to the bare skeleton of the procedure described in the literature. Such logic is devised from experimental experience with failure and has little to do with the fundamental concept underlying the algorithm, but makes it work.

Before evaluating the relative effectiveness of the various unconstrained algorithms, some remarks are appropriate concerning the criteria to use in evaluating their effectiveness.

2. Criteria for Evaluation

It is generally accepted that the primary criterion in evaluating general purpose algorithms must be whether or not the algorithm can solve most of the problems posed; that is, is the algorithm *robust*? Of course, any algorithm can be defeated by a suitably designed (pathological) problem, and even for other than pathological problems we cannot realistically ask that an algorithm solve *all* possible problems, for it is easy to pose an unconstrained non-linear programming problem that leads to negative arguments, division by zero, discontinuities, and the like. But the robustness of the algorithm is always of primary concern.

A second criterion is that of the desired degree of precision in the solution, that is, in the value of the objective function at the extremum, $f(x^*)$, and of the elements of the vector of variables x^* at the extremum. Usually the degree of precision in the solution depends upon the termination criteria used to end the computation. To provide for uniform criteria for termination, in the results described here the algorithms were adjusted so that the same relative precision in the optimal x-vector, x^*, in $f(x^*)$ and $\nabla f(x^*)$ were the joint bases for stopping the search in each code. Figure 1 indicates why both of the first two criteria must be involved in the termination procedure. If the algorithm terminates solely on the fractional change in $f(x)$ being less than some small

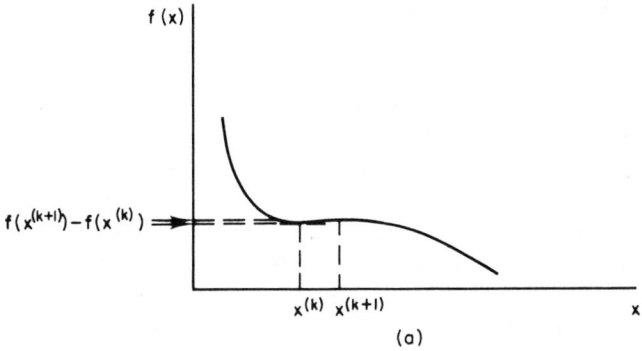

FIG. 1a. A criterion based solely on
$$\frac{f(x^{(k+1)}) - f(x^{(k)})}{f(x^{(k)})} < \epsilon$$
will terminate prematurely on a flat plateau.

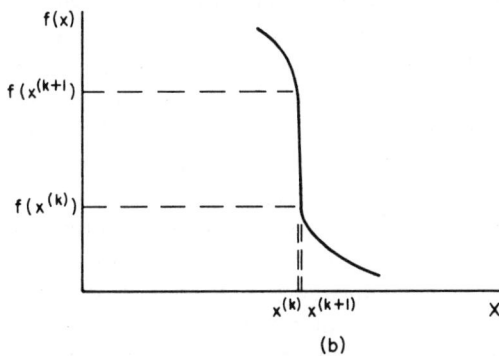

FIG. 1b. A criterion based solely on
$$\frac{x^{(k+1)} - x^{(k)}}{x^{(k)}} < \epsilon$$
will terminate prematurely on a very steep slope.

number, a flat plateau can cause premature termination. Alternatively, if the algorithm is set up to terminate solely on the fractional change in the elements of x, a steep slope can cause premature termination. Use of only the components of the gradient can lead to termination at a saddle point. One feature that should be mentioned concerning the termination criteria is that when $f(x)$ and/or x approach 0, the termination criteria must be the fractional change rather than relative change in $f(x)$ and/or x to avoid dividing by a very small number. Some authors have used as termination criteria the norm of $\nabla f(x)$, the norm of x, or the norm of the search direction d, and although these are perfectly adequate criteria under most circumstances, they do suffer from the same deficiencies identified in Fig. 1.

Assuming that uniform termination criteria are adopted, a third criterion of the effectiveness of an algorithm is the number of functional evaluations of $f(x)$ to reach the desired precision in $f(x)$ and x. Certainly this criterion is better than the number of stages, or iterations, for the number of stages varies quite widely from algorithm to algorithm and in many algorithms means something quite different than the selection of a new search direction. Nevertheless, the number of function evaluations itself is not too satisfactory a measure of efficiency for algorithms with widely differing strategies because the number of functional evaluations to devise a search direction relative to the number of functional evaluations to move in a given direction differ widely from strategy to strategy. Furthermore, how should the evaluation of the derivatives be weighted relative to the evaluation of the objective function itself in those algorithms that use derivatives so that the derivative methods include both objective function and derivative evaluations? Finally, one can reduce the number of function evaluations by all sorts of time-consuming tests, special heuristic operations, matrix operations, and so forth, so that a comparison based solely on function evaluations can easily be misleading.

Consequently, a fourth criterion, the computation time to execute an algorithm, is alternatively cited as a measure of the effectiveness of an algorithm. Although the relative time to termination is not a particularly desirable measure of the effectiveness of an unconstrained non-linear programming algorithm, in lieu of a better measure it often has to serve. Certain hazards exist in using time alone as a criterion. For simple test problems the time required to read the data and execute the print commands (we, of course, omit the printing time itself) in a code with a modestly detailed print out—say x and $f(x)$ for each stage—may prove to be two or three times the computation time experienced when these phases of the code are bypassed. In computers in which the central processing unit operates on several programs in a time-sharing recycling mode, the input–output times when added up may easily exceed by a factor of 2 or 3 the single pass time for execution. Thus, the type of computer, the care in the coding of the algorithm, and the character of the measured time all have an important bearing on the use of time of execution as a criterion. Such information is usually missing from reports in the literature describing the behaviour of a specific algorithm. In the work described below in Section 4 a CDC 6600 computer was used to determine the execution times. All printing, peripheral processing, and system manipulation times were specifically excluded from the times listed in Table 2, so that the times indeed represent the number of seconds required to execute the algorithms without interruption of any sort.

In summary, the criteria to be considered in the evaluation of the unconstrained algorithms are.

1. robustness—success in obtaining an optimal solution (to within a certain precision);
2. number of function (and derivative) evaluations;
3. computer time to termination (to within the desired degree of precision).

3. Test Problems

A number of test problems have been used by various authors of non-linear programming algorithms. Some of the test problems have been used so often that they have assumed the role of 'classics', being repetitively used to compare the performance of algorithms, usually to demonstrate that a new algorithm is as good or better than its predecessors. Table 1 lists 15 functions that were used to evaluate the respective algorithms. The functions contained only a few variables, in contrast to some least squares test functions of a large number of variables that have appeared in the literature. The latter represent a rather special type of function, however.

TABLE 1
Comparison of 15 algorithms for 15 test problems

All functions were minimized; f = failed to reach solution in 15 sec either because progress was slow or termination occurred at incorrect solution. The times shown are the execution times on a Control Data 6600. This computer has a 'standard time' of 22 sec for the standard timing (matrix inversion) program described by A. R. Colville in Techn. Report Nr. 320-2949, IBM Corp., New York Scientific Center, June 1968.

PROBLEM 1 (Zangwill, 1967)
$$f(x) = (1/15)(16x_1^2 + 16x_2^2 - 8x_1 x_2 - 56x_1 - 256x_2 + 991)$$
$$x^{(0)} = (3,8); \quad x^* = (4,9); \quad f(x^*) = -18.2$$

Algorithms	Search directions	Functional evaluations	Derivative evaluations	Time
DFP-D	1	5	4	0·006
-G	1	23	4	0·008
B-D	1	5	4	0·008
-G	1	23	4	0·009
P2-D	1	5	4	0·006
-G	1	23	4	0·008
P3-D	1	5	4	0·008
-G	1	23	4	0·008
PN-D	1	5	4	0·007
-G	1	23	4	0·007
FR-D	1	5	4	0·006
-G	1	23	4	0·010
CP-D	1	5	4	0·007
-G	1	23	4	0·009
IP-D	1	5	4	0·006
-G	1	23	4	0·005
GP-D	1	6	4	0·007
-G	1	24	4	0·007
F	2	3	6	0·004
HJ		80		0·009
NM		185		0·024
P-D	1	29		0·010
-G	1	218		0·024
R		62		0·018
S-D	1	16		0·005
-G	1	84		0·008

TABLE 1—continued

PROBLEM 2 (White and Holst, 1964)
$$f(x) = 100(x_2 - x_1^3)^2 + (1 - x_1)^2$$
$$x^{(0)} = (-1\cdot2, 1\cdot0); \quad x^* = (1,1); \quad f(x^*) = 0$$

Algorithms	Search directions	Functional evaluations	Derivative evaluations	Time
DFP-D	13	219	28	0·030
-G	14	196	30	0·022
B-D	16	230	34	0·031
-G	14	182	30	0·021
P2-D	67	789	136	0·102
-G				f
P3-D	27	297	56	0·047
-G	37	480	76	0·044
PN-D	14	857	30	0·044
-G	20	492	42	0·043
FR-D	35	1304	72	0·152
-G	28	574	58	0·047
CP-D	44	515	90	0·063
-G	32	481	66	0·043
IP-D	35	462	72	0·061
-G	141	2092	284	0·171
GP-D	10	208	58	0·032
-G	9	126	52	0·021
F	37	53	106	0·022
HJ		651		0·056
NM		359		0·035
P-D	14	284		0·038
-G	2	156		0·016
R		294		0·053
S-D	14	256		0·040
-G	14	194		0·026

TABLE 1—continued

PROBLEM 3

$$f(x) = x_1^3 + x_2^2 - 3x_1 - 2x_2 + 2$$

$$x^{(0)} = (0,2); \quad x^* = \begin{cases} (1,1), \\ (-\infty, x_2) \end{cases} \quad f(x^*) = \begin{cases} -1 \\ -\infty \end{cases}$$

Algorithms	Search directions	Functional evaluations	Derivative evaluations	Time
DFP-D	5	16	12	0·015
-G	6	64	14	0·019
B-D	5	16	12	0·012
-G	6	64	14	0·021
P2-D	8	25	18	0·020
-G	5	64	12	0·016
P3-D	6	19	14	0·017
-G	6	19	14	0·016
PN-D				a
-G				a
FR-D				a
-G				a
CP-D				a
-G				a
IP-D				a
-G				a
GP-D				a
-G				a
F	5	7	7	0·015
HJ		640		0·305
NM		190		0·173
P-D	2	24		0·014
-G	5	220		0·050
R	160	163		0·052
S-D				a
-G				a

a = converges toward global minimum, $f(x) \to -\infty$

6. UNCONSTRAINED OPTIMIZATION TECHNIQUES

TABLE 1—*continued*

PROBLEM 4 (Beale, 1958)

$$f(x) = [1\cdot5 - x_1(1 - x_2)]^2 + (2\cdot25 - x_1(1 - x_2^2)]^2 + [2\cdot625 - x_1(1 - x_2^3)]^2$$
$$x^{(0)} = (1\cdot0, 0\cdot8); \quad x^* = (3\cdot0, 0\cdot5); \quad f(x^*) = 0$$

Algorithms	Search directions	Functional evaluations	Derivative evaluations	Time
DFP-D				f
-G	10	158	22	0·022
B-D				f
-G	10	159	22	0·021
P2-D				f
-G	42	534	86	0·069
P3-D				f
-G	17	247	36	0·029
PN-D				f
-G	15	237	32	0·031
FR-D				f
-G	28	258	58	0·033
CP-D				f
-G	27	373	56	0·041
IP-D				f
-G	33	342	68	0·039
GP-D				f
-G	10	149	58	0·024
F	16	22	44	0·013
HJ		205		0·024
NM		230		0·029
P-D	27	134		0·021
-G	21	396		0·039
R		218		0·052
S-D				f
-G	9	161		0·024

TABLE 1—*continued*

PROBLEM 5 (Engvall, 1966)
$$f(x) = x_1^4 + x_2^4 + 2x_1^2 x_2^2 - 4x_1 + 3$$
Normalization of initial search vector used in the unidimensional search in all algorithms except DFP-D and B-D.
$x^{(0)} = (0\cdot5, 2\cdot0):$ $x^* = (1\cdot0, 0);$ $f(x^*) = 0$

Algorithms	Search directions	Functional evaluations	Derivative evaluations	Time
DFP-D	7	82	16	0·016
-G	8	119	18	0·017
B-D	9	120	20	0·017
-G	7	84	16	0·020
P2-D	18	106	38	0·022
-G	20	227	42	0·027
P3-D	9	65	20	0·013
-G	10	119	22	0·017
PN-D	7	57	16	0·012
-G	12	138	26	0·017
FR-D	8	58	18	0·011
-G	18	134	38	0·019
CP-D	11	79	24	0·016
-G	13	142	28	0·020
IP-D	10	71	22	0·015
-G	10	144	22	0·019
GP-D	5	51	28	0·014
-G	8	105	46	0·020
F	10	15	30	0·008
HJ		64		0·008
NM		210		0·025
P-D	6	96		0·017
-G	5	264		0·024
R		119		0·026
S-D	7	119		0·017
-G	7	137		0·016

TABLE 1—*continued*

PROBLEM 6 (Box, 1966)

$$f(x) = \sum_{i=1}^{10} [\exp(-x_1 t_i) - \exp(-x_2 t_i) - \exp(-t_i) + \exp(-10 t_i)]^2$$

$(t_i = 0 \cdot 1, 0 \cdot 2, \ldots, 1 \cdot 0)$

$x^{(0)} = (5, 0);\qquad x^* = (1, 10);\qquad f(x^*) = 0$

Algorithms	Search directions	Functional evaluations	Derivative evaluations	Time
DFP-D	12	142	26	0·277
-G	11	207	24	0·370
B-D	9	112	20	0·222
-G	9	145	20	0·262
P2-D	14	128	30	0·256
-G				f
P3-D				f
-G				f
PN-D	8	114	18	0·217
-G	14	255	30	0·442
FR-D	10	95	22	0·192
-G	33	296	68	0·587
CP-D	167	824	336	1·81
-G	11	198	24	0·351
IP-D	29	212	60	0·434
-G	292	1503	586	3·26
GP-D	6	54	34	0·138
-G	6	96	34	0·212
F	19	25	50	0·076
HJ		498		0·799
NM		268		0·436
P-D	24	161		0·275
-G	18	278		0·464
R		314		0·527
S-D	12	177		0·304
-G	17	406		0·674

TABLE 1—continued

PROBLEM 7 (Zangwill, 1967)
$$f(x) = (x_1 - x_2 + x_3)^2 + (-x_1 + x_2 + x_3)^2 + (x_1 + x_2 - x_3)^2$$
$$x^{(0)} = (100, -1, 2\cdot5); \quad x^* = (0, 0, 0); \quad f(x^*) = 0$$

Algorithms	Search directions	Functional evaluations	Derivative evaluations	Time
DFP-D	2	9	9	0·010
-G	5	99	18	0·017
B-D	3	12	12	0·009
-G	5	99	18	0·018
P2-D	7	27	24	0·012
-G	18	210	57	0·037
P3-D	5	20	18	0·011
-G	6	128	21	0·018
PN-D	3	12	12	0·008
-G	5	175	18	0·032
FR-D	3	12	12	0·014
-G	7	106	24	0·021
CP-D	19	77	60	0·021
-G	20	321	81	0·017
IP-D	28	102	87	0·020
-G	26	356	81	0·004
GP-D	3	14	30	0·010
-G	5	100	54	0·019
F	4	6	18	0·004
HJ		130		0·014
NM		810		0·096
P-D	5	84		0·014
-G	7	502		0·057
R		297		0·067
S-D	5	37		0·010
-G	6	108		0·020

TABLE 1—continued

PROBLEM 8 (Schmidt and Vetters, 1970)

$$f(x) = \frac{1}{1 + (x_1 - x_2)^2} + \sin\left(\frac{\pi x_2 + x_3}{2}\right) + \exp\left[\left(\frac{x_1 + x_2}{x_2} - 2\right)^2\right]$$

$x^{(0)} = (\tfrac{1}{2}, \tfrac{1}{2}, \tfrac{1}{2});$ $x^* = (0.78547, 0.78547, 0.78547);$ $f(x) = 3$

Algorithms	Search directions	Functional evaluations	Derivative evaluations	Time
DFP-D	8	34	27	0.034
-G	9	110	30	0.029
B-D	8	34	27	0.022
-G	10	116	33	0.033
P2-D	45	201	138	0.068
-G	45	540	138	0.125
P3-D	8	32	27	0.019
-G	12	110	39	0.038
PN-D	8	34	27	0.017
-G	16	130	51	0.038
FR-D	16	63	51	0.025
-G	17	148	54	0.040
CP-D	51	205	156	0.064
-G	26	256	84	0.059
IP-D	63	221	192	0.073
-G	90	555	273	0.143
GP-D	6	26	66	0.022
-G	8	90	90	0.038
F	9	12	36	0.015
HJ		177		0.053
NM		279		0.057
P-D	9	86		0.021
-G	11	312		0.060
R		158		0.049
S-D	8	75		0.015
-G	9	127		0.024

TABLE 1—continued

PROBLEM 9 (Engvall, 1966)

$$f(x) = \sum_{i=1}^{5} f_i^2(x), \quad f_1(x) = x_1^2 + x_2^2 + x_3^2 - 1$$

$$f_2(x) = x_1^2 + x_2^2 + (x_3 - 2)^2 - 1$$

$$f_3(x) = x_1 + x_2 + x_3 - 1$$

$$f_4(x) = x_1 + x_2 - x_3 + 1$$

$$f_5(x) = x_1^3 + 3x_2^2 + (5x_3 - x_1 + 1)^2 - 36$$

$$x^{(0)} = (1, 2, 0); \quad x^* = (0, 0, 1); \quad f(x^*) = 0$$

Algorithms	Search directions	Functional evaluations	Derivative evaluations	Time
DFP-D	17	99	54	0·028
-G	18	253	57	0·056
B-D	16	93	51	0·030
-G	18	255	57	0·044
P2-D	6317	41285	18954	7·55
-G				f
P3-D	48	275	147	0·065
-G	37	498	114	0·080
PN-D	25	169	78	0·036
-G	32	473	99	0·073
FR-D	41	283	126	0·054
-G	217	1481	654	0·154
CP-D	1219	6236	3660	1·08
-G	400	3155	1203	0·489
IP-D	2261	11374	6786	1·89
-G				f
GP-D	10	75	114	0·032
-G	12	173	138	0·044
F	22	31	93	0·022
HJ		81		0·012
NM		561		0·087
P-D	11	315		0·052
-G	13	652		0·079
R		457		0·114
S-D	16	150		0·032
-G	17	304		0·048

TABLE 1—continued

PROBLEM 10 (Fletcher and Powell, 1963)
$f(x) = 100\{(x_3 - 10\theta)^2 + [(x_1^2 + x_2^2)^{1/2} - 1]^2\} + x_3^2$,
where $2\pi\theta = \tan^{-1}(x_2/x_1)$, $-\pi/2 < 2\pi\theta < 3\pi/2$.
$x^{(0)} = (-1, 0, 0)$; $x^* = (1, 0, 0)$; $f(x^*) = 0$

Algorithms	Search directions	Functional evaluations	Derivative evaluations	Time
DFP-D	20	141	63	0·047
-G	24	380	75	0·087
B-D	21	140	66	0·047
-G	25	384	78	0·085
P2-D	1097	5795	3294	1·59
-G	4378	20476	13137	6·35
P3-D	96	451	291	0·142
-G	133	1780	402	0·402
PN-D	36	221	111	0·068
-G	42	588	129	0·134
FR-D	36	202	111	0·054
-G	149	838	450	0·222
CP-D	57	318	174	0·080
-G	121	1171	366	0·256
IP-D	1945	6659	5838	1·945
-G	661	5013	1986	1·128
GP-D	10	119	114	0·048
-G	12	200	138	0·061
F	35	42	126	0·036
HJ		1230		0·214
NM		566		0·118
P-D	4	48		0·017
-G	12	277		0·057
R		513		0·146
S-D	21	191		0·048
-G	24	430		0·090

TABLE 1—*continued*
PROBLEM 11 (Bard, 1970)

$$f(x) = \sum_{i=1}^{15} \{y_i - [x_1 + a_{1i}/(x_2 a_{2i} + x_3 a_{3i})]\}^2$$

where y_i and a_{1i}, a_{2i}, and a_{3i} are tabulated data:

y_i	a_{1i}	a_{2i}	a_{3i}	y_i	a_{1i}	a_{2i}	a_{3i}
0·14	1	15	1	0·39	8	8	8
0·18	2	14	2	0·37	9	7	7
0·22	3	13	3	0·58	10	6	6
0·25	4	12	4	0·73	11	5	5
0·29	5	11	5	0·96	12	4	4
0·32	6	10	6	1·34	13	3	3
0·35	7	9	7	2·10	14	2	2
				4·39	15	1	1

$x^{(0)} = (1, 1, 1)$; $x^* = (0·0824, 1·1330, 2·3437)$; $f(x^*) = 8·215 \times 10^{-3}$

Algorithms	Search directions	Functional evaluations	Derivative evaluations	Time
DFP-D	16	78	51	0·041
-G	11	164	36	0·058
B-D	10	55	33	0·029
-G	11	163	36	0·055
P2-D	19	93	60	0·048
-G	19	269	60	0·089
P3-D	140	495	423	0·259
-G	162	1510	489	0·541
PN-D	20	180	63	0·073
-G	23	326	72	0·109
FR-D	20	114	63	0·047
-G	47	382	144	0·171
CP-D	38	209	117	0·078
-G	42	390	129	0·135
IP-D				f
-G				f
GP-D	7	62	78	0·040
-G	7	122	78	0·056
F	38	42	126	0·047
HJ				g
NM		711		0·226
P-D	6	102		0·061
-G	6	174		0·080
R				g
S-D	18	134		0·049
-G	13	198		0·063

g = not executed.

TABLE 1—*continued*

PROBLEM 12 (Powell, 1964)
$$f(x) = (x_1 + 10x_2)^2 + 5(x_3 - x_4)^2 + (x_2 - 2x_3)^4 + 10(x_1 - x_4)^4$$
Note that the Hessian matrix of this function at its minimum is singular
$$x^{(0)} = (3, 1, 0, -1); \quad x^* = (0, 0, 0, 0); \quad f(x^*) = 0$$

Algorithms	Search directions	Functional evaluations	Derivative evaluations	Time
DFP-D	36	434	148	0·079
-G	73	526	296	0·112
B-D	38	374	156	0·071
-G	36	358	148	0·064
P2-D	1655	10317	6624	2·22
-G	1173d	15688	4696	2·27
P3-D	312	1422	1252	0·853
-G	1699	22763	6800	3·43
PN-D	30	693	124	0·084
-G	47	971	182	0·125
FR-D	104	624	420	0·092
-G	912	3774	3652	0·691
CP-D	3226	10432	12908	1·66
-G	402	4158	1612	0·536
IP-D	140	705	564	0·098
-G	7045	72318	28184	9·24
GP-D	22	149	428	0·064
-G	17	219	329	0·065
F	60	68	272	0·062
HJ		77		0·015
NM		1022		0·153
P-D	25	966		0·114
-G	37	1783		0·197
R		801		0·183
S-D	41	622		0·111
-G	112	1117		0·230

d = converged with reset only.

TABLE 1—*continued*

PROBLEM 13 (Cragg and Levy, 1969)
$$f(x) = [\exp(x_1) - x_2]^4 + 100(x_2 - x_3)^6 + \tan^4(x_3 - x_4) + x_1^8 + (x_4 - 1)^2$$
$$x^{(0)} = (1, 2, 2, 2); \quad x^* = (0, 1, 1, 1); \quad f(x^*) = 0$$

Algorithms	Search directions	Functional evaluations	Derivative evaluations	Time
DFP-D	96	424	388	0·180
-G				f
B-D	84	350	340	0·138
-G				f
P2-D	153	625	616	0·245
-G				f
P3-D	501	1753	2008	0·751
-G				f
PN-D	26	165	108	0·060
-G	31	439	128	0·110
FR-D	39	221	160	0·062
-G	72	680	292	0·156
CP-D	148	783	596	0·199
-G	157	1433	632	0·338
IP-D	121	634	488	0·164
-G	85	795	344	0·182
GP-D				f
-G	35	366	688	0·189
F	82	91	364	0·099
HJ		9283		1·52
NM		563		0·147
P-D	36	3480		0·980
-G	45	3103		0·728
R		955		0·585
S-D	128	1662		1·10
-G	339	3749		3·02

TABLE 1—*continued*

PROBLEM 14 (Wood)

$$f(x) = 100(x_2 - x_1^2)^2 + (1 - x_1)^2 + 90(x_4 - x_3^2)^2 + (1 - x_3)^3 + 10 \cdot 1(x_2 - 1)^2 + (x_4 - 1)^2 + 19 \cdot 8(x_2 - 1)(x_4 - 1).$$

The function has local minima that can cause premature termination.

$x^{(0)} = (-3, -1, -3, -1);$ $x^* = (1, 1, 1, 1);$ $f(x^*) = 0$

Algorithms	Search directions	Functional evaluations	Derivative evaluations	Time
DFP-D	57	475	232	0·085
-G	40	721	164	0·113
B-D	42	310	164	0·089
-G	40	664	172	0·092
P2-D				f
-G				f
P3-D	260	805	890	0·314e
-G				1·20e
PN-D				f
-G				f
FR-D				f
-G	189	3288	760	0·330
CP-D	136	704	548	0·188
-G	186	2745	748	0·410
IP-D	231	9202	928	2·42
-G	639	4264	2560	1·59
GP-D	16	367	250	0·079
-G	15	268	364	0·089
F	60	61	244	0·024
HJ		836		0·152
NM		797		0·154
P-D	25	276		0·041
-G	39	850		0·098
R		1043		0·378
S-D	38	715		0·171
-G	54	905		0·244

e = with reset gave best time.

TABLE 1—continued

PROBLEM 15

$$f(x) = \sum_{i=1}^{5} \frac{a_i + x_i^2}{1 + x_i^2}$$

where a_i is a random number in the interval (0·1, 1·0) *on each iteration.*

$x^{(0)} = (1, -1, 1, -1, 1);\qquad x^* = (0, 0, 0, 0, 0);\qquad f(x^*) = 0$

Algorithms	Search directions	Functional evaluations	Derivative evaluations	Time
DFP-D	21	800	110	0·308
-G	17	294	90	0·127
B-D	27	4511	140	1·56
-G	26	361	135	0·155
P2-D	15	207	80	0·092
-G	12	179	65	0·077
P3-D	18	1744	95	0·636
-G	16	162	85	0·076
PN-D				f
-G	27	450	140	0·195
FR-D				f
-G				f
CP-D	20	242	105	0·097
-G	18	228	95	0·086
IP-D	14	216	75	0·084
-G	10	123	55	0·051
GP-D				f
-G				f
F				f
HJ				f
NM				f
P-D				f
-G				f
R				f
S-D				f
-G				f

4. Evaluation of Unconstrained Algorithms

A number of comparisons of the effectiveness of algorithms have appeared in the literature (Bard, 1970; Box, 1965, 1966; Curtin, 1968; Fletcher, 1965; Huang and Levy, 1970; Jones, 1970; Leon, 1966; Murtagh and Sargent, 1970; Pearson, 1969; Pierson and Rajtora, 1970; Wortman, 1969). However, examination of the test results presented so far can only give a fragmentary picture of the relative effectiveness of the unconstrained algorithms because each study used different unidimensional search methods, different termination criteria, and different methods of counting equivalent function evaluations. A more desirable state of affairs would be to conduct an evaluation using a uniform set of standards and test problems.

To this end the following algorithms were tested on the problems listed in Table 1.

Analytical derivatives

(1) DFP: Davidon-Fletcher-Powell, rank 2 (Fletcher and Powell, 1963).
(2) B: Broyden (rank 1) (1965).
(3) P2: Pearson No. 2 (1969) without reset.
(4) P3: Pearson No. 3 (1969) without reset.
(5) PN: Projected Newton (Pearson, 1969).
(6) FR: Fletcher-Reeves (1964), reset each $n + 1$ iterations.
(7) CP: Continued Partan (Shah *et al.*, 1964).
(8) IP: Iterated Partan (Shah *et al.*, 1964).
(9) GP: Goldstein-Price (1967).
(10) F: Fletcher (1970).

Derivative free

(11) HJ: Hooke-Jeeves (1961).
(12) NM: Nelder-Mead (1965).
(13) P: Powell (1964).
(14) R: Rosenbrock (1960).
(15) S: Stewart (DFP with numerical derivatives) (1967).

Details of the logic of these algorithms can be found in the cited references, and also in Himmelblau (1972). The first 10 algorithms use analytical first derivatives, whereas the last five algorithms do not require analytical derivatives.

Two unidimensional searches were employed for each algorithm in which a unidimensional search was required: the Golden Section (G) and the DSC-Powell (designated D) search (Box *et al.*, 1967). Consequently, in effect 26 different algorithms were tested. The termination criteria for the unidimensional searches were all the same and the termination criteria for the algorithms were all the same being as follows:

1. termination of the unidimensional searches: The square of the norm of the vector between the current point to the middle point was than 10^{-6};
2. termination of the algorithm itself:
   (a) Relative change in $f(x) \leqslant 10^{-5}$ [or absolute change in $f(x) \leqslant 10^{-5}$ if $f(x) \to 0$];
   (b) relative change in $x_i \leqslant 10^{-5}$, $i = 1, \ldots, n$ (or absolute change in $x_i \leqslant 10^{-5}$ if $x_i \to 0$);
   (c) absolute change in elements of $\nabla f(x) \leqslant 10^{-4}$.

Of particular interest in the study are answers to the following questions:

1. which are the better and which are the poorer algorithms?;
2. how does the nature of the problem, that is the degree of non-linearity, the number of variables, and so forth affect the performance of an algorithm?;
3. how do the derivative-free algorithms compare with those that use derivatives?

Table 1 lists the times in seconds to execute each algorithm on a CDC 6600 computer together with the corresponding number of function evaluations, derivative evaluations, and iterations for the 15 test problems. Footnotes to the table designate the instances in which an incorrect answer was obtained, excessive time was required, and so forth. The blank spaces for the search algorithms exist because the derivative evaluation and search direction counts sometimes have no meaning comparable to the other algorithms. The term 'derivative evaluation' refers not to the number of gradient calls but to the number of derivatives evaluated so that if a gradient had three elements, one gradient call would yield three derivative evaluations. A derivative count is a better measure than a gradient call count because derivatives can vary widely in complexity. The term 'search directions' is equivalent to number of iterations in most variable metric and conjugate methods, but means little in the many of the derivative free methods, and hence the count has been omitted for some of the algorithms.

Each algorithm can be examined in relation to the character of the minimization problem by scanning the headings of Table 1. Not enough problems with many variables are included in this study to warrant an evaluation of the effect of dimensionality on the algorithms, but the effect of non-linearity can be observed. General problems such as 6 and 13 require more time for all the algorithms than the lower order functions, but nothing startling is observed.

6. UNCONSTRAINED OPTIMIZATION TECHNIQUES

Table 2, which lists the number of failures by algorithm in order of increasing number, shows the order of reliability or robustness with respect to the test problems. Because problem 15 was a stochastic problem, a count excluding problem 15 is also listed. For either case, a large group of algorithms exists of the same degree of robustness.

TABLE 2
Number of failures on 15 problems

No. of failures excluding Problem 15	Algorithm	No. of failures including Problem 15	Algorithm
0	FR-G	0	CP-G
	CP-G	1	DFP-D
	GP-G		DFP-G
	F		B-D
	HJ		B-G
	NM		PN-G
	P-D		FR-G
	P-G		CP-D
	R		GP-G
	S-G		F
1	DFP-D		HJ
	DFP-G		NM
	B-D		P-D
	B-G		P-G
	PN-G		R
	CP-D		S-G
	S-D	2	P2-D
2	P2-D		P3-D
	P3-D		P3-G
	P3-G		IP-D
	PN-D		IP-G
	FR-D		S-D
	IP-D	3	PN-D
	IP-G		FR-D
	GP-D		GP-D
>2	P2-G	>3	P2-G

To provide some insight as to the relative performance of the group of algorithms with less than two failures, the algorithms were ranked by their respective execution times. Table 3 gives the relative performance of the algorithms for the 14 deterministic problems. A value of 1·00 designates the algorithm with the smallest execution time, and the other entries in one column represent the respective times relative to 1·00 in increasing order.

TABLE 3
Ranking by Relative times

PROBLEM 1		PROBLEM 2		PROBLEM 3	
Relative time	Algorithm	Relative time	Algorithm	Relative time	Algorithm
1·00	F	1·00	P-G	1·00	B-D
1·50	DFP-D	1·31	B-G	1·17	P-D
1·75	PN-G	1·31	GP-G	1·25	DFP-D
1·75	CP-D	1·37	DFP-G	1·25	F
1·75	GP-G	1·37	F	1·58	DFP-G
2·00	DFP-G	1·63	S-G	1·75	B-G
2·00	B-D	1·87	DFP-D		
2·00	S-G	1·94	B-D		
2·25	B-G	2·19	NM		
2·25	CP-G	2·50	S-D		
2·25	HJ	2·69	PN-G		
2·50	FR-G	2·69	CP-G		
2·50	P-D	2·88	P-D		
4·50	R	2·94	FR-G		
6·00	NM	3·31	R		
6·00	P-G	3·50	HJ		
PROBLEM 4		PROBLEM 5		PROBLEM 6	
1·00	F	1·00	F	1·00	F
1·62	B-G	1·00	HJ	2·79	GP-G
1·62	P-G	2·00	DFP-D	2·92	B-D
1·69	DFP-G	2·00	CP-D	3·43	B-G
1·85	GP-G	2·00	S-G	3·62	P-D
1·85	HJ	2·13	DFP-G	3·65	DFP-D
1·85	S-G	2·13	B-D	4·63	CP-G
2·23	NM	2·13	PN-G	4·87	DFP-G
2·39	PN-G	2·13	P-D	5·74	NM
2·54	FR-G	2·38	FR-G	5·82	PN-G
3·00	P-G	2·50	B-G	6·10	P-G
3·15	CP-G	2·50	CP-G	6·95	R
4·00	R	2·50	GP-G	7·25	FR-G
		3·00	P-G	8·88	S-G
		3·13	NM	10·05	HJ
		3·25	R	23·8	CP-D

TABLE 3—*continued*

Relative time	Algorithm	Relative time	Algorithm	Relative time	Algorithm
PROBLEM 7		PROBLEM 8		PROBLEM 9	
1·00	F	1·00	F	1·00	HJ
2·25	B-D	1·40	P-D	1·83	F
2·50	DFP-D	1·47	B-D	2·33	DFP-D
3·50	HJ	1·60	S-G	2·50	B-D
3·50	P-D	1·93	DFP-G	3·67	B-G
4·25	DFP-G	2·20	B-G	3·67	GP-G
4·25	CP-G	2·26	DFP-D	4·00	S-G
4·50	B-G	2·53	PN-G	4·33	P-D
4·75	GP-G	2·53	GP-G	4·67	DFP-G
5·00	S-G	2·67	FR-G	6·08	PN-G
5·25	FR-G	3·27	R	6·58	P-G
5·25	CP-D	3·53	HJ	7·25	NM
8·00	PN-G	3·80	NM	9·50	R
14·2	P-G	3·93	CP-G	12·8	FR-G
16·7	R	4·00	P-G	40·7	CP-G
24·0	NM	4·27	CP-D	90·0	CP-D
PROBLEM 10		PROBLEM 11		PROBLEM 12	
1·00	P-D	1·00	B-D	1·00	HJ
2·12	F	1·41	DFP-D	4·13	F
2·76	DFP-D	1·62	F	4·27	B-G
2·76	B-D	1·90	B-G	4·33	GP-G
3·55	P-G	1·93	GP-G	4·73	B-D
3·59	GP-G	2·00	DFP-G	5·27	DFP-D
4·71	CP-D	2·10	P-D	7·47	DFP-G
5·00	B-G	2·17	S-G	7·60	P-D
5·12	DFP-G	2·69	CP-D	8·33	PN-G
5·29	S-G	2·76	P-G	10·2	NM
6·94	NM	3·76	PN-G	12·2	R
7·88	PN-G	4·66	CP-G	13·1	P-G
8·59	R	5·90	FR-G	15·3	S-G
12·6	HJ	7·79	NM	35·7	CP-G
13·1	FR-G			46·0	FRG
15·1	CP-G			110	CP-D

TABLE 3—continued

Relative time	Algorithm	Relative time	Algorithm
PROBLEM 13		PROBLEM 14	
1·00	F	1·00	F
1·11	PN-G	1·71	P-D
1·39	B-D	3·54	DFP-D
1·48	NM	3·71	B-D
1·57	FR-G	3·71	GP-G
1·82	DFP-D	3·83	B-G
1·91	GP-D	4·08	P-G
2·01	CP-D	4·71	DFP-G
3·41	CP-G	6·33	HJ
5·91	R	6·42	NM
7·34	P-G	7·83	CP-D
9·90	P-D	10·7	S-G
15·4	HJ	13·8	FR-G
30·5	S-G	15·8	R
		17·1	CP-G

In order to reduce the mass of data and reach some reasonably comprehensible conclusions, the algorithms were ranked according to their relative execution times, and the rankings averaged. If the algorithm failed, it was assigned a ranking of 14. Table 4 gives the overall ranking for the algorithms, each problem being weighted equally. It could be argued that the more difficult problems (in some sense) should be weighted more heavily than the easy problems, but this was not done. Although the absolute values of the function evaluations or execution times may have little meaning, the relative ranking of the algorithms should be a quite reasonable quantitative mechanism for evaluation. It was difficult to decide how to rank the algorithms that failed to solve a problem and those that failed to solve problem 15, but it was decided to include all the problems but No. 15, and to list the algorithms not solving a problem at the bottom of the ranking.

A surprising result of the ranking was that the algorithms tended to cluster into groups. Consequently, Table 4 lists the algorithms in Table 3 in increasing rank order according to execution times classed by the qualitative terms superior, good, fair, etc., a procedure that seems to be more meaningful for the limited set of test problems used than a continuous classification. Within each classification the order is that of the computed ranking.

Fletcher's algorithm was significantly better than all the others. As expected, the search algorithms were slower than many of the algorithms that used derivatives, but what is of interest is the high ranking of Powell's algorithm. Problem 15 seemed to cause the derivative-free algorithms, and Fletcher's algorithms to fail, the significance of which is not fully understood. The algorithms not included in Table 3 or 4 are not generally recommended because they could terminate prematurely or could be ineffective because of unduly excessive use of computer time. In the derivative methods, slow oscillations in the search indicated that the direction matrix becomes nearly singular. Incorporation of an appropriate restart procedure restoring the direction

TABLE 4
Evaluation of unconstrained algorithms from execution times
(Numbers in parenthesis indicate average rank based on Table 3)

Classification	Algorithm
Superior	Fletcher (1·8)
Very Good	DFP-D (4·7)
	B-D (4·7)
Good	GP-G (5·6)
	B-G (6·1)
	P-D (6·1)
	DFP-G (6·8
Fair	S-G (8·8)
	HJ (9·6)
	PN-G (10·0)
	P-G (10·0)
	NM (10·0)
	FR-G (10·1)
	CP-G (11·2)
	R (11·9)
	CP-D (12·0)

matrix to a positive definite form can improve some of the variable metric methods, but this step will not lift the algorithm into the superior or very good class.

Attempts to compare quite different algorithms on the basis of the number of function and derivative evaluations can be less satisfactory than the use of execution times, particularly if the times are to be determined on the same computer using common subroutines and the problems are solved to the same degree of precision. The main obstacle to the use of the number of function and derivative evaluations is that these two quantities need to somehow be combined into a suitable single measure of 'equivalent evaluations'. It would be possible to evaluate the time to compute each derivative relative to each function, and use the relative times as the weighting factors for the amalgamation, but even this procedure is subject to question. For example, the gradient components are evaluated much more often in the Goldstein-Price algorithm than in the Fletcher-Reeves algorithm leading to some relative distortion between these two algorithms. Consequently, ranking by 'equivalent function' evaluation count has not been carried out in this study but can be done by the interested reader from the data in Table 1. If the analysis were carried out it is fairly clear that the ranking of superior and very good algorithms would remian about the same as in Table 4.

5. Conclusions

Fifteen algorithms (26 when two different unidimensional searches are included) have been ranked according to execution time and robustness in solving fourteen test problems. The Fletcher algorithm was clearly superior to all the others, followed by the Davidon-Fletcher-Powell (rank 2) and Broyden (rank 1) algorithms using the DCS-Powell unidimensional search. Of particular interest to the user from the viewpoint of the program preparation time is the high ranking of the derivative free algorithm by Powell (1964) with the DCS-Powell unidimensional search.

Acknowledgement

The author is indebted to Michael Andenberg for his helpful and illuminating discussions and for his significant contribution in the preparation of the computer codes used in this study.

References

Bard, Y. (1970). Comparison of gradient methods for the solution of nonlinear parametric estimation problems. *SIAM J. Numer. Anal.* **7**, 159–186.

Beale, E. M. L. (1958). 'On an Iterative Method of Finding a Local Minimum of a Function of More than One Variable'. Technical Report No. 25, Statistical Techniques Research Group, Princeton University.

Box, M. J. (1965). A new method of constrained optimization and a comparison with other methods. *Comput. J.* **8**, 42–52.

Box, M. J. (1966). A comparison of several current optimization methods, and the use of transformations in constrained problems. *Comput. J.* **9**, 67–77.

Box, M. J., Davies, D., and Swann, W. H. (1969). 'Nonlinear Optimization Techniques'. ICI Monograph No. 5, Oliver and Boyd, London.

Broyden, C. G. (1965). A class of methods for solving nonlinear equations. *Math. Comput.* **19**, 577–584.

Cragg, E. E., and Levy, A. V. (1969). Study on a supermemory gradient method for the minimization of functions. *J. Optim. Theory Applns* **4**, 191–205.

Curtin, J. F. (1968). 'The Application of Direct Search and Descent Methods of Function Minimization to Parameter Estimation'. Dept. of Supply, Weapons Res. Establishment, Salisbury, Australia.

Engvall, J. L. (1966). 'Numerical Algorithm for Solving Over-determined Systems of Nonlinear Equations'. NASA Document N70-35600.

Fletcher, R. (1965). Function minimization without evaluating derivatives—a review. *Comput. J.* **8**, 33–41.

Fletcher, R. (1970). A new approach to variable metric algorithms. *Comput. J.* **13**, 317–322.

Fletcher, R., and Powell, M. J. D. (1963). A rapidly convergent descent method for minimization. *Comput. J.* **6**, 163–168.

Fletcher, R., and Reeves, C. M. (1964). Function minimization by conjugate gradients. *Comput. J.* **7**, 149–154.

Goldstein, A. A., and Price, J. F. (1967). An effective algorithm for minimization. *Num. Math.* **10**, 184–189.

Himmelblau, D. M. (1972). 'Applied Nonlinear Programming'. McGraw-Hill, New York.

Hooke, R., and Jeeves, T. A. (1961). 'Direct search' solution of numerical and statistical problems. *J. Ass. comput. Mach.* **8**, 212–221.

Huang, H. Y., and Levy, A. V. (1970). Numerical experiments on quadratically convergent algorithms for function minimization. *J. Optim. Theory Applns* **6**, 269–282.

Jones, A. (1970). Spiral—A new algorithm for nonlinear parameter estimation using least squares. *Comput. J.* **13**, 301–308.

Leon, A. (1966). A Comparison among eight known optimizing procedures. *In*: 'Recent Advances in Optimization Techniques' (A. Lavi and T. P. Vogl, eds.), pp. 23–42. John Wiley, New York.

Murtagh, B. A., and Sargent, R. W. H. (1970). Computational experience with quadratically convergent minimization methods. *Comput. J.* **13**, 185–194.

Nelder, J. A., and Mead, R. (1965). A simplex method for function minimization. *Comput. J.* **7**, 308–313.

Pearson, J. D. (1969). Variable metric methods of minimization. *Comput. J.* **12**, 171–178.

Pierson, B. L., and Rajtora, S. G. (1970). Computation experience with the Davidon method applied to optional control problems. *IEEE Trans.* **SSC4**, 240–242.

Powell, M. J. D. (1964). An efficient method for finding the minimum of a function of several variables without calculating derivatives. *Comput. J.* **7**, 155–162.

Rosenbrock, H. H. (1960). An automatic method for finding the greatest or least value of a function. *Comput. J.* **3**, 175–184.

Schmidt, J. W., and Vetters, K. (1970). Ableitungsfreie Verfahren für nichtlineare Optimierungsprobleme. *Numer. Math.* **15**, 263–282.

Shah, B. V., Beuhler, R. J., and Kempthorne, O. (1964). Some algorithms for minimizing a function of several variables. *J. Soc. ind. appl. Math.* **12**, 74–92.

Stewart, G. W. (1967). A modification of Davidon's minimization method to accept difference approximations to derivatives. *J. Ass. comput. Mach.* **14**, 72–83.

White, B. F., and Holst, W. R. (1964). Paper submitted at the 1946 Spring Joint Computer Conf., Washington, D.C.

Wood, C. F., Westinghouse Research Laboratories, Pittsburg, Pa.

Wortman, J. D. (1969). NLPROG, Ballistic Research Laboratories Memorandum Report No. 1958, Aberdeen Proving Ground, Md.

Zangwill, W. I. (1967). Minimizing a function without derivatives. *Comput. J.* **10**, 293–296.

Zwart, P. (1970). Unimodal functions. *J. Optim. Theory Applns* **6**, 155–156.

7. A Consideration of Non-gradient Algorithms for the Unconstrained Optimization of Functions of High Dimensionality

J. M. PARKINSON AND D. HUTCHINSON
Department of Computational Science,
The University of Leeds, Leeds, England

Summary

Consideration is given initially to the efficiency and limitations of three currently favoured minimization algorithms when applied to representative functions of high dimensionality. Numerical results are presented for several sets of quadratic and non-quadratic test functions, and for each set the dimensionality varies over a range 10 to at least 100. Primary computer storage is shown to be a serious limitation on the feasible dimensionality of the objective function, and the effect of introducing secondary storage is analysed. Finally, a new algorithm is presented, based on the Nelder and Mead simplex method, which has a substantially lower storage requirement; indeed, necessary storage is proportional to dimensionality n rather than to n^2 as holds for all three original methods. Results are given for the new algorithm when applied to the same sets of test functions, showing clearly its advantage over the other algorithms when primary computer storage is a limiting factor.

1. Introduction

There is at present little published data on the relative efficiency of algorithms for unconstrained optimization when applied to functions of medium to high dimensionality. Such problems do occur in practice, for example in process control, but are treated in an *ad hoc* fashion, so that the experience gained is unsystematic and difficult to quantify. Box (1966), Kowalik and Osborne (1968), and Spendley (1969) have all published limited comparisons for functions of dimensionality as high as 20, but emphasized the need for further investigation. In this paper results are given for three selected methods when applied to sets of test functions whose dimensionality varies over the range 10 to 150. It is shown that primary computer storage is a severe limitation on the possible dimensionality of the objective function for many current methods, and that the introduction of backing store relieves this limitation at the expense of a drastic increase in total run time. The only algorithm which could easily be modified to substantially reduce the necessary storage proved to be the Nelder and Mead (1965) adaptation of the simplex search technique. Thus, a new algorithm is presented, based on the simplex method and retaining its well known robustness, which reduces the necessary storage though at the expense of efficiency for low dimensionality.

2. Choice of Methods

The methods chosen for the initial investigation were:

(i) the method of Powell (1964) using conjugate directions (hereafter termed P64) which was chosen because it is probably the most widely quoted algorithm, particularly in comparisons with new methods, and also the most commonly used non-gradient method;

(ii) the Stewart-Davidon method (1967) (hereafter termed SD), being an appropriate choice since it is a representative gradient method, even though differences of function values are used to estimate the gradient vectors. Moreover, in his original paper Stewart claims that this method is more efficient than P64, the superiority becoming more marked as the dimensionality is raised;

(iii) the simplex method of Nelder and Mead (1965) (hereafter termed NMS), which our own investigations (Parkinson and Hutchinson, 1971) have shown in a more favourable light, even though in several published comparisons it is suggested that the method becomes less competitive as the dimensionality of the objective function is raised.

3. Choice of Test Functions

Published results have made it clear that all current methods can suffer from severe hold-ups, or even breakdown, when applied to complex test functions, even though the dimensionality is as low as two. Moreover, the likelihood of failure clearly rises with the dimensionality. However, the determination of relative robustness for the three algorithms was not our primary concern so that it seemed best to concentrate on relatively straightforward functions for high dimensions, typical of those which may well be considered in practice.

The functions considered all have their unique minimum at $\mathbf{x} = [0, 0, \ldots, 0]$, though the function value at the minimum of $F1, F2, F3, F4$ is zero in each case, and for $F5$ or $F6$ it is unity. The starting point in each case is defined by $\mathbf{x} = [1, 1, \ldots, 1]$, and the corresponding function values for the test functions $F1$ to $F6$ are $(n^2 + n)/2$, $(n^2 + n + 2)/2$, $(n^2 + 3n)/2$, $(n^2 + n)$, $\cosh 1$, $1 + \cosh 1$, respectively, where n is the dimensionality of the function. The functions, with the corresponding Hessian matrix H at the minimum, are:

$$F1(\mathbf{x}) = \sum_{i=1}^{n} ix_i^2, \qquad H = \text{diag}\,[2i],$$

$$F2(\mathbf{x}) = \sum_{i=1}^{n} ix_i^2 + x_1^4, \qquad H = \text{diag}\,[2i],$$

$$F3(\mathbf{x}) = \sum_{i=1}^{n} ix_i^2 + nx_n^4, \qquad H = \text{diag}\,[2i],$$

$$F4(\mathbf{x}) = \sum_{i=1}^{n} i(x_i^2 + x_i^4), \qquad H = \text{diag}\,[2i],$$

$$F5(\mathbf{x}) = \frac{1}{n} \sum_{i=1}^{n} \cosh x_i, \qquad H = \text{diag}\,[1/n],$$

$$F6(\mathbf{x}) = \frac{1}{n} \sum_{i=1}^{n} (\cosh x_i + x_i^4), \qquad H = \text{diag}\,[1/n].$$

The first function $F1$ is a simple quadratic form, typical of a 'sums of squares' function, which would however be expected to flatter those algorithms which terminate for quadratic functions, subject only to the limitations of the linear search. The next two functions $F2$, $F3$ are non-quadratic, but may be viewed as simple perturbations of $F1$; the Hessian is no longer constant though it is the same diagonal matrix at the minimum of $F1$, $F2$, or $F3$. The function $F4$ differs more substantially from $F1$ and should reduce the relative effectiveness of the algorithms with the property of quadratic convergence. Functions $F5$ and $F6$ are similar non-quadratic functions where the latter may be considered a perturbation of the former; in each case the Hessian is a diagonal matrix at the minimum which tends to the null matrix as $n \to \infty$, thus making numerical optimization more difficult. As n is increased for the first four test functions the contrary applies; more precisely, the Hessian form is unchanged but the starting point becomes less favourable because the corresponding function value increases.

4. Initial Investigation

The programs used in the investigation were, so far as possible, definitive versions of each algorithm and taken from the original papers. Algorithm P64 uses the Box ALGOL version of Powell's original FORTRAN listing; algorithm SD uses the program published as DAPODMIN, Algorithm 46, in the *Computer Journal* (1970), and the simplex program of NMS was written from the flow diagram given in the original Nelder and Mead paper. However, the performance of each algorithm depends in practice on the values of certain preset parameters as well as on the precise coding, e.g. parameters associated with the linear search or initial conditions; where recommended, the original parameters have been retained. In the NMS algorithm the arbitrary initial simplex corresponds to the choice of scale parameters $istep_i = 0.5$, all i, which

TABLE 1

A comparison of the required function evaluations (FE) and corresponding times against dimensionality n, for test function $F1 = \sum_{i=1}^{n} ix_i^2$ where the function is minimized to accuracy of order 10^{-8} by three standard algorithms and the new algorithm LSS.

Dimension-ality	P64 FE	P64 Time (sec)	SD FE	SD Time (sec)	NMS FE	NMS Time (sec)	LSS FE	LSS Time (sec)
10	120	2	115	2	499	6	531	8
20	230	8	430	12	1129	28	1203	34
30	341	18	943	41	1846	69	2068	87
40	449	32	1638	95	2602	129	2957	165
50	561	50	2197	159	3205	199	3879	271
60	673[a]	72[a]	2902	252	4025[a]	300[a]	5185	435
70	780	98	3436[a]	348[a]	4774	416	7109	696
80	890	127	4329	500	5502	547	10063	1127
90	1003	161	5198	676	6151	689	13688	1724
100	1112	198	5897	852	6905	859	17757	2485
110	b	b	6631	1054	b	b	22375	3490
120	b	b	7421	2287	b	b	27357	4595
130	b	b	8203	1541	b	b	32661	5941
140	b	b	9013	1823	b	b	38202	7352
150	b	b	b	b	b	b	43881	9113

[a] The primary store requirement already exceeds 8K, and for all dimensionalities above this.
[b] The primary store limit of 16K of 48 bit words is exceeded.

is by no means the optimum. Moreover, relative performance depends on the initial point and the accuracy required. However, under these constraints each method was employed to minimize each test function with dimensionality varied over the range 10 to 150 in steps of 10. A subset of the results obtained are presented in Tables 1 to 4, inclusive. Terminating procedures were adjusted to give results to the same accuracy in the objective function.

In Table 1 it was further assumed that available storage is restricted to the nominal 16K of 48 bit words; hence the incomplete columns which emphasize the respective storage requirements of the algorithms. Tables 1 and 2 present results for the very simple functions $F1$ and $F4$, whereas Tables 3 and 4 apply

7. NON-GRADIENT ALGORITHMS

to the non-quadratic functions $F5$ and $F6$. A direct comparison of the results for SD and NMS shows that NMS is much less efficient for the lower dimensionalities with all the test functions, but that NMS becomes more competitive as the dimensionality is raised, and indeed is directly comparable for the very simple functions with dimensionality greater than 100. This contradicts published speculation. Direct comparison of SD with P64 shows the former

TABLE 2

A comparison of the required function evaluations against dimensionality n, for test function $F4 = \sum_{i=1}^{n} i(x_i^2 + x_i^4)$, where the function is minimized to accuracy of order 10^{-8} by three standard algorithms

Dimensionality	Required function evaluations		
	P64	SD	NMS
10	223	203	523
20	441	624	1175
30	665	1114	1898
40	890	1813	2683
50	1110	2396	3291
60	1330	3132	4103
70	1550	3743	4863
80	1770	4652	5591
90	1990	5529	6239
100	2210	6513	6994
—	—	—	—
150	3312	10102	11037

to be more efficient for dimensionality below 20, approximately, varying somewhat with the test function: the order is reversed for dimensionality greater than 20, and the clear advantage of P64 rises appreciably with the dimensionality. However, the advantage illustrated in Table 1 is partly spurious for the following reasons: (i) the P64 algorithm performs a sequence of linear searches with a quadratic fit in each of the coordinate directions initially, together with another linear search which completes the first iteration; the solution is so related to the initial point for $F1$ that it may, ideally, be determined in one iteration; and (ii) P64 incorporates two convergence (terminating) criteria, but the results presented were obtained using only the less stringent criterion since invoking the second also caused repeated failures. Thus, the apparently excellent performance of P64 in Table 1 is surprisingly poor when one considers that the process has quadratic convergence, the function is quadratic, and the initial point is particularly favourable—this reflects on the efficiency of the linear search. The results for functions $F2$ and $F3$, which are perturbations of $F1$, are not presented in detail here, but are generally within one or two per cent of those quoted in Table 2 for SD and NMS whereas for P64 they are rather worse. The function $F4$ is a more substantial modification of $F1$, and comparison of the results in Tables 1 and 2 shows that P64 now requires about twice as many function evaluations, SD perhaps 10% more, but NMS only one or two per cent more.

TABLE 3

A comparison of the required function evaluations (FE) and storage requirement against dimensionality n, for test function $F5 = \frac{1}{n}\sum_{i=1}^{n} \cosh x_i$, where the function is minimized to accuracy of order 10^{-8} by three standard algorithms and the new algorithm LSS

Dimen-sionality	P64 FE	P64 Storage (words)	SD FE	SD Storage (words)	NMS FE	NMS Storage (words)	LSS FE	LSS Storage (words)
10	581	4900	175	5233	542	4554	575	5368
20	883	5260	862	5488	2253	4944	2323	5568
30	1162	5820	1483	5843	3872	5534	4007	4768
40	1710	6580	2030	6298	5398	6284	5984	5968
50	2015	7540	2673	6853	6623	7314	7743	6168
60	2230	8700	3201	7508	8325	8504	8567	6368
70	2603	10060	3878	8263	9636	9824	11971	6568
80	2931	11620	4282	9118	11673	11484	16233	6768
90	3181	13380	4611	10073	13260	13274	20797	6968
100	3516	15340	5072	11128	14393	15264	25745	7168

With the assumed limit on primary storage of a nominal 16K of 48 bit words P64 and NMS can only minimize the test functions up to dimensionality just over 100, whereas SD is effective for dimensionality just over 140. The storage limit is appropriate to a KDF9 computer, on which most of the investigation was completed, but the word length on the ICL 1906A, also used in the investigation, is 24 bits so that the corresponding limit would be about 32K words where double length arithmetic is employed. This is equivalent to the primary store of a small/medium computer, as used in process control for instance, where dimensionality much greater than 100 is common.

It is worth noting that the robustness of NMS was confirmed for the higher dimensions since no failure occurred, whereas SD had an isolated failure and a few surprisingly poor results; P64 had the difficulty previously mentioned, failing about 50% of the time with the simple functions when the second criterion was incorporated, but giving no failures when this was dropped.

Finally, execution times are included in Table 1 for completeness; the times are those observed for minimizations performed on the KDF9 rounded to the nearest second, but times approximately a twentieth of those quoted were obtained on the ICL 1906A. The function evaluations and execution times

TABLE 4

A comparison of the required function evaluations (FE) and corresponding additional time against dimensionality n, where the primary store is arbitrarily limited to 8K of 48 bit words but magnetic disc secondary storage is available; the test function is $F6 = \frac{1}{n}\sum_{i=1}^{n}(\cosh x_i + x_i^4)$, and is minimized to accuracy of order 10^{-8} by three standard algorithms and the new algorithm LSS.

Dimen-sionality	P64		SD		NMS		LSS	
	FE	Additional time (sec)	FE	Additional time (sec)	FE	Additional time (sec)	FE	Additional time (sec)
10	682	N.R.	181	N.R.	655	N.R.	687	N.R.
20	1105	N.R.	882	N.R.	1978	N.R.	2048	N.R.
30	1317	N.R.	1603	N.R.	3652	N.R.	3885	N.R.
40	2011	N.R.	2380	N.R.	4438	N.R.	4702	N.R.
50	2272	N.R.	3253	N.R.	6472	N.R.	7034	N.R.
60	2433	N.R.	4061	N.R.	7532	1936	8656	N.R.
70	2809	EX.	4888	444	8742	2813	11077	N.R.
80	3039	EX.	5662	789	10683	4022	15244	N.R.
90	3333	EX.	6328	954	12168	5385	19605	N.R.
100	3691	EX.	7073	1137	13588	6768	24540	N.R.

N.R. = secondary store Not Required as primary store limit of 8K not exceeded.
EX. = primary store limit Exceeded, but this method not readily adaptable to the use of secondary storage.

parallel one another quite closely, but show that the 'overheads' due to the computation of the algorithm is least for NMS and greatest for P64, though the precise comparison depends on the program details.

5. Storage Considerations

5.1. *Relation of Dimensionality to Necessary Primary Storage*

The limitation of 16K of 48 bit words has been clearly demonstrated in the tables, notably in Tables 1 and 3. However, the actual core storage used was recorded for each algorithm when applied to function $F1$ for the full range in dimensionality, and the results are presented in Fig. 1. The limitation of 8K is

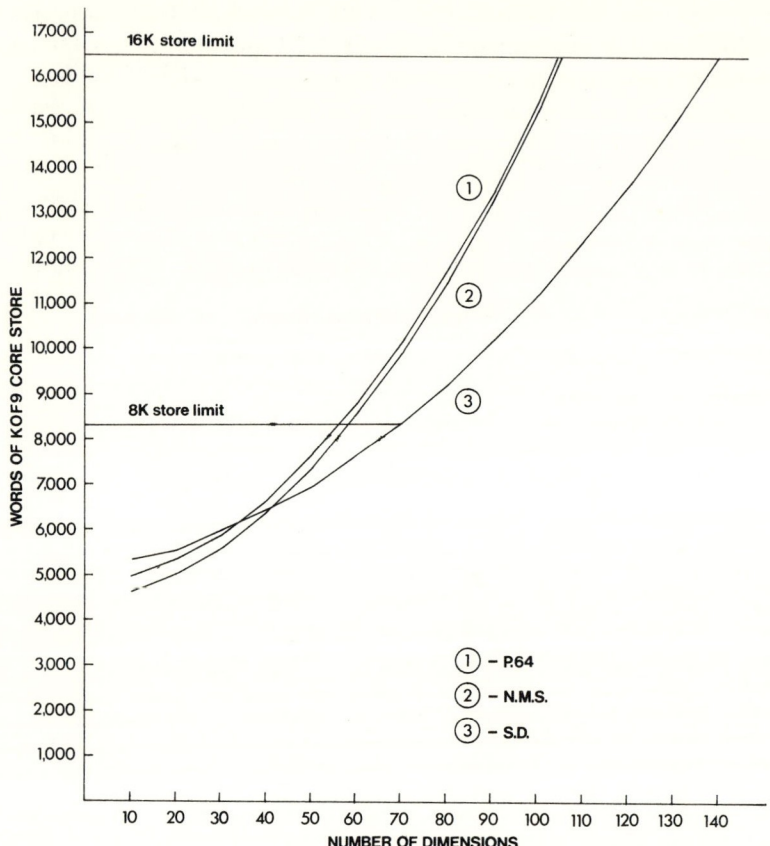

FIG. 1. The relation between required primary storage and the number of dimensions for each of the three standard algorithms—observed values for KDF9 using function $F1$.

also shown in the tables and again in Fig. 1. The algorithms were then considered in detail with a view to reducing their storage requirements, either by more efficient data handling or by modification of the algorithm. In each case it was found that a large array dependent on n^2 (n being the dimensionality of the object function) was involved, related to the n coordinates of n points in n dimensions, and that this array became the dominant part of the required storage as n increased. Moreover, the particular relationships between dimensionality and required storage were derived for the KDF9 for each algorithm:

SD, required storage $= 5078 + (n^2 + 21n)/2$ words
P64, required storage $= 4740 + n^2 + 6n$ words
NMS, required storage $= 4364 + n^2 + 9n$ words

For n greater than 40, approximately, the algorithms are quoted in order of increasing demands on computer store.

Efficient data handling did not offer general prospects of substantial improvements, though packing the coordinates with reduced accuracy (e.g. two coordinate values per word) in NMS is advantageous provided the objective function is such that the method remains relatively insensitive to small perturbations of each simplex. The algorithms SD and P64 are much more sensitive to this modification, which is therefore not recommended.

5.2. *The Use of Secondary Storage*

Consideration was given, both practically and theoretically, to the use of secondary store of either magnetic tape or disc, both for storing the large array in segmented form and for segmenting the program. The latter provided little relaxation in the dimensionality which may be tackled, since the program soon becomes a minor proportion of the required storage as the dimensionality is raised: the former proved to be feasible, as far as implementation is concerned, in the cases of SD and NMS. However, P64 though evidently the best

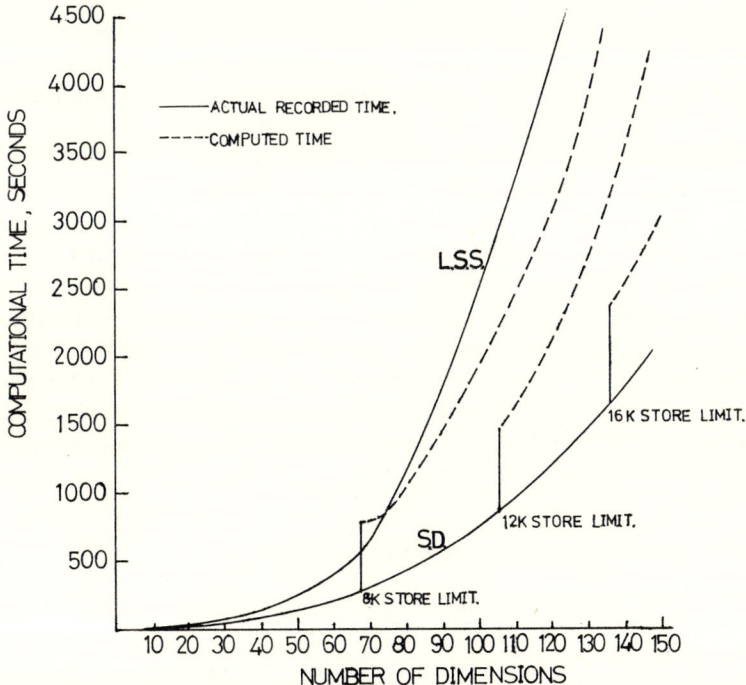

FIG. 2. Typical results showing the effect of primary storage limits on computational time as the number of dimensions is raised—algorithms SD and LSS on function $F1$.

prospect was immediately ruled out, since the large array present holds several different parameters and the array elements are accessed frequently, either singly or in widely differing groups. In contrast, the previous two methods generally required all the array elements each time the corresponding large array is accessed. After validation trials attention was concentrated on theoretical analysis, since the initial test runs were heavy on computer time. This analysis, based on the access time of the devices, number of possible accesses per iteration, and so on, is discussed further in the next section; typical results are presented in Fig. 2.

5.3. *Theoretical Estimates of Additional Time Required by Secondary Storage*

For both NMS and SD the corresponding large array which leads to a violation of the limit on primary storage is assumed stored completely on secondary storage. At each access of the array it is brought into core in a series of blocks, each of which is operated on, and then returned to secondary storage. The blocks are as large as possible, subject to being integer multiples of n. The extra time required for a complete minimization is based on (i) the total number of iterations, (ii) the average number of array accesses per iteration, and (iii) the average time to read into core and return to secondary store the complete array. For SD the number of accesses per iteration is actually 4, with another access each time a particular matrix (related to the Hessian) is set up; for NMS the number of accesses per iteration averages close to 2.

The average times for transfers involving magnetic tape and disc were based on KDF9 peripherals, the computed times being based on (i) initial access time, (ii) transfer rate to core, (iii) transfer rate to secondary store, and (iv) either track switching or time to complete a rewind, depending on the type of storage. A number of computer trials gave close agreement with the computed effect of secondary storage. Figure 2 shows clearly that the introduction of secondary storage increases the required time for a minimization quite substantially for the KDF9. A similar analysis for the ICL 1906A makes the same point even more dramatically, since operations within core took about $\frac{1}{20}$ of the time for the KDF9, but operations involving secondary storage were comparable for the two computers.

6. New Algorithm Using Reduced Primary Storage

The crippling effect on efficiency of the necessity for secondary storage suggested that new or modified algorithms might be preferable at high dimensionality. Unfortunately, P64 offered little prospect of modification because of the type of array access, and SD could only be made less demanding through the doubtful expedient of data packing. NMS could make most reliable use of data packing, but offered best prospects through modification of the algorithm. The outcome was a new method for unconstrained minimization designed for minimizing functions of relatively high dimensionality in limited primary storage. The algorithm (Fig. 3), has as its central part a variation of the simplex method, but the large array proportional to n^2 (which in the case of the simplex method holds the position vectors in n dimensions of $n + 1$ points), has been eliminated. In the original method the initial simplex is set up by taking an initial point and generating the other n points by projecting through the initial point along lines parallel to the n coordinate axes. However, by storing only the initial point and the n respective displacements (corresponding to the n coordinate directions), the simplex is defined and the respective function values may be determined, as may the centroid of the simplex also. A list of the *ten* highest functions values and their coordinates (an arbitrary but satisfactory practical choice) is then built up in descending order, and the basic simplex algorithm is initiated, using the 'top' point in the list as the potential reflection point. At the end of each simplex iteration the elements of the list are shifted up one location, thus deleting the top point, and the new point is tested to see whether its function value lies within the scope of the current list; if so, the list is re-arranged by inserting the new point in the correct position. This process continues until the list is exhausted, whereupon a complete restart is made about the point of lowest function value so far encountered; the latter has been stored for this purpose, with the algorithm continuing until convergence is indicated. Thus, at worst, a restart will be needed every ten iterations, and at best the process will keep on inserting new points into the list, acting therefore exactly like the ordinary simplex method.

7. Results for New Algorithm

The performance of the new algorithm LSS is illustrated in Tables 1, 3, and 4. For each test function LSS is only slightly less efficient than NMS at low dimensionality, where efficiency is measured by 'required function evaluations', but falls away as the dimensionality is increased. However, a trade-off of speed against storage was expected, and is confirmed in the low storage requirements quoted in Table 3. The actual formula for the storage requirement was obtained:

LSS, required storage $= 5168 + 20n$ words

so that 8K, and 16K will suffice for dimensionalities 150 and 560, respectively. The comparable dimensionalities for SD, the strongest competitor, are about 70 and 140, respectively. If primary storage is limited then efficiency is better measured by execution time, and Table 4 shows the impact of secondary storage on the required times. Figure 2 also illustrates the effect of secondary storage

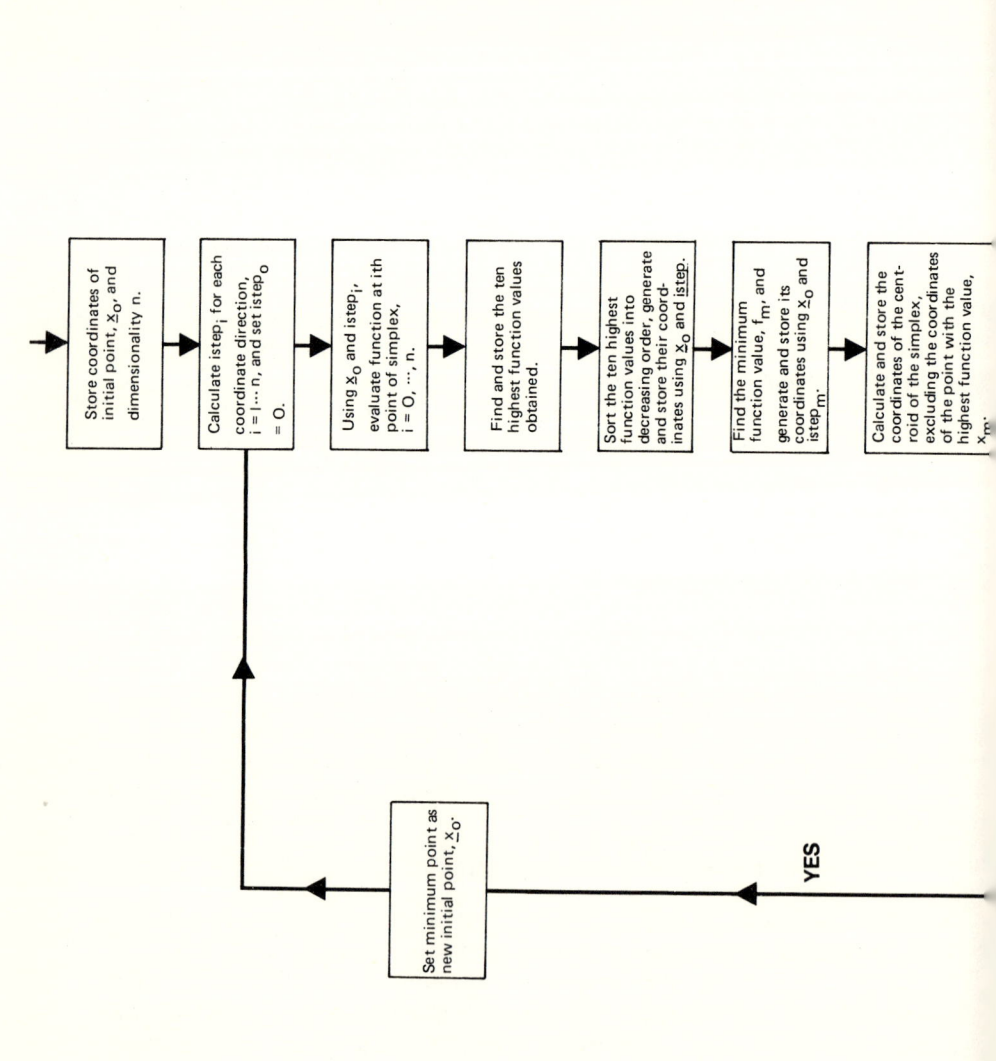

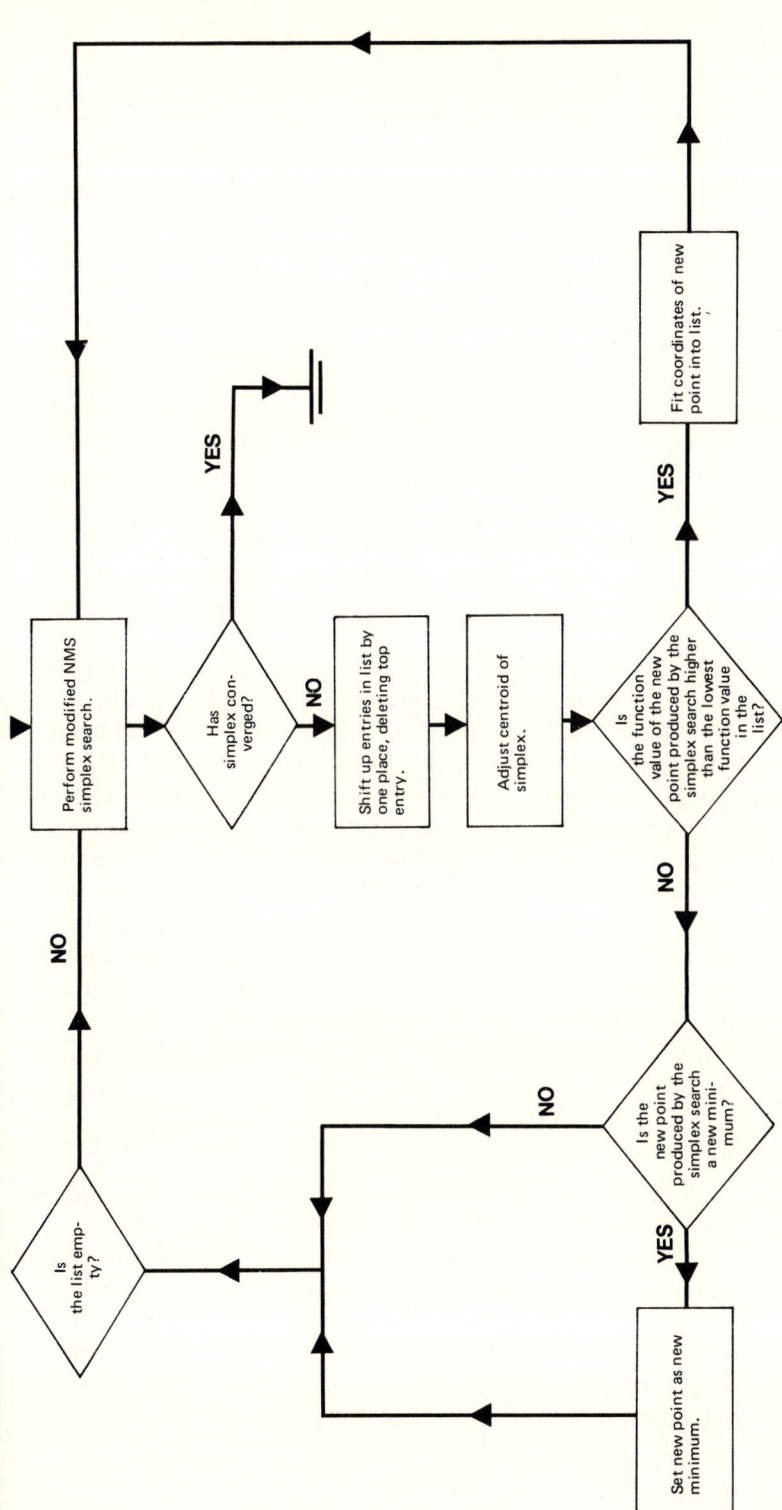

FIG. 3. Flow diagram for the new algorithm LSS.

(magnetic tape in this case) on execution time, showing that LSS becomes much more competitive with SD under these conditions, and increasing the dimensionality or reducing the available core store will favour LSS.

8. Failures in Standard Algorithms

During the investigations no failures were experienced with either NMS or LSS, but frequent failures with the original version of P64 and an occasional difficulty with SD have already been noted. In the case of SD a real overflow occurred when minimizing function $F1$ with $n = 40$, but this was eliminated by reducing the required accuracy to the order 10^{-8}, and this accuracy was taken as standard for the complete investigation. With functions $F2$ and $F3$ an occasional result still requires, say, 20% more function evaluations than anticipated, but this effect is also more marked if higher accuracy is required. The actual failure has not been explained, but the occurrence is so rare that the algorithm is clearly highly robust, though its performance is much more erratic than the other algorithms, even for the very regular test functions considered.

The failures with P64 were always programmed exits caused by an inability to satisfy criterion 2. Criterion 1 gives 'convergence' when the elements of the correction vector δx are sufficiently small. If the latter solution is denoted by s1, then invoking criterion 2 generates a new starting point by taking a specific step δu from s1 and a new solution s2 is determined by applying criterion 1; a third minimum s3 is sought on the line joining s1 and s2, and convergence is accepted if all three solutions are sufficiently close. If not, the restart process is repeated in a modified form, but continued failure causes the programmed exit previously mentioned. However, the latter procedure is unlikely to terminate satisfactorily if criterion 1 does not give a good estimate of the solution, and this is probably the cause of the trouble in the present instance. Hence, it is suggested that criterion 2 be dropped, and that convergence be tested by comparing results for two or more starting points but without any automatic restart.

9. Conclusions

The initial investigations show that for these relatively simple test functions algorithm SD is marginally better than P64 for dimensionality below 20, but that algorithm P64 becomes increasingly superior as the dimensionality is raised. Algorithm NMS is a poor third at low dimensionality but improves and becomes comparable with SD when the dimensionality exceeds 100. The conclusions so far apply on the assumption that primary storage is unlimited, and contradict published predictions on the relative effect of dimensionality. Algorithm P64, however, is somewhat flattered by the particular choice of starting points, and if higher accuracy were required the relative advantage of P64 would be reduced. Furthermore, the second convergence criterion in P64 is best omitted as it can lead to frequent failures.

The new algorithm LSS will, in general, require far less storage than the three standard algorithms, since storage is proportional to the dimensionality n rather than to n^2 as is the case for P64, SD, and NMS. Hence, as the dimensionality is raised with a reasonable limit on primary storage, P64 and NMS are the first algorithms to need secondary storage, followed by SD, but LSS follows very much later; for example, if available primary store is 16K of 48 bit words the algorithms P64, NMS, SD, and LSS will just call on secondary storage for dimensionalities 105, 106, 151, and 561, respectively. However, the fastest 'in core' algorithm, P64, is not readily modified to make even tolerably efficient use of secondary storage if the primary store limit is reached. The algorithms SD and NMS can be so modified, but execution time rises rapidly if any type of secondary storage is employed. Hence, though LSS is the least efficient on the basis of function evaluations it may become competitive on the basis of execution time, and even be fastest at high dimensionality, as well as using by far the least store.

Finally, the robustness of NMS (and LSS) is confirmed over the full range in n, so that NMS may be preferred to P64 where storage is not a prime consideration, but robustness is likely to prove an advantage.

Acknowledgements

Mr. Parkinson's work was carried out under the sponsorship of the Science Research Council.

References

Box, M. J. (1966). A comparison of several current optimization methods, and the use of transformations in constrained problems. *Comput. J.* **9**, 67–77.
Kowalik, J., and Osborne, M. R. (1968). 'Methods for Unconstrained Optimization Problems.' American Elsevier Publishing Co., New York.
Lill, S. A. (1970). A modified Davidon method for finding the minimum of a function, using difference approximation for derivatives'Algorithm 46.*Comput. J.***13**,111–113.
Nelder, J. A., and Mead, R. (1965). A simplex method for function minimization. *Comput. J.* **7**, 308–313.
Parkinson, J. M., and Hutchinson, D. (1971). 'An Investigation into the Efficiency of Variants on the Simplex Method.' Paper presented at the Conference on Numerical Methods for Optimization, Dundee.
Powell, M. J. D. (1964). An efficient method for finding the minimum of a function of several variables without calculating derivatives. *Comput. J.* **7**, 155–162.
Spendley, W. (1969). Nonlinear least squares fitting using a modified simplex minimization method. *In*: 'Optimization' (R. Fletcher, ed.), pp. 259–270. Academic Press, London.
Stewart, G. W., III (1967). A modification of Davidon's minimization method to accept difference approximations to derivatives. *J. Ass. comput. Mach.* **14**, 72–83.

8. An Investigation into the Efficiency of Variants on the Simplex Method

J. M. PARKINSON AND D. HUTCHINSON
Department of Computational Science,
The University of Leeds, Leeds, England

Summary

The Nelder and Mead version of the simplex method is widely accepted as more robust, though rather less efficient, than several of the more recent methods for unconstrained optimization. However, most assessments and comparisons have been based on relatively few results. The investigations described here may be considered in two main parts: the effect on efficiency of (1) selection of initial parameters, (2) variants on the basic algorithm.

In the first part, the effect of the initial scale, shape, and orientation are investigated, and as a result several automatic procedures are proposed which may be employed to set up an appropriate initial simplex. Secondly, a number of internal modifications to the basic iteration are considered, and the most beneficial of these are recommended. Finally, a short section is included which discusses the usual convergence criteria and their limitations.

1. Introduction

The simplex search technique published in 1962 by Spendley *et al.*, was designed for specialized applications, but became significant for general unconstrained optimization problems with the publication by Nelder and Mead (1965) of their powerful adaptation of the original algorithm, (hereafter denoted NMS). Published numerical comparisons of current algorithms are scarce and often of doubtful generality. However, some work at Leeds (Schnabel, 1966) showed NMS as the most reliable of the currently favoured methods in the presence of noise, as well as being reasonably efficient in any event. Kowalik and Osborne (1968) confirm these results, and Box *et al.* (1969) quote NMS as the 'most efficient of all current sequential techniques', though Box (1966) suggested that NMS became relatively less competitive as the number of dimensions were increased. However, a number of important parameters in the NMS have not previously been properly explored, and the investigations described here include a consideration of improved versions of the algorithm.

Numerical comparisons are notoriously difficult to generalize, involving as they do such factors as choice of test function, starting position, convergence criterion and initial step length, as well as the computer software and hardware characteristics. The number of function evaluations is used here as a measure of work done, and the numerical results presented explicitly are a small subset of those obtained. Comprehensive results will be published in due course (J. M. Parkinson, Ph.D. Thesis, 1972).

2. Notation and Basic Algorithm

A proper simplex is defined as a set of $n + 1$ points in $[n]$ which span the space. If the points are mutually equidistant the simplex is described as regular (e.g. an equilateral triangle in [2]).

The following notation is used:

V_i, $i = 0, 1, \ldots, n$, are vertices of the simplex in $[n]$ with co-ordinates $\mathbf{x}_i$;

$istep_i$, $i = 1, 2, \ldots, n$, are scalars associated with the n co-ordinate directions;

$\mathbf{X}_i$, $i = 1, 2, \ldots, n$, are unit vectors in the n co-ordinate directions;

$f(\mathbf{x})$ is the function to be minimized, and $f_i \equiv f(\mathbf{x}_i)$;

$f_h = \max_i f_i$, and the corresponding vertex has co-ordinates $\mathbf{x}_h$;

$f_l = \min_i f_i$, and the corresponding vertex has co-ordinates $\mathbf{x}_l$;

$\bar{\mathbf{x}} = \dfrac{1}{n} \sum_{\substack{i=0 \\ i \neq h}}^{n} \mathbf{x}_i$, the centroid of all $\mathbf{x}_i$, excluding $\mathbf{x}_h$;

$(\mathbf{x}_i, \mathbf{x}_j)$ = distance between the points $\mathbf{x}_i$ and $\mathbf{x}_j$.

Throughout the paper we consider minimization of the function $f(\mathbf{x})$, as opposed to its maximization.

In the original simplex search (1) the initial simplex is regular, and this feature is preserved as each new simplex is formed by replacing $\mathbf{x}_h$ by $(2\bar{\mathbf{x}} - \mathbf{x}_h)$. In the NMS the initial simplex is defined by $\mathbf{x}_0$ and $\mathbf{x}_i = \mathbf{x}_0 + istep_i \mathbf{X}_i$, $i = 1, 2, \ldots, n$, where $istep_i$ determines the distance of V_i from V_0. In [2] this simplex is a right-angled triangle for rectangular co-ordinates, but this feature is rapidly destroyed as the successive iterations introduce three fundamental operations; namely, reflection, expansion, and contraction. The corresponding new points $\mathbf{x}_r$, $\mathbf{x}_e$, $\mathbf{x}_c$, and their selection, is defined below:

(1) Reflection: $\mathbf{x}_h$ is replaced by $\mathbf{x}_r = (1 + \alpha)\bar{\mathbf{x}} - \alpha \mathbf{x}_h$ provided that $f_i > f_r > f_l$ for at least one i other than $i = l$ or $i = h$; α is the reflection coefficient.

(2) Expansion: $\mathbf{x}_h$ is replaced by $\mathbf{x}_e = \gamma \mathbf{x}_r + (1-\gamma)\bar{\mathbf{x}}$ provided that (i) $f_r < f_e$, and (ii) $f_e < f_l$. If (i) holds but not (ii), then $\mathbf{x}_h$ is replaced by $\mathbf{x}_r$; γ is the expansion coefficient.

(3) Contraction: $\mathbf{x}_h$ is replaced by $\mathbf{x}_c = \beta \mathbf{y} + (1-\beta)\bar{\mathbf{x}}$ where $\mathbf{y}$ is $\mathbf{x}_h$ or $\mathbf{x}_r$ according as $f_r \geq f_h$ or $f_r < f_h$; β is the contraction coefficient.

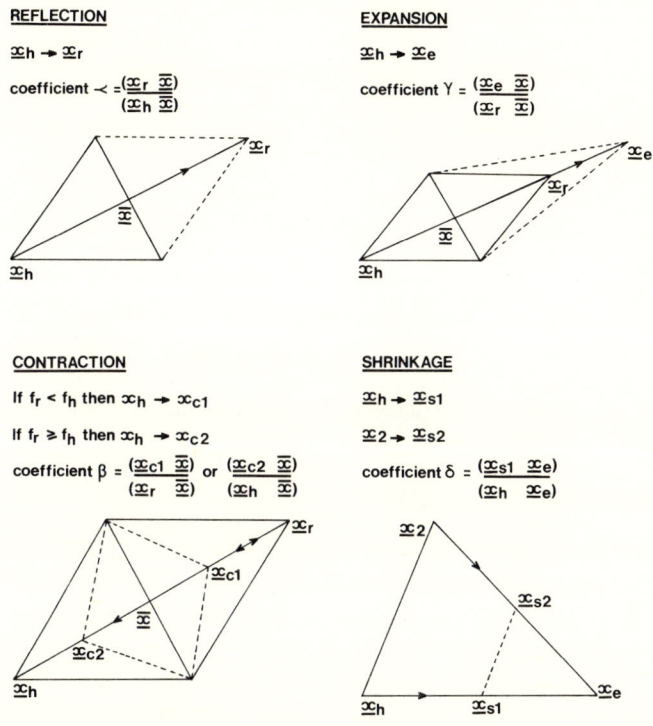

FIG. 1. The basic operations of the simplex method.

A further operation, shrinkage, can be employed within an iteration: if $f_c \geq f_h$, then the new simplex is defined by $\mathbf{x}_l$ and $\frac{1}{2}(\mathbf{x}_i + \mathbf{x}_l)$, all i except $i = l$.

However, this operation can be generalized, the above formula becoming $\delta \mathbf{x}_i + (1-\delta)\mathbf{x}_e$, where normally $\mathbf{x}_e = \mathbf{x}_l$ and we term δ the shrinkage coefficient. The convergence criterion in NMS is that

$$\frac{1}{n}\sum_{i=0}^{n}(f_i - \bar{f})^2 < eps,$$

where *eps* is a parameter which is set according to the accuracy required, and $\bar{f}$

is the mean of f_i over all i; a complete discussion of the algorithm is given by Nelder and Mead (1965) in their paper.

At this stage we also define orientation: a new initial simplex may be generated by a sequence of rotations θ_1, θ_2, ... of the NMS initial simplex about $\mathbf{x}_0$ in the respective co-ordinate planes. In [2] there is only one co-ordinate plane, and a single rotation $0 \leqslant \theta_1 < 2\pi$ uniquely determines the orientation: this is illustrated in Fig. 2.

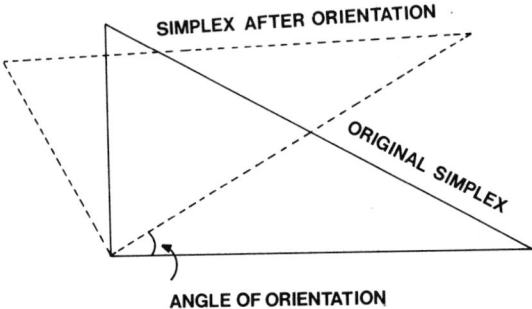

FIG. 2. Orientation of the initial simplex in a plane.

3. Possible Modifications of NMS

The progress of a simplex search which involves reflection only is illustrated in Fig. 3, showing that there are two cases, and for $2n$ iterations progress is $n\sqrt{2}\cos\theta$ or $n\cos\phi$, respectively, where the simplex chosen has sides 1, 1, $\sqrt{2}$. Fastest and slowest progress here is $n\sqrt{2}$ and n, respectively, so that in general, progress depends on the orientation of the initial simplex as well as its magnitude. The investigation therefore included the effect on efficiency of the selection of the initial parameters of scale and orientation. Automatic procedures for generating these important parameters were developed.

The choice of α, β, γ has been termed a strategy, and Nelder and Mead recommended $(1, \frac{1}{2}, 2)$ as a good, safe one on the basis of a relatively small number of trials. Thus further consideration was given to possible strategies and also to the variation of δ, the shrinkage coefficient. Another plausible variation is to repeat the expansion process within an iteration as long as the next function value gives a new minimum; this is simply a linear search which is best terminated by translating the current simplex to the new position (discussed later) rather than to permit gross distortion of this simplex by replacing only one point.

As well as modifications within iterations, one may insert tests or re-setting procedures between iterations. However, in order to preserve the essential simplicity of the algorithm, all but one were ultimately rejected, that retained being aimed at reducing the incidence of numerous successive contractions. In our experience the latter phenomenon occurs with NMS when a drastic rescaling of the simplex is needed in order to change substantially the direction of search, and can even give rise to apparent convergence at a false minimum.

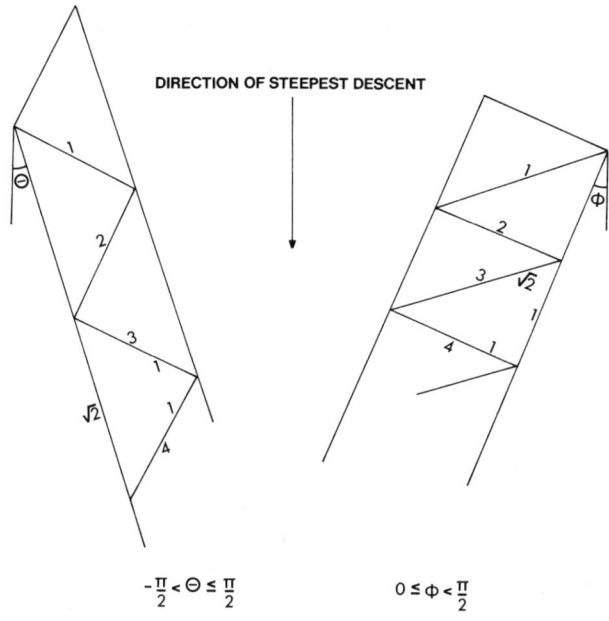

FIG. 3. Rate of progress down an inclined plane with reflection only.

Finally, a number of stopping criteria were compared, though these should normally be selected according to the objective required, e.g. position or value of the minimum. The simplest solution to the problem of false convergence seems simply to step away from the point of apparent convergence, restart, and compare the two solutions, repeating as necessary until agreement is obtained.

4. The Test Functions

The test functions selected were those most commonly used in the literature for comparative purposes, and generally these are of low dimensionality but designed to strain the efficiency of optimization algorithms. A simple sum of squares function was included since several modifications would be expected to be really effective only in such regular cases. The starting positions are specified in each case, and it must be borne in mind that these affect comparisons, particularly since they are normally relatively close to the solution.

(i) Rosenbrock's 'parabolic valley':
$f = 100(x_2 - x_1^2)^2 + (1 - x_1)^2$ and $f_{min} = 0$ at $(1, 1)$.
Starting point $(-1.2, 1.0)$.

(ii) Box's exponential function:
$$f = \sum_y \{[\exp(-x_1 y) - \exp(-x_2 y)] - x_3[\exp(-y) - \exp(-10y)]\}^2,$$
where $y = 0.1(0.1)1.0$ and $f_{min} = 0$ at $(1, 10, 1)$. Starting point $(0, 20, 20)$.

(iii) Beale's function:
$$f = \sum_{i=1}^{3} [C_i - x_1(1 - x_2^i)]^2,$$
where $C_1 = 1.5$, $C_2 = 2.25$, and $C_3 = 2.625$. $f_{min} = 0$ at $(3, 0.5)$.
Starting point $(0, 0)$.

(iv) Rosenbrock's 'cube function':
$f = 100(x_2 - x_1^3)^2 + (1 - x_1)^2$ and $f_{min} = 0$ at $(1, 1)$.
Starting point $(-1.2, 1.0)$.

(v) Wood's function:
$f = 100(x_2 - x_1^2)^2 + (1 - x_1)^2 + 90(x_4 - x_3^2)^2$
$\quad + (1 - x_3)^2 + 10.1((x_2 - 1)^2 + (x_4 - 1)^2)$
$\quad + 19.8(x_2 - 1)(x_4 - 1),$
and $f_{min} = 0$ at $(1, 1, 1, 1)$. Starting point $(-3, -1, -3, -1)$.

(vi) Sum of squares function:
$$f = \sum_{i=1}^{n} i x_i^2 \; (n = 5, 10) \quad \text{and} \quad f_{min} = 0 \text{ at } \mathbf{x} = 0.$$

Starting point $(1, 1, \ldots, 1)$.

Note: In all cases accuracy was obtained to $eps = 10^{-8}$.

5. The Effect of Initial Parameters

The distribution of the $(n + 1)$ points of the initial simplex determines its 'shape', which may vary widely. Preliminary experiments varying the positions of V_i, $i = 1, 2, \ldots, n$ with $(\mathbf{x}_0, \mathbf{x}_i)$ constant showed that shape is relatively unimportant, e.g. a regular initial simplex with NMS is comparable with the usual NMS initial simplex. Similar results were obtained when varying the shape of the NMS initial simplex by altering $istep_i$ for

$$\sum_{i=1}^{n} istep_i = \text{constant}$$

8. EFFICIENCY OF VARIANTS

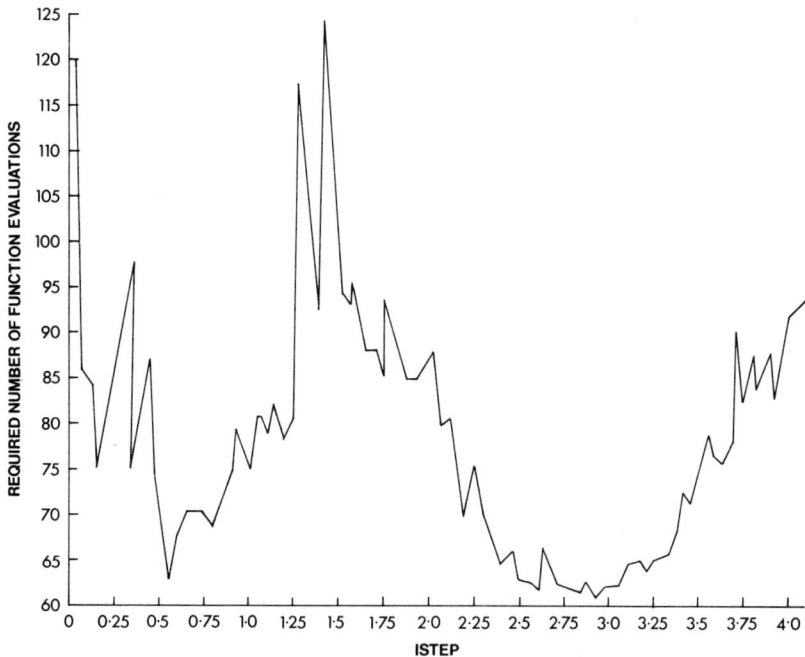

FIG. 4. Plot of *ISTEP* against required number of function evaluations to minimize Beale's function ($eps = 10^{-8}$).

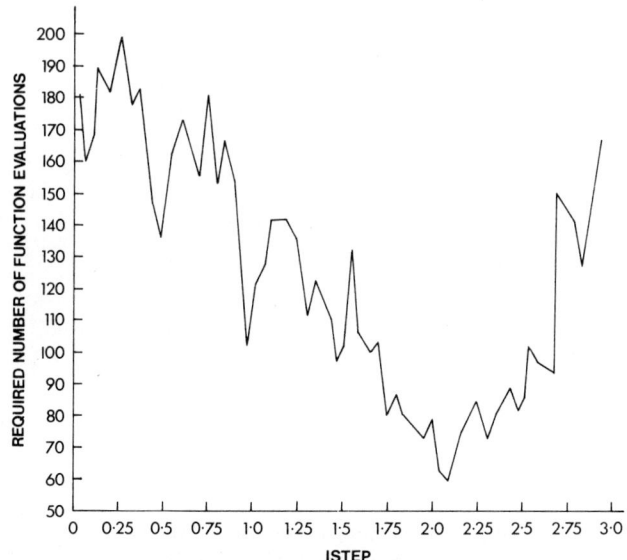

FIG. 5. Plot of *ISTEP* against required number of function evaluations to minimize Rosenbrock's cube function ($eps = 10^{-8}$).

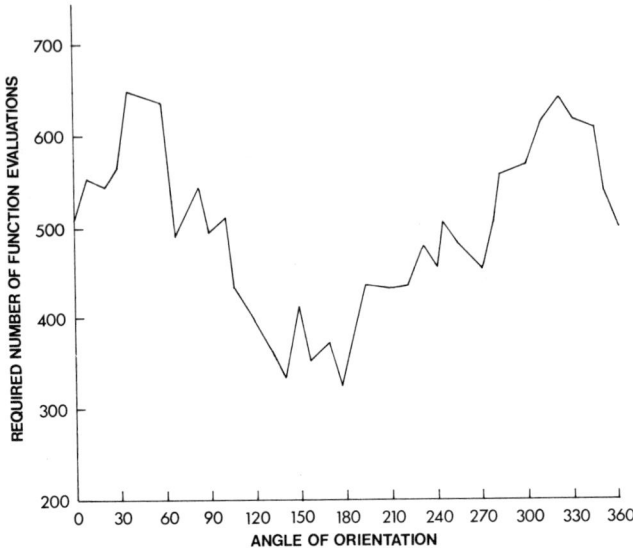

FIG. 6. Plot showing variation in function evaluations as the initial simplex is rotated through 360° in steps of 10° for sum of squares function (*ISTEP* = 1, *eps* = 10^{-8}).

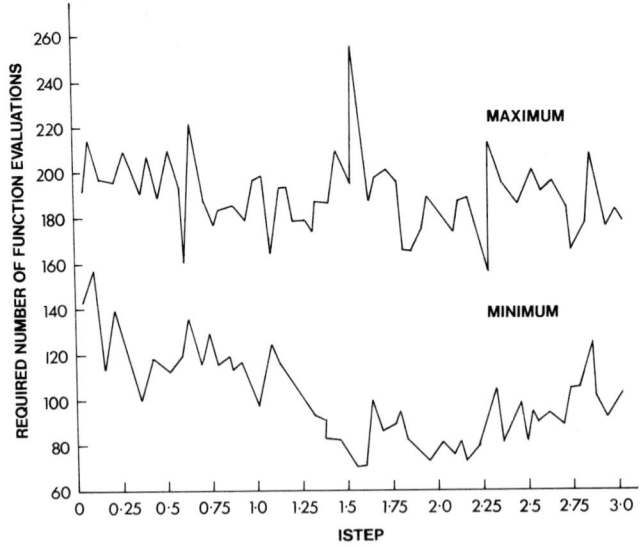

FIG. 7. Plot showing maximum and minimum number of required function evaluations, against increasing *ISTEP*, where the orientation of the initial simplex is varied through 360° in steps of 45°, for Rosenbrock's function (*eps* = 10^{-8}).

This may be anticipated since the initial simplex is rapidly modified by the action of the algorithm. Hence in subsequent investigations $istep_i$ was kept independent of i, quoted as *ISTEP*, giving a symmetric initial simplex. Varying the 'scale' of the initial simplex (i.e. the value of *ISTEP*), however, produced more significant variations in the number of function evaluations required to minimize each function. In Figs 4 and 5 typical results are shown. It may again be anticipated that a good initial simplex would have $istep_i$ ($i = 1, 2, ..., n$), and

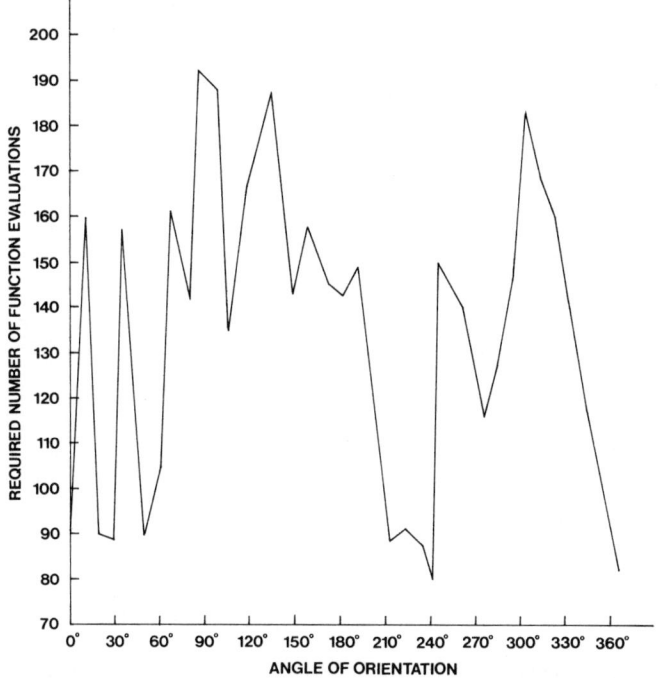

FIG. 8. Plot showing variation in required function evaluations as initial simplex is rotated through 360° in steps of 10° for Rosenbrock's function (*ISTEP* = 2, $eps = 10^{-8}$).

so its scale, comparable with the corresponding co-ordinate distances from the starting point to the solution. Since the single parameter *ISTEP* is considered (i.e. each $istep_i = ISTEP$) it is reasonable to assume that as *ISTEP* increases, and the simplex passes through these distances in the respective co-ordinate directions a number of minima will appear on a graph of *ISTEP* against function evaluations required, for any particular function. Figures 4 and 5 confirm this, Fig. 4 showing minima around 0·5 and 3·0, which are the co-ordinate distances of the solution of Beale's function from the starting point at (0, 0); Fig. 5 shows a minimum around 2·2, which is the distance of the solution (1, 1) from the starting point (−1·2, 1) in the x_1 co-ordinate direction for Rosenbrock's 'cube' function. Only one minimum shows up here, since the starting point and final solution both have the same x_2 co-ordinate.

Turning to the orientation of the initial simplex, it was found that varying this had a dramatic effect for all the test functions. For the sum of squares function, the number of function evaluations required was found to be surprisingly sensitive to orientation, though as anticipated, varying essentially periodically as the initial simplex was rotated through 2π (see Fig. 6). Rotation

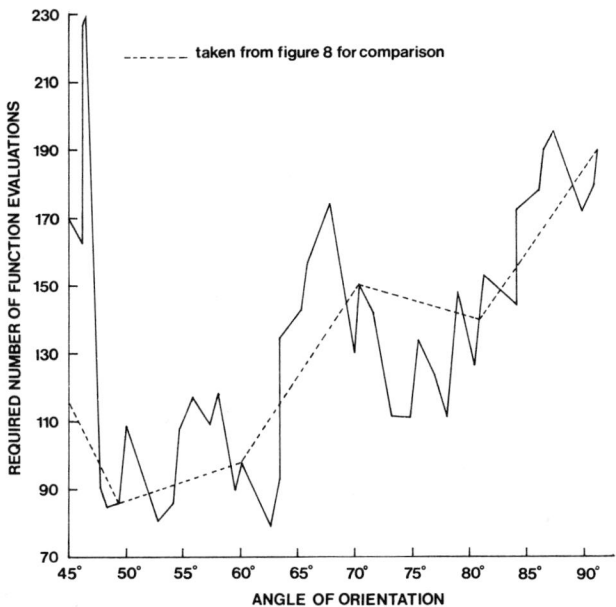

FIG. 9. Plot showing variation in required function evaluations as initial simplex is rotated through 45° in steps of 1° for Rosenbrock's function (*ISTEP* = 2, *eps* = 10^{-8}).

through 2π gave a variation of ±35% in the required number of function evaluations. For the more complex functions, however, the sensitivity to orientation becomes much greater, and the periodicity is largely concealed, as can be well seen in Figs 7, 8, and 9. The latter show the persistent variation in the number of required function evaluations as the increment in the angle of orientation is decreased; indeed, the phenomenon is still marked when the increments are further reduced by a factor of 100. Similar results were obtained for all the complex test functions; a change of only 1° may change the required number of function evaluations by ±45%. Thus it appears that the deliberate choice of an initial orientation holds an element of risk for all but very regular functions, due to the wide variations in the number of required function evaluations necessitated by even small changes in orientation. The difficulty is compounded by raising the dimensionality of the objective function since this widens the choice of orientation parameters in direct proportion.

6. Automatic Setting of the Initial Parameters

Several automatic procedures were developed which would determine appropriate initial parameters for scale and orientation, but only two gave sufficient and regular gains over random selection to merit normal use, and both are concerned solely with the scale of the simplex.

The sensitivity of the test functions to orientation makes very accurate prediction of these parameters desirable, and this is not possible at present. However, the most obvious automatic selection of the initial orientation is straightforward, being based on the localized 'steepest descent' deduced from the function values on a trial simplex (i.e. the vector sum of the estimated gradient vectors in the co-ordinate directions). For example, the 'steepest descent' vector would be used to set $\theta = 0$ in Fig. 2. Experience has shown that this procedure is advantageous for very regular functions such as sums of squares, but is otherwise uncertain. Moreover, in its present form the estimate of the steepest descent vector depends on the scale ($ISTEP$) of the trial simplex.

In the first method for setting scale (see Fig. 10) a simple linear search is used to determine an approximate minimum in each co-ordinate direction, such points being taken with the initial point as defining the initial simplex. Hence, for an objective function which is relatively regular the first reflection of the simplex should be to a point in the general region of the required minimum. In fact the linear searches are performed until a minimum is obtained or preset limits are violated—$istep_i$ is then set to the distance of the terminal point from the initial point for the corresponding i co-ordinate.

In the second method (see Fig. 11) two estimates are obtained for the non-local steepest descent directions by setting up two trial simplices of substantially different scale. Then the positions of the minima in these directions are determined by linear search, and a further search is employed on the line joining the previous two minima: the latter search yields an overall minimum, and $ISTEP$ is then set to the distance of this minimum from the initial point (and $istep_i = ISTEP$, all i). However, the linear searches are again subject to preset limits.

In both cases the limits are employed to restrict the range of values of $istep_i$ for all i, and arbitrary upper and lower limits are both included. Such limits are desirable to avoid numerical difficulties or pathological situations, but the actual values employed (0·1, 15, respectively) are based on experience with these particular test functions.

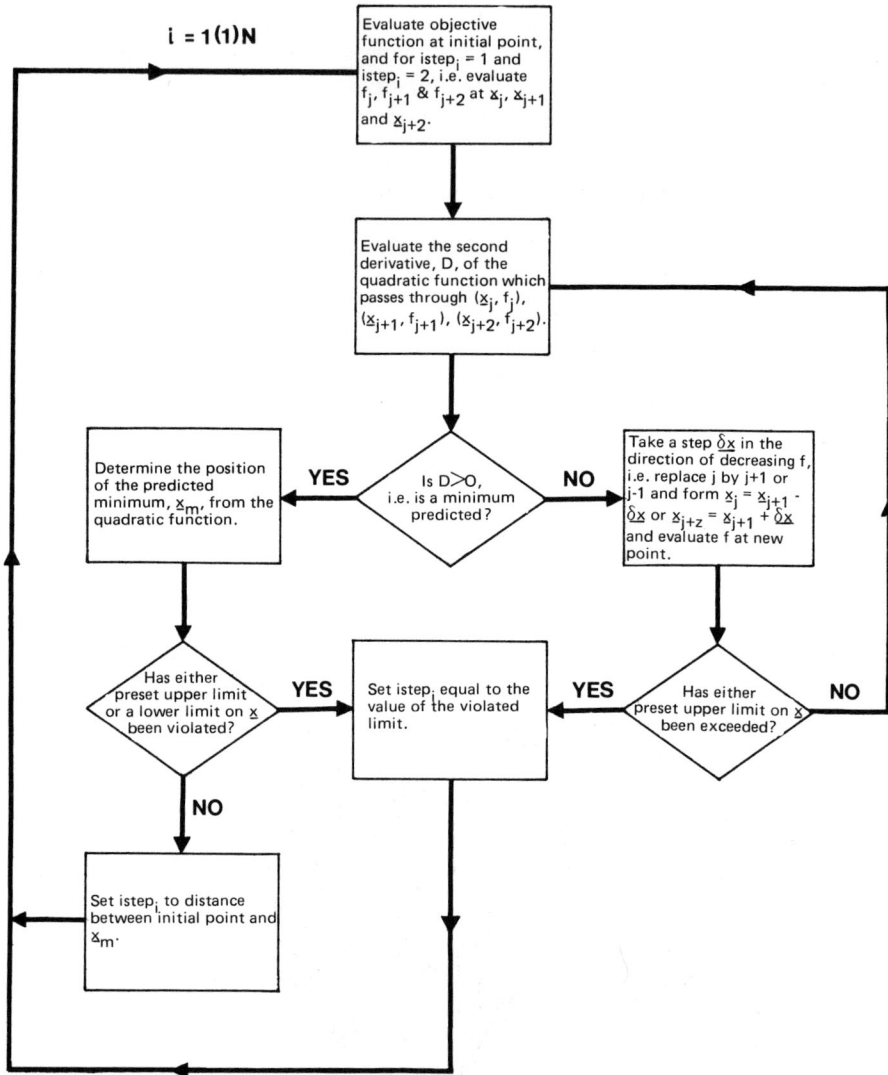

FIG. 10. Flow diagram for 'linear search' method of setting the scale of the initial simplex.

The gain in efficiency is naturally most marked for the quadratic functions, say 40% reduction in function evaluations over the average for a set of random scale parameters, though the comparison is affected by the range over which the 'random' selection is made. However, the results for the other functions (Table 1) still show worthwhile gains over random selection: for example, for

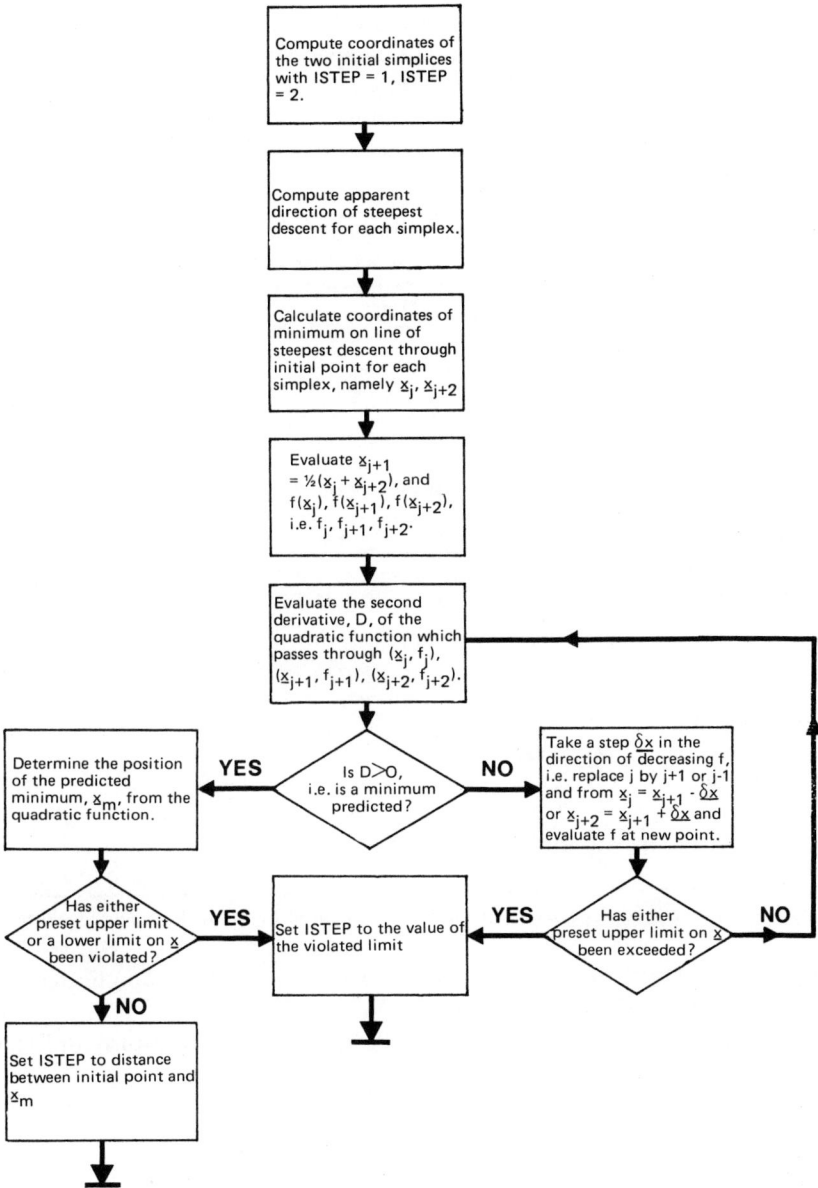

Fig. 11. Flow diagram for method based on 'steepest descents' for setting the scale of the initial simplex.

Rosenbrock's function the average number of function evaluations required is 173 for *ISTEP* varied over the set 0·1(0·1)10 as against 165 or 140 for the two automatic setting procedures. The comparison for Rosenbrock's function is little changed if the average is computed over the set 0·1(0·05)5, though compressing the range further finally eliminates the possibility of gain provided that the optimum value of *ISTEP* is included in it. On balance the procedure

TABLE 1
A comparison of two automatic scale setting algorithms, one based on linear search, the other on 'steepest descent'

Function			Method based on linear search		Method based on steepest descent		
		$istep_i$	Extra function evaluations	Total function evaluations	*ISTEP*	Extra function evaluations	Total function evaluations
Rosenbrock's parabolic valley	(1)	1·642	7	165	1·690	8	140
	(2)	0·440					
Beale's function	(1)	1·125	6	71	1·749	8	76
	(2)	0·405					
Box's function	(1)	−2·29					
	(2)	5·000	12	230	11·515	12	117
	(3)	1·200					
Rosenbrock's cubic function	(1)	1·639	6	184	3·129	8	148
	(2)	0·480					
Wood's function	(1)	3·570	16	495	5·458	16	529
	(2)	1·763					
	(3)	3·101					
	(4)	2·285					

Note: 'Extra function evaluations' are those required by the scale setting algorithms.

based on estimates of steepest descent is best, though more complicated than that based only on linear searches.

Other automatic procedures were employed but with little return for further sophistication.

7. The Effect of Modifications within a Basic Iteration

The most beneficial of all such modifications was found to be the inclusion of unlimited expansion. Initially, this was followed by a standard shrinkage towards the new low point, with δ, the shrinkage coefficient, set equal to the number of times the expansion was repeated. However, if a large number of

8. EFFICIENCY OF VARIANTS

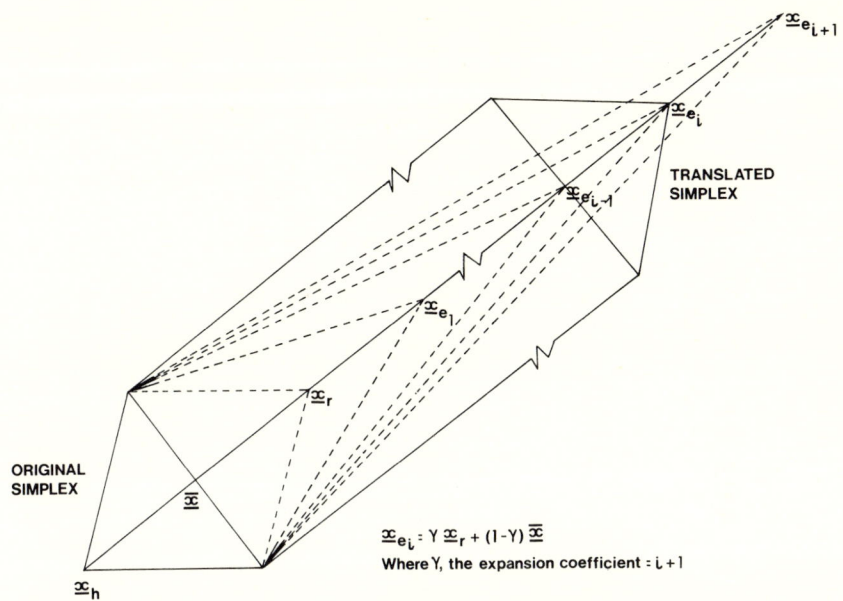

FIG. 12. Unlimited expansion of simplex followed by translation in two dimensions.

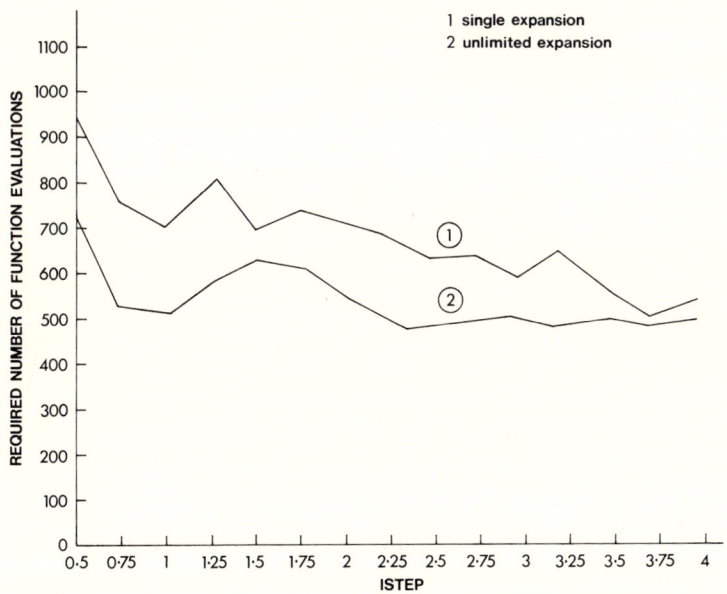

FIG. 13. Plot showing number of function evaluations required to minimize sum of squares function in 10 dimensions with and without unlimited expansion ($eps = 10^{-8}$).

TABLE 2
A comparison of the number of function evaluations required to minimize Rosenbrock's 'parabolic valley'; with and without unlimited expansion

ISTEP	Normal NMS expansion	Unlimited expansion
1·0	125	170
1·2	143	168
1·4	138	151
1·6	175	156
1·8	201	106
2·0	85	120
2·2	74	79
2·4	129	136
2·6	148	174
2·8	176	93
3·0	152	105
3·2	158	186
3·4	234	133
3·6	192	137
3·8	166	142
4·0	194	145
4·2	204	138
4·4	183	160
4·6	184	151
4·8	205	140
5·0	191	148
5·2	197	144
5·4	179	120
5·6	200	121
5·8	200	102
6·0	179	193
6·2	179	196
6·4	171	197
6·6	201	192
6·8	192	126
7·0	181	147

Mean function evaluations with normal NMS expansion = 172
Mean function evaluations with unlimited expansion = 145

expansions take place in one iteration, then the simplex produced as a result of such shrinkage becomes very small, giving rise to the possibility of false convergence. Thus a new operation was incorporated which we term 'translation'. This involves moving the simplex towards the new lowest point

8. EFFICIENCY OF VARIANTS

predicted by unlimited expansion, in such a way that the scale and orientation are unaltered (Fig. 12).

Incorporating unlimited expansion followed by translation of the original simplex is clearly most advantageous for regular functions, as illustrated in

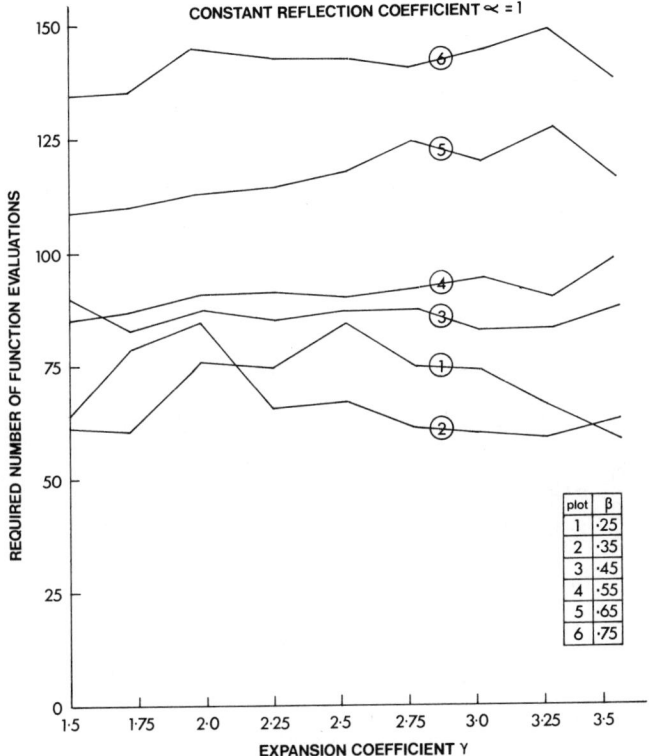

FIG. 14. Plot of required function evaluations (averaged over a range of *ISTEP*) against the expansion coefficient (γ) as the contraction coefficient (β) is increased, with the reflection coefficient (α) constant, $\alpha = 1$ ($eps = 10^{-8}$).

Fig. 13, but the gain must be related to the associated values of *ISTEP* and the distance of the initial point from the minimum.

With more complex functions, and so fewer clear 'down hill runs,' the gain is less obvious, although a significant improvement can be seen for Rosenbrock's 'parabolic valley' function in Table 2, over a range for *ISTEP*.

Considering now the strategy of (α, β, γ), the recommendation made by Nelder and Mead was based on trials with roughly 100 combinations. We have systematically investigated several thousand combinations and, as may be anticipated, there is no evident general strategy which gives the best results for all the test functions. However, typical results (Figs 14 and 15), which are for Beale's function, indicate that the expansion coefficient, γ, has least effect on efficiency with no clearly defined optimum value, whereas the contraction

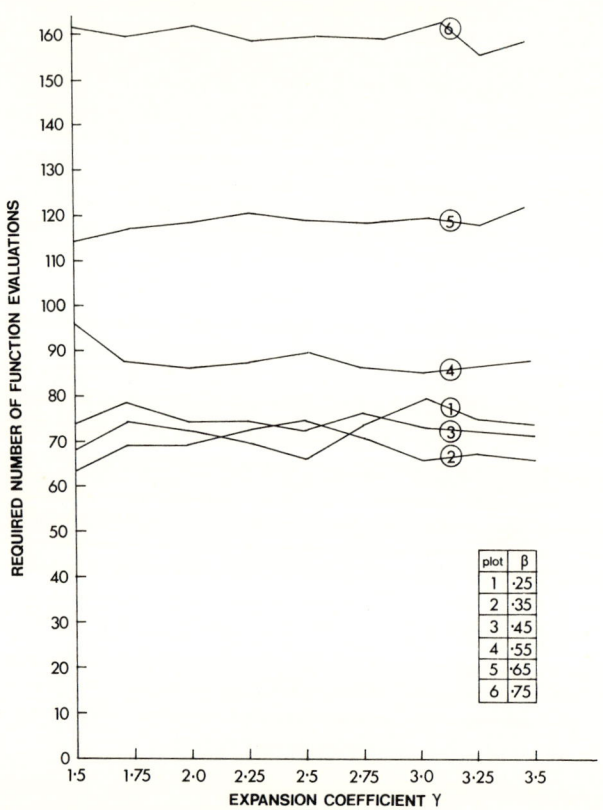

FIG. 15. Plot of required function evaluations against the expansion coefficient (γ) as the contraction coefficient (β) is increased, with the reflection coefficient (α) constant, $\alpha = 0.8$ ($eps = 10^{-8}$).

coefficient, β, has most effect, and clear optima exist (at about 0.35 in these cases). The reflection coefficient, α, is similar to β in the magnitude of its effect and the existence of optima; though the effect of changes in either α or β must be negligible in the neighbourhood of their respective optima, changes of 10% produce changes of ±10% in necessary function evaluations when α or β are remote from their respective optima by 40%.

The same effects were found for all the other test functions, and generally of the same magnitude though the contraction coefficient became less significant in the case of the sum of squares function. It should be noted that the results are based on a fixed orientation though averaged over a range of values for *ISTEP*. Overall, the strategy (2, 0·25, 2·5) was somewhat superior to (1, 0·5, 2) for the set of test functions (average gain of 15%) although this may partly reflect the special properties of these functions, which are so difficult to minimize.

Numerous trials have shown that repeated shrinkage is usually associated with convergence, but repeated contractions indicate a hold-up while the direction of search is inappropriate. The most satisfactory treatment of the former involved counting the shrinkage operations and successively halving the shrinkage coefficient on repetition; this simply speeds up final convergence. On the other hand, a count of repeated contractions was used to trigger a suspension of the contraction operation if the same point is contracted repeatedly; this encouraged 'break out' in oscillatory situations at the bottom of a 'steep sided valley'. The gain, even on the complex functions, was not great (usually <5%), but it did eliminate all but one of the cases of false convergence, though only six such failures were obtained during thousands of trials.

8. Convergence Criteria

It is clear that in general terminating criteria do not guarantee convergence, and that different criteria will cause termination at different positions and for different function values. The simplest terminating criteria are based on corrections becoming sufficiently small, or gradients becoming small, or function values becoming close, or perhaps a quadratic fit giving an 'interior' minimum. Combinations of these are frequently employed, and they can also be associated with other restrictions, e.g. in NMS the test may only follow shrinkage. Nelder and Mead used the condition that the standard deviation over the simplex is small, namely:

$$\frac{1}{n}\sum_{i=0}^{n}(f_i-f)^2 < eps$$

and the test is performed after each iteration; this test is not stringent, but has been satisfactory for all but six cases out of several thousand runs with the unmodified NMS. In each of these cases the failure was apparent convergence to a false solution. In no case did the algorithm fail to converge. However, systematic trials were performed with different criteria substituted in the standard NMS, the most stringent being the requirements that

$$(f_h - f_l) < eps_1$$

and

$$\frac{1}{n} \sum_{i=0}^{n} (\mathbf{x}_i^{(k+1)}, \mathbf{x}_i^{(k)})^2 < eps_2$$

where $\mathbf{x}_i^{(k)}$, $\mathbf{x}_i^{(k+1)}$ are successive positions for V_i, eps_1, and eps_2 are scalars, which restrict the range in f and the corrections to x_i, all i, respectively. There was little to choose between the criteria examined, though the above test gave the most reliable relation between the convergence parameters and the accuracy attained. It is therefore preferred, but in common with all criteria tested it repeated the six failures with Beale's function.

As previously mentioned it is desirable to include a restart to try and detect false convergence. Also, the critical of the two criteria in the test above should be identified so that the user can satisfy his own requirements by a further run with eps_1 or eps_2 reduced, if necessary.

9. Conclusions

The principal investigations involved introducing a variety of modifications into the basic NMS but on an individual basis. Then a final algorithm, termed PHS, was developed which incorporated the most useful of these modifications. The first important modification introduced is that which automatically sets the *scale* of the initial simplex using trials based on steepest descent directions. The alternative, based on linear searches is less satisfactory (see Table 1), but both can give gains of 20%. The next modification is the introduction of so-called *unlimited expansion*. (This is only restricted in practice by the limit to the total number of function evaluations.) The merits of the modification are clearly illustrated in Table 2 and Fig. 13 and can frequently give gains of 20%, though the more complex the objective functions the less predictable is the gain. The next most significant factor affecting efficiency is the choice of *strategy*, which may already be adjusted in NMS though normally set to (1, 0·5, 2); the use of (1, 0·25, 2·5) as a global strategy gave typical gains of about 10% with these test functions. Finally, the *shrinkage* and *contraction checks* which were also incorporated accelerate final convergence and reduce instances of false convergence, respectively, with only marginal gains in efficiency.

The orientation of the initial simplex has a significant effect on efficiency, but the relationship can be too sensitive for an automatic predictor to provide sufficient accuracy at this time (see Figs 6, 7, 8 and 9). Nevertheless, trials with automatic selection for regular functions were encouraging.

The full investigation confirmed the robustness of NMS and indeed all its variants, but the new improved version PHS is substantially more efficient for all the test functions considered. Results for a subset of these functions are quoted in Table 3 to correspond with a set of published results for NMS, but similar gains were obtained with Wood's function and the sum of squares function. In conclusion PHS makes the simplex method highly competitive with currently more favoured methods, particularly for complex functions of low dimensionality.

TABLE 3

A comparison of results for the new algorithm PHS with those for NMS taken from Kowalik and Osborne

Function	Kowalik and Osborne simplex (NMS)		Modified simplex (PHS)	
	Number of function evaluations	Accuracy	Number of function evaluations	Accuracy
Rosenbrock's parabolic valley	200	6.6×10^{-8}	135	5.8×10^{-8}
Beale's function	100	2.01×10^{-8}	67	3.6×10^{-8}
Box's function	290	3.8×10^{-5}	117	6.7×10^{-8}
Rosenbrock's cubic function	206	3.79×10^{-8}	148	3.7×10^{-8}

Acknowledgements

Mr. Parkinson's work was carried out under the sponsorship of the Science Research Council.

References

Box, M. J. (1966). A comparison of several current optimization methods, and the use of transformations in constrained problems. *Comput. J.* **9**, 67–77.
Box, M. J., Davies, D., and Swann, W. H. (1969). 'Non-linear Optimization Techniques'. I.C.I. Monograph No. 5, Oliver and Boyd, Edinburgh.
Kowalik, J., and Osborne, M. R. (1968). 'Methods for Unconstrained Optimization Problems'. American Elsevier Publishing Co., New York.
Nelder, J. A., and Mead, R. (1965). A simplex method for function minimization. *Comput. J.* **7**, 308–313.
Schnabel, B. K. (1966). 'An Investigation into the Effects of Random Error on a Selection of Current Minimization Methods'. M.Sc. Dissertation, University of Leeds.
Spendley, W., Hext, G. R., and Himsworth, F. R. (1962). Sequential application of simplex designs in optimisation and evolutionary operation. *Technometrics* **4**, 441–461.

9. An Experimental Comparison of Three Random Searches

GÜNTHER SCHRACK AND NICK BOROWSKI
*The University of British Columbia, Department of Electrical Engineering,
Vancouver, British Columbia, Canada*

Summary

Although random searches are inefficient as compared with gradient searches, a number of promising algorithms from this area of Monte Carlo methods have recently been proposed. Experiments were conducted on a digital computer in order to evaluate three random searches against each other (Matyas; Schumer *et al.*; and Kjellström). Experimental results are reported on a number of test functions considering the algorithms' behaviour with respect to the rate of average convergence, scaling, dimensionality, and sensitivity to noise.

1. Introduction

This paper presents the results of numerical experiments which were carried out in order to test the practical suitability of the sequential random search algorithms proposed by Matyas (1965), Schumer *et al.* (1968), and Kjellström (1969) as well as to gain some knowledge of their relative merits. The algorithms were chosen for the promise of potential improvement of the search (Matyas), for the claim of possible suitability to problems of high dimension (Schumer *et al.*), and for the novelty of approach (Kjellström).

All algorithms are programmed to solve the unconstrained minimization problem: given the starting point x^0, find x^* such that $f(x^*)$ is a minimum where x is an N-vector and f is the objective function.

The term *random search* is used to denote the solution of a deterministic optimization problem by Monte Carlo techniques. As no derivatives enter into the algorithms, it is a direct search. Constraints can be incorporated without difficulty although the algorithms are usually devised for the unconstrained optimization problem. Both simultaneous and sequential random searches have been proposed, where the terms simultaneous and sequential have the usual connotation and implication as to their suitability to analogue and digital computers.

First, brief descriptions of the algorithms will be given, to be followed by a discussion of the results as obtained for a number of test functions. The final paragraphs will summarize the conclusions.

2. The Random Searches Under Investigation

2.1. *Matyas' Adaptive Random Optimizer (ARS)*

Let x, d, b, u, and ξ be N-vectors; let t, r_s, r_f, c_s, c_f, α_s, α_f be scalars. In the algorithm (defined in Fig. 1), x^0 is the given starting point and ξ is a non-normalized, uniformly distributed random N-vector, generated by the Brown-Muller algorithm, see Muller (1959).

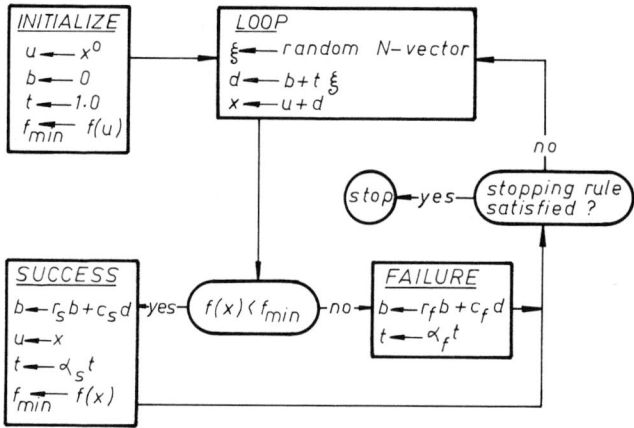

FIG. 1. Matyas' adaptive random search.

Each step d is determined from a random element $t\xi$ and a bias b. The bias b represents the past experience and is a linear combination of the previous step and the previous bias (cf. Fig. 2). The coefficients r_s, r_f control the contribution of the bias b^{k-1} from the previous step to the present bias; they are termed *rejection coefficients*. The coefficients c_s, c_f control the contribution of the previous step d^{k-1} to the present bias; they are called *detection coefficients*. According to Matyas, these coefficients must observe the following inequalities: $0 \leqslant r_s, r_f \leqslant 1$, $c_s > 0$, $c_f \leqslant 0$, $r_s + c_s > 1$, $|r_f + c_f| < 1$. The coefficients α_s, α_f increase and decrease the stepsize control t^k and thus influence the rate of change of the step size of the random part of the search. In the experiments conducted, the following values were used: $r_s = r_f = 0{\cdot}75$, $c_s = 0{\cdot}50$, $c_f = -0{\cdot}25$, $\alpha_s = 1{\cdot}10$, $\alpha_f = 0{\cdot}90$. These values do not seem to be optimal.

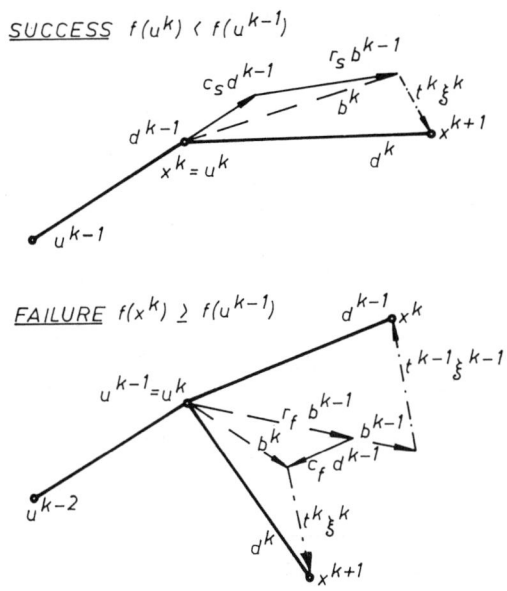

Fig. 2. Determination of a trial point in ARS.

2.2. *The Adaptive Step Size Random Search (ASSRS) of Schumer and Steiglitz*

Let x, u, y, and ξ be N-vectors; let t, α_s, α_f be scalars; and let M be an integer. In Fig. 3, which defines the algorithm, x^0 is the given starting point and ξ is a normalized, uniformly distributed random N-vector. The number M is the maximum number of successive failures after which the step size is decreased. Each step is entirely composed of the random vector $t\xi$ and the past experience is incorporated in the stepsize t only. The step size is controlled by the coefficients α_s and α_f which are given by

$$\alpha_s = 1\cdot 0 + \alpha,$$
$$\alpha_f = 1\cdot 0/(1\cdot 0 + \alpha),$$

with $0 < \alpha < 1\cdot 0$; used were $M = 3N$, $\alpha = 0\cdot 618$.

The concept of ASSRS is based on a theoretical investigation of this type of algorithm with the hypersphere as the objective function.

2.3. *Kjellström's Random Search (KRS)*

This algorithm is based on statistical concepts, and thus represents an entirely different approach compared to the other two searches. The characteristic feature of this method is its use of a random walk in a suitably defined region

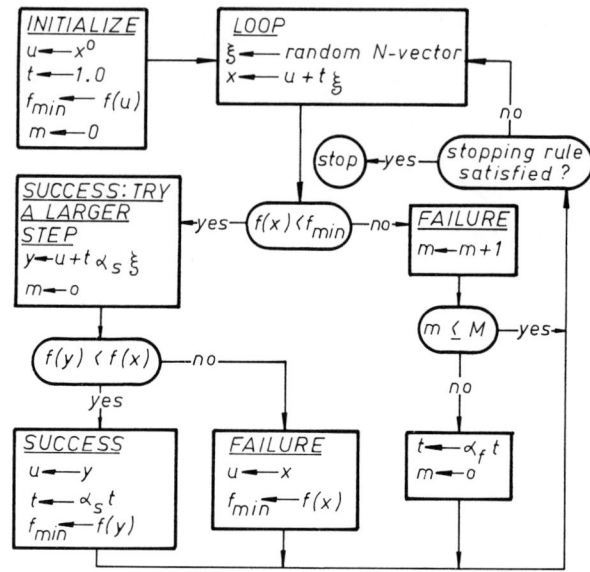

FIG. 3. Adaptive step size random search.

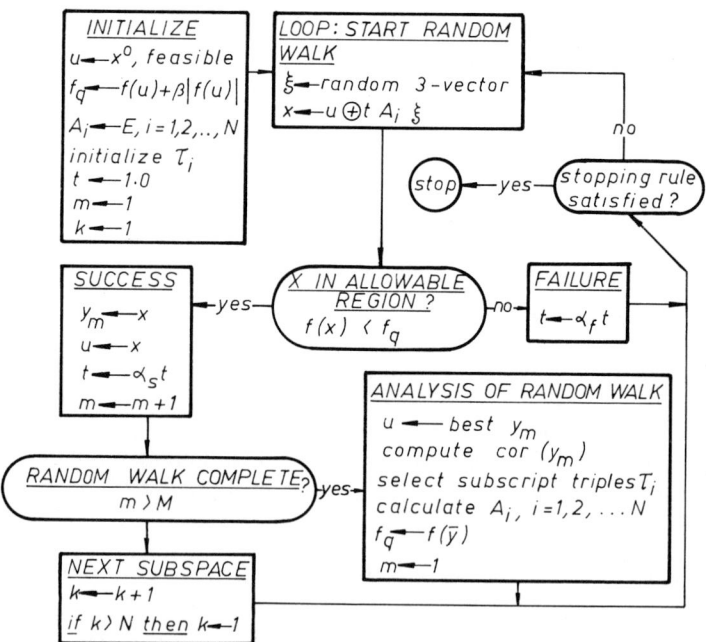

FIG. 4. Kjellström's random search.

9. THREE RANDOM SEARCHES 141

of the parameter space in order to determine the correlation between parameters. The information thus gained is analysed and then used to guide the three-dimensional random walk in the succeeding iteration.

Let x, $\{y_j\}$, and u be N-vectors; let ξ be a 3-vector; let t, α_s, α_f, β be scalars; let $\{A_j\}$, E be 3×3 matrices, E the unit matrix; let M be an integer, and $\{\tau_i\}$ triples of integers. In the algorithm, presented in Fig. 4, x^0 is the given starting point and ξ is a non-normalized, uniformly distributed random 3-vector. The operation $\oplus$ adds the 3-vector $tA_i\xi$ to the three elements of u as defined by the subscript triple τ_i and leaves the remaining elements of u unchanged. Initially $\{\tau_i\} = \{(1,2,3), (2,3,4), \ldots, (N,1,2)\}$. The vectors y_m store the successful steps, upon which the analysis of the random walk is based. This analysis is carried out by first calculating the $N \times N$ correlation matrix of the x_i with the y_j's, selecting for each x_i, $i = 1, 2, \ldots, N$, the two closest correlated x's, thus defining the subscript triples τ_i. These triples define the subspaces in which the random walks are conducted. Furthermore, the scaling of the random walks is controlled by the A_i matrices which are obtained by the use of the concentration ellipsoid obtained from the correlation matrix and the corresponding τ_i. Finally, all $f(y_m)$ are sorted and the median $\bar{y}$ selected to define $f_q \leftarrow f(\bar{y})$. The step size control constants α_s, α_f were determined experimentally by Kjellström and have the values $\alpha_s = 1\cdot 115$, $\alpha_f = 0\cdot 933$. The scalar β defines the 'ceiling' of the allowable region within which the random walks are conducted. The number of steps in each random walk is given by M; used were $\beta = 0\cdot 01$ and $M = 3N$.

3. Numerical Experiments and Results

Each author of the investigated algorithms reported numerical results for a certain objective function. A close duplication of these experiments proved to be difficult, however, as in each case some of the design parameters were not reported. Nonetheless, a comparison of the results obtained here with those reported seem close enough to justify the assumption that the algorithms were implemented as intended by their authors.

All algorithms are coded in FORTRAN IV and tested on an IBM 360-67 running under the Michigan Terminal System. The following characteristics were studied: rate of average convergence for several different types of objective functions, effect of scaling, effect of dimension, and the effect of Gaussian and uniform noise in the objective functions on the rate of average convergence.

3.1. *Rate of Average Convergence*

As will be seen below, the optimum seeking capability of the searches vary widely, depending on the test function used. It was not possible to reduce the function value of some of the test functions to a predetermined value such as, say, 10^{-4} (all test functions have the optimum value of $f(x^*) = 0$). Therefore, all comparisons are based on the function value obtained for a given number of function evaluations n (successes and failures). Chosen were the values $n = 100, 200$, and 300.

The rate of convergence of a random search depends heavily, of course, on the sequence of random numbers used to generate the random vectors ξ. In order to average out this dependence, each optimization was conducted r times sequentially, thus representing runs with different random number

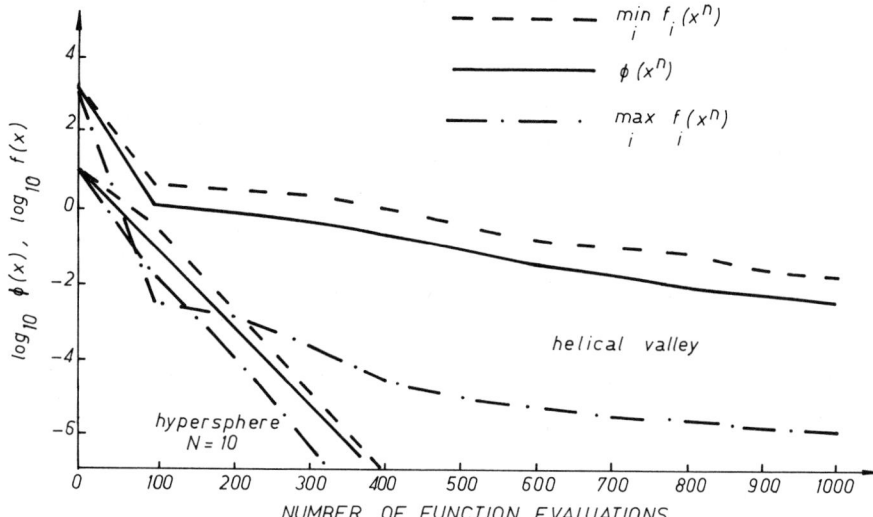

FIG. 5. Deviations from $\phi(x)$ for the adaptive random search.

generator initializations. Let $f_i(x^n)$ be the value of $f(x)$ at the nth function evaluation of the ith run, $i = 1, 2, \ldots, r$, and let $\phi(x^n)$ be the average value of the objective function from r runs, i.e.

$$\phi(x^n) = \frac{1}{r} \sum_{1}^{r} f_i(x^n).$$

Then the *rate of average convergence* is defined as the rate of convergence of $\phi(x)$. All results reported here are based on $r = 10$. The experimental results for each test function are displayed in tables giving the value of $\phi(x^n)$ for $n = 0, 100, 200$, and 300. $\phi(x^0) = f(x^0)$ is the value at the starting point. In the tables, the notation 1·21 E-3 denotes $1·21 \times 10^{-3}$.

A plot of the numerical results for two of the objective functions discussed below is reproduced in Fig. 5, displaying $\phi(x^n)$, $\min_i f_i(x^n)$, and $\max_i f_i(x^n)$ versus n, $n \leq 900$. As can be seen, the deviation of a single optimization from $\phi(x^n)$ can be considerable.

3.2. Basic Test Functions

The first four test functions were selected to represent objective functions with varying degrees of difficulty.

Function 1: the hypersphere of dimension $N = 5$,

$$f(x) = \sum_1^N x_i^2, \qquad x^0 = (1, 1, \ldots, 1).$$

TABLE 1
Test function 1: hypersphere, $N = 5$

n	0	100	200	300
ARS	5·00	1·21 E-3	9·52 E-8	6·47 E-12
ASSRS	5·00	1·41 E-3	3·80 E-7	4·09 E-11
KRS	5·00	6·60 E-1	2·24 E-1	1·45 E-1

TABLE 2
Test function 2: helical valley, $N = 3$

n	0	100	200	300
ARS	2·50 E3	4·23	2·88	1·29
ASSRS	2·50 E3	2·19	1·69	1·20
KRS	2·50 E3	1·47 E1	4·52	3·00

The results are given in Table 1. The algorithms ARS and ASSRS perform about equally well, showing a reduction of $\phi(x)$ to $\simeq 10^{-12}$ with $n = 300$ function evaluations. The algorithm KRS, however, was considerably slower: after 300 function evaluations the value of $\phi(x)$ was reduced to only 0·145.

Function 2: Fletcher and Powell's (1963) helical valley with starting point as suggested by its authors. For this function, the behaviour of the three searches were similar, see Table 2. Initially, for approximately the first 100 function evaluations, good progress was made, ASSRS being best, followed by ARS. However, the searches showed a change in average convergence thereafter, slowing down considerably (cf. Fig. 5, where the results for the helical valley for the ARS are shown graphically for $n \leqslant 900$).

Function 3: Powell's quartic of dimension $N = 4$ with starting point as given in Fletcher and Powell (1963). The rate of average convergence for ARS and ASSRS are similar (see Table 3) and approximately equal to that for function 2, whereas KRS was again considerably slower.

TABLE 3
Test function 3: Powell's quartic, $N = 4$

n	0	100	200	300
ARS	2·15 E2	7·27 E-1	6·49 E-2	9·40 E-3
ASSRS	2·15 E2	6·97 E-1	1·02 E-1	2·74 E-2
KRS	2·15 E2	9·48	5·12	4·68

Function 4: the moon problem (see Kjellström, 1969) of dimension $N = 6$;

$$f(x) = \begin{cases} \sum_{1}^{N} (x_i - 0\cdot5)^2 - N/4 & \text{if } g(x) \geqslant N, \\ \infty, & \text{if } g(x) < N, \end{cases}$$

where

$$g(x) = \sum_{1}^{N} x_i^2, \qquad x^0 = (-1\cdot2, -1\cdot2, \ldots, -1\cdot2).$$

TABLE 4
Test function 4: moon problem, $N = 6$

n	0	100	200	300
ARS	1·58 E1	4·45	3·61	3·32
ASSRS	1·58 E1	6·67	6·35	6·21
KRS	1·58 E1	4·24	2·38	1·58

This problem is representative for constrained optimization. The forbidden region is the interior of a hypersphere with radius $\sqrt{N}$, the minimum is at $x^* = (1, 1, \ldots, 1)$. The starting point x^0 is on the antipodal point of x^* but off the surface such that the search path is forced to follow around the 'moon'. The results (Table 4) show that the problem is difficult for all three searches. However, this is the only function where KRS was the best, a notable fact in view of the difficulty of the problem.

3.3. *Effect of Scaling*

Function 5: to demonstrate the extent to which the algorithms depend on the scaling of the objective function, the following simple hyperellipsoid of dimension $N = 5$ served as test function:

$$f(x) = 0 \cdot 1 x_1^2 + \sum_2^N x_i^2, \qquad x^0 = (1, 1, \ldots, 1).$$

TABLE 5
Test function 5: hyperellipsoid, $N = 5$

n	0	100	200	300
ARS	4·10	2·05 E-2	6·84 E-4	5·76 E-6
ASSRS	4·10	2·38 E-2	6·28 E-4	1·40 E-5
KRS	4·10	5·46 E-1	2·44 E-1	1·46 E-1

Again, ARS and ASSRS behaved similarly and KRS was much slower (Table 5). A comparison of the results with those for the hypersphere of the same dimension (function 1), however, shows that KRS does not differ significantly in behaviour whereas the other two were affected by the change of scale, being much slower for function 5.

3.4. *Effect of Dimensionality*

Function 6: hypersphere of dimension $N = 10$. The effect of dimensionality of a function was tested by comparison of the rate of average convergence for the hyperspheres of dimensions $N = 5$ and $N = 10$. From these experiments it is apparent (see Table 6) that the efficiency of the three searches is influenced in a similar fashion by an increase in dimensions: for a given reduction of the function value, a doubling of the dimension doubles the number of function evaluations required for ARS, more than doubles this number for ASSRS, but quadruples these for KRS (furthermore, see Fig. 5 for the results for the hypersphere of dimension $N = 10$, obtained for ARS for $n \leqslant 900$).

TABLE 6
Test function 6: hypersphere, $N = 10$

n	0	100	200	300
ARS	1·00 E1	1·18 E-1	8·48 E-4	6·12 E-6
ASSRS	1·00 E1	2·20 E-1	4·50 E-3	7·09 E-5
KRS	1·00 E1	3·71	2·64	2·39

3.5. *Effect of Gaussian Noise Perturbations to the Objective Function*

Test functions 7 and 8 below are not deterministic, they are stochastic functions. Nevertheless, random searches can be applied to such functions, as the following results show.

Function 7: the hypersphere of dimension $N = 5$ perturbed by a Gaussian noise level of 1% of the value of the objective function:

$$f(x) = \left(\sum_1^N x_i^2\right)(1 \cdot 0 + 0 \cdot 01 \xi_1), \qquad x^0 = (1, 1, \ldots, 1),$$

TABLE 7
Test function 7: hypersphere with 1% Gaussian noise, $N = 5$

n	0	100	200	300
ARS	5·01	6·03 E-4	5·56 E-8	2·13 E-12
ASSRS	5·01	1·04 E-3	1·10 E-7	2·70 E-11
KRS	5·01	5·74 E-1	2·25 E-1	1·63 E-1

where ξ_1 is normally distributed with zero mean and unit variance.

All three random searches performed nearly identically (Table 7) when compared with the deterministic hypersphere of the same dimension, i.e. they do not seem to be affected by such random variations of the objective function.

3.6. *Effect of Uniform Noise Added to the Objective Function*

Function 8: for this function, the noise added is independent of the value of the objective function. The test function is (with $N = 5$):

$$f(x) = \left|\sum_1^N x_i^2 + 0 \cdot 05 \xi_2\right|, \qquad x^0 = (1, 1, ,\ldots, 1),$$

where ξ_2 is uniformly distributed in $(-1, 1)$. The algorithms stop progressing when the noise level is reached (Table 8). However, up to that level, the rate of average convergence was not appreciably affected.

TABLE 8
Test function 8: hypersphere with 1% uniform noise, $N = 5$

n	0	100	200	300
ARS	4·96	1·18 E-2	1·10 E-2	1·07 E-2
ASSRS	4·96	1·78 E-2	1·68 E-2	1·64 E-2
KRS	4·96	5·49 E-1	2·16 E-1	1·15 E-1

4. Conclusion

All three random searches exhibited a linear or nearly linear average convergence behaviour. In the early stages of a search, Matyas' adaptive random search (ARS) and the adaptive step size random search (ASSRS) show a much higher rate of average convergence than Kjellström's random search (KRS), usually more than twice as much with the exception of function 4. For all eight test functions, ARS and ASSRS show a similar rate of average convergence, ARS being slightly better with the exception for function 2. Algorithm KRS is considerably slower than the other two except for the moon problem (function 4). The algorithms ARS and ASSRS are seriously affected by the scaling of a function, a characteristic KRS does not display. All three searches are affected by the dimensionality of a function; KRS, however, more so than the other two. Gaussian noise added to a function does not affect the searches.

The searches generally showed a superior rate of average convergence during the first 100 function evaluations. This suggests the use of random searches as a starting algorithm for other optimization algorithms. Also, the improvement of a random search should concentrate on an improvement of its behaviour near the optimum.

This study clearly shows that a further investigation of random searches is warranted and desirable.

References

Fletcher, R., and Powell, M. J. D. (1963). A rapidly convergent descent method for optimization. *Comput. J.* **6**, 163–168.

Kjellström, G. (1969). Network optimization by random variation of component values. *Ericsson Technics* **25**, 133–151.

Matyas, J. (1965). Random optimization. *Automat. Rem. Contr.* **26**, 224–251.

Muller, M. E. (1959). A note on a method for generating points uniformly on N-dimensional spheres. *Comm. ACM* **2**, 19–20.

Schumer, M. A., and Steiglitz, K. (1968). Adaptive step size random search. *IEEE Trans.* **AC-13**, 270–276.

10. The Choice of Step Length, a Crucial Factor in the Performance of Variable Metric Algorithms

L. C. W. Dixon

The Numerical Optimization Centre, The Hatfield Polytechnic, Hatfield, Hertfordshire, England

Summary

A number of different variable metric algorithms have recently been proposed and many reports containing numerical examples of their performance are now available. In many of these reports different choices of step length have been used, thus complicating the comparison of their relative performance. At first sight this might appear to be an effect of secondary importance. It is the aim of this paper to show that it is crucial.

Many of these algorithms belong to a family introduced by Huang (1970), who showed that, provided the step length is determined by an accurate line search, then all these algorithms produce an identical sequence of points on a quadratic function. Dixon (1971) gives conditions for a subset of this family to generate identical points on more general functions. It will be shown that the members of the Broyden (1967) family satisfy these conditions. As most implementations do not undertake accurate line searches, it follows that the reported differences in performance of members of this family are crucially dependent on the choice of step length.

1. Introduction

The suggestion by Davidon (1959) that the inverse Hessian matrix of a function can be constructed iteratively during the solution of an n-dimensional minimization problem provides the basis of many optimization algorithms. Since the well-known implementation by Fletcher and Powell (1963) was found to fail under certain circumstances (Bard, 1968) a considerable number of alternative updating formulae have been suggested.

Recently, it has been shown that these formulae can be combined into families, and the properties of these families have been theoretically investigated. Regrettably, most of these theoretical studies only apply to quadratic functions. Two families will be discussed in this paper: the one parameter family introduced by Broyden (1967) and discussed in detail by Broyden (1970), Fletcher (1970), Shanno (1970) and Goldfarb (1970), and the more general family introduced by Huang (1970).

Many of the above papers include numerical data gathered by testing implementations on standard test functions. In these comparisons many different updating formulas have been used in conjunction with different strategies for determining the step length. In this paper some theoretical properties of the behaviour of these algorithms on general functions are presented. The published numerical data is then examined, with emphasis on these theoretical properties of the different strategies and supplemented by a consistent set of data.

2. Theoretical Properties

2.1. *Properties of Huang's Family*

In Huang (1970) a family of updating formulae is introduced given by

$$H_i = H_{i-1} + \rho_{i-1} \frac{\Delta x_{i-1}(C_1 \Delta x_{i-1} + C_2 H_{i-1}^T \Delta g_{i-1})^T}{(C_1 \Delta x_{i-1} + C_2 H_{i-1}^T \Delta g_{i-1})^T \Delta g_{i-1}} + \\ - \frac{H_{i-1} \Delta g_{i-1}(K_1 \Delta x_{i-1} + K_2 H_{i-1}^T \Delta g_{i-1})^T}{(K_1 \Delta x_{i-1} + K_2 H_{i-1}^T \Delta g_{i-1})^T \Delta g_{i-1}}, \qquad (1)$$

where the parameters ρ_{i-1}, $C^{(i-1)} = C_1/C_2$, $K^{(i-1)} = K_1/K_2$ provide three degrees of freedom at each iteration.

He considered the following implementation. At a point x_i calculate f and g_i and hence the direction

$$p_i = H_i^T g_i. \qquad (2)$$

Choose a step length α_i so that

$$x_{i+1} = x_i + \Delta x_i \qquad (3)$$

and

$$\Delta x_i = -\alpha_i p_i, \qquad (4)$$

where α_i is chosen such that

$$g_{i+1}^T p_i = 0. \qquad (5)$$

The matrix H is then updated using (1) where

$$\Delta g_{i-1} = g_i - g_{i-1}, \qquad (6)$$

and the process repeated until a point is found at which

$$g_i^T g_i < \epsilon_1. \qquad (7)$$

10. THE CHOICE OF STEP LENGTH

Huang (1970) showed theoretically that all members of family (1) generate the same sequence of points x_i when applied to a quadratic function. Dixon (1971) extended this result by considering the behaviour on any differentiable function and showed that

$$p_i = -bu_i \qquad (8)$$

where

$$u_i = \left\{ \prod_{j=0}^{i-1} \left(I - \frac{\Delta x_j \Delta g_j^T}{\Delta x_j^T \Delta g_j} \right) \right\} H_0^T g_i +$$

$$+ \sum_{j=0}^{i-2} \left\{ \prod_{k=j+1}^{i-1} \left(I - \frac{\Delta x_k \Delta g_k^T}{\Delta x_k^T \Delta g_k} \right) \right\} \left(\rho_j \frac{\Delta x_j^T g_i}{\Delta x_j^T \Delta g_j} \right) \Delta x_j \qquad (9)$$

and

$$b = + \frac{(\alpha_{i-1} K_1 + K_2) \Delta g_{i-1}^T H_{i-1}^T g_{i-1}}{(K_1 \Delta x_{i-1} + K_2 H_{i-1}^T \Delta g_{i-1})^T \Delta g_{i-1}}. \qquad (10)$$

Suppose that at any iteration

$$\alpha_{i-1} K_1 + K_2 \neq 0 \quad \text{and} \quad (K_1 \Delta x_{i-1} + K_2 H_{i-1}^T \Delta g_{i-1})^T \Delta g_{i-1} \neq 0. \qquad (11)$$

Let us extend the definition (5) of α_i, so that α_i is defined uniquely on non-convex functions. Then all members of the family having the same values of ρ_i generate identical sequences x_i on any general function, provided conditions (11) are not satisfied. (The proof remains valid if ρ_i varies from iteration to iteration.)

The unique value of α_i can be obtained by accepting the first local minimum in the downhill direction $\pm p_i$.

It will be noted that these necessary and sufficient conditions agree with the numerical data in Huang and Levy (1970) who tested four algorithms with $\rho_i = 1$ and three with $\rho_i = 0$ and noted that these produced two identical sequences of points.

2.2. *Properties of Broyden's Family*

In the notation given above the original Davidon-Fletcher-Powell (DFP) formula is given by

$$H_{\text{DFP}} = H + \frac{\Delta x \Delta x^T}{\Delta x^T \Delta g} - \frac{H \Delta g \Delta g^T H}{\Delta g^T H \Delta g} \qquad (12)$$

which is a member of Huang's family with $\rho = 1$, $C = \infty$, $K = 0$ and was included in the Huang and Levy test series.

In Broyden (1967) the family of formulae given by

$$H_i = H_{\text{DFP}} + \beta \left(\frac{H\Delta g \, \Delta g^T H}{\Delta g^T H \Delta g} (\Delta g^T \Delta x) - H\Delta g \Delta x^T - \Delta x \Delta g^T H + \frac{\Delta x \Delta x^T}{\Delta x^T \Delta g} \Delta g^T H \Delta g \right) \qquad (13)$$

is introduced. In Broyden (1970) the properties of this family are examined for a quadratic function, and for reasons of stability in the convergence of the sequence H_i to G^{-1}, the value

$$\beta = \frac{1}{\Delta x^T \Delta g} \qquad (14)$$

is recommended. In this report the combination of (13) and (14) will be denoted as the Broyden-Fletcher-Shanno (BFS) formula H_{BFS}.

The same family is considered in Fletcher (1970) but in the form

$$H_i = (1 - \phi) H_{\text{DFP}} + \phi H_{\text{BFS}} \qquad (15)$$

and also in Shanno (1970), where it is written as

$$H_i = H_{i-1} + t \frac{\Delta x \Delta x^T}{\Delta x^T \Delta g} + \frac{[(1-t)\Delta x - H\Delta g][(1-t)\Delta x - H\Delta g]^T}{[(1-t)\Delta x - H\Delta g]^T \Delta g}. \qquad (16)$$

An important member of this family is the symmetric rank-one (R1) formula

$$H_{\text{R1}} = H + \frac{(\Delta x - H\Delta g)(\Delta x - H\Delta g)^T}{(\Delta x - H\Delta g)^T \Delta g} \qquad (17)$$

which corresponds to $\rho = 1$, $C = -1$, $K = -1$ and was also included in the Huang and Levy test series. Wolfe (1967) showed that with this formula it is unnecessary to satisfy condition (5) if the sequence $\{H_i\}$ is to tend to G^{-1} in a finite number of steps on a quadratic function.

For this reason the author prefers to consider the family in the form

$$H = H_{\text{R1}} - avv^T \left(\frac{(\Delta x^T \Delta g)(\Delta g^T H \Delta g)}{(\Delta x - H\Delta g)^T \Delta g} \right) \qquad (18)$$

where

$$v = \frac{\Delta x}{\Delta x^T \Delta g} - \frac{H\Delta g}{\Delta g^T H \Delta g} \qquad (19)$$

leads to Broyden's family but is not the only possible choice of v. This form emphasizes the rank-two nature of the update and also demonstrates that it is the presence of the terms avv^T, introduced for stability reasons, that introduces

10. THE CHOICE OF STEP LENGTH

TABLE 1

Algorithm		Broyden	Fletcher	Shanno		Huang's Family	
	a	β	ϕ	t	ρ	C	K
RANK 1	0	$\dfrac{1}{(\Delta x - H\Delta g)^T \Delta g}$	$\dfrac{\Delta x^T \Delta g}{(\Delta x - H\Delta g)^T \Delta g}$	0	1	-1	-1
H_{DFP}	1	0	0	1	1	∞	0
H_{BFS}	$\dfrac{\Delta g^T H \Delta g}{\Delta x^T \Delta g}$	$\dfrac{1}{\Delta x^T \Delta g}$	1	∞	1	∞	∞
H	a	$\dfrac{1-a}{(\Delta x - H\Delta g)^T \Delta g}$	$\dfrac{(1-a)\Delta x^T \Delta g}{(\Delta x - H\Delta g)^T \Delta g}$	$\dfrac{a(\Delta x - H\Delta g)^T \Delta g}{(a\Delta x - H\Delta g)^T \Delta g}$	1	$\dfrac{-(\Delta x + H\Delta g)^T \Delta g}{\Delta x^T \Delta g}$ $\dfrac{(\Delta x - aH\Delta g)^T \Delta g}{(a-1)\Delta x^T \Delta g}$	$\dfrac{-(a-i)\Delta g^T H \Delta g}{(a\Delta x - H\Delta g)^T \Delta g}$

the need for condition (5). The relationship between these expressions is given in Table 1.

Fletcher (1970) demonstrated that for stability reasons a must lie at or between the limits corresponding to the DFP and BFS algorithms, and recommended using H_{DFP} if

$$\Delta g^T H \Delta g > \Delta x^T \Delta g$$

or otherwise switching to H_{BFS}.

It follows from the table that all members of the Broyden (1967) family are members of Huang's family with $\rho_i = 1$, and hence belong to a subset. If α_i were chosen to satisfy (5) and (11) they would all generate identical sequences $\{x_i\}$. The fact that this had never been observed prior to Huang and Levy (1970) led the author to undertake an analysis of the published data on this family. Before proceeding to this analysis some comments will be made on the likely effect of common safeguards on these identical sequences.

2.3. The Predicted Step Length

From formula (9) it follows that members of Huang's family with identical ρ_i generate the same directions and hence when combined with conditions 5 and 12 the same sequence $\{x_i\}$. However, the initial prediction

$$p_i = -bu_i \tag{20}$$

depends on the value of b. As

$$b = + \frac{(\alpha K + 1)\Delta g_{i-1}^T H_{i-1}^T g_{i-1}}{(K \Delta x_{i-1} + H_{i-1}^T \Delta g_{i-1})^T \Delta g_{i-1}} \tag{21}$$

this obviously depends upon K, and hence is different for each member of Broyden's family. It will be noted that b can take positive, negative, zero and infinite values, and while b positive corresponds to a downhill direction, the others are not so convenient.

If $b < 0$ then the initial slope is uphill and the matrix must be non-positive definite. Theoretically this cannot occur with the H_{DFP} or H_{BFS} updating formula, but it is possible with H_{R1} when most implementations reject direction (19) and reset H. This policy destroys the sequence $\{x_i\}$.

If $\alpha_i = -1/K$, then $b = 0$, and the next step identically vanishes. This result has been noted before for the rank-one formula ($K = -1$).

If $\Delta x^T \Delta g = -\Delta g^T H \Delta g / K$, then division by zero would have to be prevented.

Neither of these difficulties can arise when $K = 0$ or ∞, which correspond to the DFP and BFS formulae.

Of the three commonly used members of the family only the rank-one needs

10. THE CHOICE OF STEP LENGTH

to be protected for the last two difficulties. In the Sargent and Murtagh (1970) implementation the safeguard for $\alpha = 1$ is to retain

$$H_i = H_{i-1} \tag{22}$$

and the safeguard for division by zero is to replace the formulae by

$$H_i = H_{i-1} + \frac{(\Delta x - H\Delta g)(\Delta x - H\Delta g)^T}{(\Delta x - H\Delta g)^T(\Delta x - H\Delta g)} \tag{23}$$

which is not a member of Huang's family. The use of (22) or (23) destroys the sequence $\{x_i\}$ and to obtain a consistent sequence the author found it necessary to treat b and u_i separately in (20) restricting extreme values of b and accepting uphill directions. It is not implied that this necessarily improves the efficiency.

3. Numerical Evidence

In order to confirm that the above behaviour occurred, some detailed runs were undertaken. In these runs a parabolic interpolation process was used. As it was necessary to find the minimum exactly, a highly safeguarded search was employed. The fact that simple parabolic routines could terminate at points other than the minimum has been reported in an earlier paper (Dixon, 1969). The procedure used, first evaluated the function at

$$|\alpha^{(0)}| = \min(1, s)$$

where s was a safety limit, the step being taken in the downhill direction. A parabola was then fitted to the known function value and slope at $\alpha = 0$ to give a further prediction $\alpha^{(1)}$.

If $|\alpha^{(1)}| > |\alpha^{(0)}|$ and a bracket had not been formed, then the outerpoint was extended by a constant factor 5 until a bracket occurred. If $|\alpha^{(1)}| < |\alpha^{(0)}|$ and a bracket had not been found, then $\alpha^{(0)} := \alpha^{(1)}$ and a parabola was fitted again to the function value and slope at $\alpha = 0$ until a bracket occurred. Once a bracket $L < \alpha < M$ has been found, parabolic interpolation was employed; if the prediction $\alpha^{(E)} = L + E(M - L)$ and $0.25 < E < 0.75$ then $\alpha^{(E)}$ was accepted, otherwise $E = 0.25$ or 0.75 was preferred. If this point was an improvement then the new bracket was established and the process repeated. If the point was not an improvement then the far end point was also rejected in favour of a point midway between it and the current best point; the bracket enclosing the best of these points was then accepted. This process ensures that on a skew function one quarter of the bracket round the minimum is removed at each interpolation.

The point was accepted when either the new predicted value $|\alpha^{(E)}|$ agreed with the current value $|\alpha^{(c)}|$ within $\epsilon_2|\alpha^{(c)}|$, or the bracket itself was reduced to $\epsilon_2|\alpha^{(c)}|$ in size. Using this procedure with $\epsilon_2 = 10^{-7}$ on Powell's function gave

TABLE 2
Behaviour on Powell's function: accurate line search

Iteration number	DFP	BFS	Fletcher Switch	Rank-one	Rank-one safeguarded
1	30·8302	30·8302	30·8302	30·8302	30·8302
2	18·5408	18·5408	18·5408	18·5408	18·5408
3	10·4095	10·4095	10·4095	10·4095	10·4095
4	2·93573 E-2	2·93559 E-2	2·93573 E-2	2·93556 E-2	2·93556 E-2
5	2·31547 E-2	2·31539 E-2	2·31547 E-2	2·31537 E-2	2·31537 E-2
6	7·06969 E-3	7·06806 E-3	7·06418 E-2	7·05737 E-3	7·05737 E-3
7	4·47354 E-3	4·4725 E-3	4·47069 E-3	4·46561 E-3	4·46561 E-3
8	2·13856 E-3	2·13865 E-3	2·13918 E-3	2·14008 E-3	2·14008 E-3
9	1·99420 E-3	1·99509 E-3	1·99756 E-3	2·00075 E-3	2·00075 E-3
10	1·32662 E-3	1·33148 E-3	1·34282 E-3	1·35640 E-3	1·94696 E-3
11	1·28077 E-4	1·19872 E-4	1·19219 E-4	1·18368 E-4	6·95484 E-5
12	4·47069 E-5	4·23202 E-5	4·21829 E-5	4·19831 E-5	1·75158 E-5
13	6·15553 E-6	5·54279 E-6	5·42274 E-6	5·32000 E-6	4·83724 E-7
14	1·63994 E-6	1·48582 E-6	1·45576 E-6	1·42964 E-6	9·79914 E-8
15	2·20355 E-7	1·87257 E-7	1·79436 E-7	1·72515 E-7	1·46662 E-9
16	5·86361 E-8	5·00679 E-8	4·79858 E-8	4·61418 E-8	2·39827 E-10
17	7·15174 E-9	5·62893 E-9	5·22315 E-9	4·88093 E-9	1·84082 E-12
18	1·89154 E-9	3·94280 E-9	3·62250 E-9	3·33838 E-9	C
19	1·81931 E-9	1·36482 E-9	1·27209 E-9	1·20459 E-9	
20	C	2·65064 E-10	2·62303 E-10	2·88125 E-10	

the function values in Table 2; it will be noted that the first four columns which should theoretically be identical are remarkably consistent. The amount of computer rounding error is very small. The fifth column gives the numbers for a rank-one process in which uphill directions were rejected and steepest descent used instead; this occurred at iteration 10 and in this case lead to improved convergence. When the runs were repeated with less exact line searches, $\epsilon_2 = 10^{-4}$ for example, the similarity in the first four columns was no longer obvious. As no implementation (known to the author) attempts to find the minimum on a non-quadratic function as accurately as this, the fact that this property had not been observed before is not too surprising.

4. Alternative Strategies for Choosing the Step Length

4.1. *Acceptability Criteria*

In Section 2 implementations of the variable metric idea were considered that assumed that the step length α_i was chosen to satisfy condition (5) namely

$$g_{i+1}^T p_i = 0.$$

Very few implementations have attempted to satisfy this condition in practice, and of those that have, e.g. Broyden (1970), there is little indication that the tolerances used were similar to those discussed in Section 3.

Most implementations are satisfied if an improved point has been found by an interpolation process, either cubic or parabolic. Such an implementation would, of course, find the true minimum along the line on a quadratic function. The implementations due to Fletcher and Powell (1963) and Fletcher (1966) are of this form.

More recently, following Goldstein and Price (1967), it has been suggested that a point may be accepted if it gives an acceptable reduction in function value. The acceptability conditions are often a combination of the following conditions.

I. The following conditions are often used to prevent the step size from being too large:

(1) $f(x_i) - f(x_{i+1}) > 0$. This has now been shown to be unsatisfactory, Fletcher (1970b).

(2) $f(x_i) - f(x_{i+1}) > \epsilon_3 \alpha g_i^T H_i g_i$.

(3) $f(x_i) - f(x_{i+1}) > \epsilon_3 \delta f_Q$;

where $\alpha g_i^T H_i g_i$ would be the reduction in f if the function were truly linear and δf_Q the reduction if it were truly quadratic. As these conditions must be satisfied as $\alpha \to 0$, their failure indicates that the step tried is probably too large. In the implementations leading to the results discussed in Section 5.2, condition (2) was used with $\epsilon_3 = 0.1$.

II. If
$$(1 - \epsilon_3)\alpha g_i^T H_i g_i < f(x_i) - f(x_{i+1}) < (1 + \epsilon_3)\alpha g_i^T H_i g_i$$
then the move taken has been effectively linear and should be increased. This is a slight modification of a test given in Goldstein and Price (1967). The value $\epsilon_3 = 0\cdot 1$ was used to obtain the values given in Section 5.2.

III. Positive-definite step. In most implementations positive definite matrices are required and it is usual only to accept a point that is consistent with this condition, i.e. is such that $\Delta x^T \Delta g > 0$.

Alternative versions of conditions I and II above, together with a discussion of their convergence properties, is given in Wolfe (1969).

4.2. Search Procedure

Implementations will be given a four-digit code. The *first* digit will be A if the line search is nominally accurate, B if a bracketed point is accepted and C if a combination of conditions I–III is used.

Three main aspects of the search procedure will be distinguished in the rest of the code: the initial step, any subsequent interpolation, and any subsequent extrapolation.

The *second* letter in the code will denote the *initial step*.

A will denote
$$\alpha^{(0)} = 1.$$
B will denote
$$\alpha^{(0)} = \min(1, s),$$
where s is a safety limit. Commonly used values of s have been
$$s = 2(f - est)/g^T Hg, \text{ if } f > est,$$
$$s = 1 \text{ otherwise,}$$
where *est* is the estimated value of the function value at the minimum; or
$$s = 2|\Delta x_{i-1}|/|Hg|,$$
using the previously successful step to limit the subsequent initial step.

C will denote a more general system where the initial value of α is determined by the previous behaviour.

The *third* letter of the code will denote the *interpolation procedure*. The main feature will be coded where alternatives are possible.

A will denote constant factor reduction.
B will denote parabolic interpolation.
C will denote cubic interpolation.
E will include any other policy.

The *fourth* and final letter of the code will denote the *extrapolation procedure*.
A will denote constant factor extrapolation.
B will denote parabolic extrapolation.
E will include any other policy.

As an example of the use of this code, the algorithm described in Fletcher and Powell (1963) is B, BCA. It accepts a bracketed point, applies a safety limit to the first step, and uses cubic interpolation and constant factor extrapolation.

4.3. *Published Implementations*

In Table 3 the codings of a number of published implementations are listed. This table is no doubt incomplete but indicates the distribution of categories that have been tried.

In preparing Table 3 it became obvious that there were notable gaps; there were no DFP or BFS results in which the acceptability criteria were applied, whilst the switching policy had only been reported in combination with these

TABLE 3
Codings of published algorithms

Implementation	Line search A	Bracket B	Conditions C
DAVIDON-FLETCHER-POWELL FORMULA			
Fletcher-Powell	63		BCA
Fletcher	66		CCA
Stewart[a]	67	BBB	
Fletcher-Lill[a]	68		CBB
Pearson	69	EA[b]	
Shanno	70		ACA
Greenstadt	70	BBA	BBA
Broyden	70	BBB	
Bard	70	CBB	
Murtagh and Sargent	70	BCA	
Biggs	70	CCA	CCA, ACB
Huang and Levy	70	[b]	
Fletcher	70		CCA
This report		BBA	BBA, ACB	BBA
BROYDEN-FLETCHER-SHANNO FORMULA			
Shanno	70		ACA
Broyden	70	BBB	
Shanno and Kettler	70		ACB
Biggs	70		ACB, CCA
This report		BBA	BBA, ACB	BBA

TABLE 3—*continued*

Implementation	Line search A	Bracket B	Conditions C
RANK-ONE			
Powell	69		Special
Shanno	70	ACA	
Bard	70	CBB	CBB
Murtagh and Sargent	70	BCA	ACE, BCE
Huang and Levy	70	[b]	
This report		BBA	BBA, ACB	BBA
SWITCHING			
Fletcher	70		ACA
This report		BBA	BBA, ACB	BBA

The first letter of the code has been transferred to the head of the column.
[a] Indicates the use of approximate derivatives.
[b] Indicates that the relevant information was not available to the author when writing this paper.

criteria. There were very few comparisons of the effect of parabolic as distinct from cubic interpolation and no direct comparison of all four updating procedures. It was therefore decided to complete the table by including results obtained on all twelve combinations using the parabolic approach and all four updating policies using the cubic interpolation bracketing policy. Before introducing a more complete comparison based on these results it seemed desirable to see how consistent the published data was. For consistency comparisons only the most commonly tested functions could be used and only three seemed to be sufficiently mutual for conclusions to be drawn. These were the Rosenbrock, Powell and Wood quartic functions.

5. Numerical Results

5.1. *Accurate Line Search Routines and Comparison of Iterations Required*

The published data giving the number of iterations required for convergence when accurate line searches are undertaken is given in Table 4.

New data derived especially for this report is also included. The iteration numbers are remarkably consistent considering the different interpretations of 'accurate line search' and the different termination criteria that are, no doubt, combined there. It is known that the number of iterations required for convergence on Powell's quartic is very sensitive to the termination criteria whilst the other functions are relatively insensitive and this no doubt explains the

10. THE CHOICE OF STEP LENGTH

higher figures required by Huang and Levy on this function. The other variations in the table are all well within the scatter found by the author during tests with different values of ϵ_2. For instance with $\epsilon_2 = 10^{-5}$, on Rosenbrock's function, the small differences in starting points on iteration 11 when using the DFP and BFS formulae were enough to produce very different unidirectional minima, separated by a region of high function values, and the subsequent convergence favoured the BFS formulae by four iterations (a very similar result to that reported by Broyden, 1970).

TABLE 4
Comparison of iteration numbers with accurate line search routines

	Algorithm details			Function details		
Formula	Author		Code	Rosenbrock	Powell	Wood
DFP	Stewart	67	A, BBB[a]	24	19	
	Pearson	68	A, EA[b]	19		40
	Greenstadt	70	A, BBA	26	18	
	Broyden	70	A, BBB	23	18	
	Huang and Levy	70	A,[b]	22	32	40
	This report		A, BBA	19	20	34
BFS	Broyden	70	A, BBB	19	26	
	This report		A, BBA	19	21	40
Fletcher switch	This report		A, BBA	19	21	40
Rank-one	Huang and Levy	70	A,[b]	22	32	40
	This report		A, BBA	19	21	40

[a] Indicates the use of approximate derivatives.
[b] Indicates that the relevant information was not available to the author, when writing this paper.

In conclusion, it can be stated that, when the iteration is undertaken sufficiently accurately, then the numerical evidence confirms that the iteration pattern is independent of updating formulae in this family.

5.2. *Systematic Comparison*

To evaluate the significance of the different choices of updating formula and step-length, a series of tests was undertaken using all sixteen permutations described above on twelve test functions. These functions include some standard test functions, a series of exponential functions, and three new functions, a skew penalty function of the type generated in S.U.M.T., and two new functions with singular matrices at the minimum, one ROS(8) having a flat minimum formed by raising both brackets in the usual Rosenbrock function to power 8 and the other RECIP having a cusp at the minimum. These twelve functions are divided into two groups for discussion purposes: Group A (nine functions) on which most methods were successful and Group B (three functions) on which the picture was more complex. The details of the functions are given below:

1. ROS(2), introduced by Rosenbrock (1960):
$$f(x) = 100(x_2 - x_1^2)^2 + (1 - x_1)^2.$$
Starting approximation: $x = (-1, 2, 1)$.

2. POW, introduced by Powell (1962):
$$f(x) = (x_1 + 10x_2)^2 + 5(x_3 - x_4)^2 + (x_2 - 2x_3)^4 + 10(x_1 - x_4)^4.$$
Starting approximation: $x = (3, -1, 0, 1)$.

3. WOOD, introduced by Wood (cited in Colville, 1968):
$$f(x) = 100(x_2 - x_1^2)^2 + (1 - x_1)^2 + 90(x_4 - x_3^2)^2 + (1 - x_3)^2$$
$$+ 10 \cdot 1[(x_2 - 1)^2 + (x_4 - 1)^2] + 19 \cdot 8(x_2 - 1)(x_4 - 1).$$
Starting approximation: $x = (-3, -1, -3, -1)$.

4. BOX(2), introduced by Box (1966):
$$f(x) = \sum_{i=1}^{10} [\exp(-x_1 z_i) - \exp(-x_2 z_i) - \exp(-z_i) + \exp(-10 z_i)]^2,$$
$(z_i = i/10)$.

Starting approximation: $x = (5, 0)$.

5. EXP(2), introduced by Biggs (1971):
$$f(x) = \sum_{i=1}^{10} [\exp(-x_1 z_i) - 5\exp(-x_2 z_i) - \exp(-z_i) + 5\exp(-10 z_i)]^2,$$
$(z = i/10)$.

Starting approximation: $x = (1, 2)$.

6. EXP(3), introduced by Biggs (1971):
$$f(x) = \sum_{i=1}^{10} [\exp(-x_1 z_i) - x_3 \exp(-x_2 z_i) - \exp(-z_i) + 5\exp(-10 z_i)]^2,$$
$(z_i = i/10)$.

Starting approximation: $x = (1, 2, 1)$.

10. THE CHOICE OF STEP LENGTH

7. EXP(4), introduced by Biggs (1971):

$$f(x) = \sum_{i=1}^{10} [x_3 \exp(-x_1 z_i) - x_4 \exp(-x_2 z_i) - \exp(-z_i) + 5\exp(-10 z_i)]^2,$$

($z_i = i/10$).

Starting approximation: $x = (1, 2, 1, 1)$.

8. PEN:

$$f(x) = (x_1 - 5)^2 + x_2^2 + 10^{-4}/(x_2 - x_1^2), \quad \text{if } x_2 > x_1^2.$$

Starting approximation: $x = (2, 5)$.

9. ROS(8):

$$f(x) = 100(x_2 - x_1^2)^8 + (1 - x_1)^8.$$

Starting approximation: $x = (-1 \cdot 2, 1)$.

10. EXP(5), introduced by Biggs (1971):

$$f(x) = \sum_{i=1}^{11} [x_3 \exp(-x_1 z_i) - x_4 \exp(-x_2 z_i) + 3\exp(-x_5 z_i) - \exp(-z_i)$$

$$+ 5\exp(-10 z_i) - 3\exp(-4 z_i)]^2, \quad (z_i = i/10).$$

Starting approximation: $x = (1, 1, 1, 1, 1)$.

11. WEIBULL, introduced by Shanno (1970):

$$f(x) = \sum_{i=1}^{99} \left[\exp\left(-\frac{(y_i - x_3)^{x_2}}{x_1} \right) - z_i \right]^2,$$

where

$$y_i = 25 + [50 \log_e (1/z_i)]^{2/3}, \quad (z_i = i/100).$$

Starting approximation: $x = (250, 0 \cdot 3, 5)$.

12. RECIP:

$$f(x) = (x_1 - 5)^2 + x_2^2 + x_3^2/(x_2 - x_1^2), \quad \text{if } x_2 > x_1^2.$$

Starting approximation: $x = (2, 5, 1)$.

The detailed runs on the first nine easier functions are presented in Table 5, in which the upper of each pair of numbers represents the number of equivalent function evaluations required and the lower the number of iterations. The total number of equivalent function evaluations required is given in the final column.

TABLE 5
Comparison of performance on nine test functions

Function		ROS(2)	Powell	Wood	BOX(2)	EXP(2)	EXP(3)	EXP(4)	PEN	ROS(8)	Total
					ACCURATE LINE SEARCH						
DFP	A, BBA	267 / 19	352 / 20	564 / 34	179 / 11	85 / 6	151 / 11	316 / 19	191 / 12	135 / 8	2240
BFS		261 / 19	334 / 21	599 / 40	111 / 8	88 / 6	151 / 11	308 / 20	170 / 12	111 / 7	2133
Switch		260 / 19	329 / 21	619 / 40	159 / 11	86 / 6	147 / 11	288 / 19	184 / 12	111 / 7	2183
R1		266 / 19	290 / 18	553 / 39	108 / 8	87 / 6	152 / 11	296 / 19	190 / 13	130 / 8	2072
					PARABOLIC BRACKETING TECHNIQUE						
DFP	B, BBA	160 / 33	256 / 33	885 / 122	90 / 17	64 / 13	79 / 13	416 / 57	249 / 55	134 / 25	2333
BFS		132 / 29	136 / 19	320 / 49	53 / 11	53 / 11	67 / 11	152 / 22	121 / 26	71 / 12	1105
Switch		138 / 30	203 / 28	341 / 50	53 / 11	54 / 11	73 / 12	182 / 26	181 / 38	90 / 16	1315
R1		155 / 32	291 / 40	407 / 60	49 / 10	52 / 11	73 / 12	132 / 19	267 / 56	98 / 18	1524

10. THE CHOICE OF STEP LENGTH

ACCEPTABLE POINT TECHNIQUE

	C, BBA		F							
DFP	138, 39	1035, 180	F	361, 110	172, 54	155, 36	439, 84	208, 53	511, 157	3019F
BFS	103, 31	134, 25	298, 56	62, 17	56, 16	78, 18	172, 33	112, 28	134, 42	1149
Switch	121, 36	189, 35	370, 70	67, 18	56, 16	87, 19	147, 27	228, 56	138, 43	1403
R1	164, 46	178, 32	407, 72	41, 10	47, 13	64, 14	164, 30	270, 67	129, 40	1464

CUBIC BRACKETING TECHNIQUE

	B, ACB			F						
DFP	210, 21	280, 14	705, 43	F	63, 6	128, 11	305, 20	267, 40	150, 12	2108F
BFS	183, 21	270, 14	445, 36	114, 7	96, 8	128, 11	285, 19	309, 40	129, 9	1959
Switch	234, 23	340, 20	475, 23	114, 7	63, 6	128, 11	255, 17	261, 40	144, 11	2014
R1	138, 13	336, 16	490, 26	96, 6	63, 6	128, 11	270, 18	363, 40	102, 9	1986

TABLE 6
Comparison of two difficult test-functions

Updating	EXP(5) function				Weibull function			
	A, BBA	B, BBA	C, BBA	B, ACB	A, BBA	B, BBA	C, BBA	B, ACB
DFP	3540 / 204	F	837 / 133	F	F	F	F	F
BFS	1366 / 82	349 / 44	204 / 30	792 / 41	767 / 47	772 / 102	1673 / 316	F[a]
Switch	1996 / 118	F	682 / 102	828 / 42	F	1003 / 155	F	F
R1	1205 / 75	502 / 63	1058 / 147	882 / 50	1016 / 63	F	2882 / 546	F

[a] Shanno and Kettler (1970) report convergence in 444 e.f.e. for their implementation. This difference is apparently due to different safeguards being applied when the function could not be evaluated due to overflow errors.

The following conclusions can be drawn:

(1) on this set of functions, with an accurate line search, the updating formula chosen is not particularly important. The number of iterations on each function is very consistent;
(2) in a comparison based on equivalent function evaluations, the use of a cubic bracketing technique provides little improvement over an efficient function value line search routine, even on functions for which n is small. For larger functions it might be anticipated that the accurate parabolic line search would be more efficient;
(3) the use of a parabolic bracketing technique, or an acceptable point technique gives a much improved efficiency when used in conjunction with the Broyden-Fletcher-Shanno formula. Of the two strategies the parabolic bracketing technique seems slightly more efficient. This is similar to the algorithm listed in Fielding (1971), but the first bracketed point having a reduced function value is accepted.

On the three more difficult functions the behaviour is rather different. The behaviour on EXP(5) and WEIBULL is shown in Table 6.

(1) With these more difficult functions, the identity of performance with the accurate line search routines is lost and a study of the print out reveals that the DFP formula leads to a sequence $\{x_i\}$ that is effectively confined to a subspace. These results are presumably due to an accumulation of round off errors in both the line searches and updating strategies.
(2) These results again indicate that the Broyden-Fletcher-Shanno formula combined with the parabolic bracketing technique is both efficient and reliable.

When these 16 routines were run on the final function RECIP none succeeded in finding the correct answer. Many methods have been tried on this problem, Rosenbrocks (1961) technique reached 16·502 in 560 function evaluations, Hooke and Jeeves (1959) reached 16·5634 in 192 function evaluations whilst the BFS, B, BBA algorithm stopped with 16·76 after 93 function evaluations. The function is included to show that there are situations in which the matrix updating approach should and does fail.

5.3. *A Comparison with Dominant Degree Results*

In Biggs (1971) the principle of modifying the updating formulae to include information obtained about their non-quadratic nature was proposed. The algorithms that result from that proposal are part of Huang's family but have variable ρ_i and therefore do not form part of the same subset. In his original paper results were published that used the modified updating formula and a dominant degree line search routine that, as in the cubic routine, calculates derivatives at each point; these results have been extended to include the remaining functions given above. The total equivalent to those given in Table 5 is 1138, emphasizing that his combination is a marked improvement over cubic interpolation when function and gradient calculations are always combined.

6. Conclusions

(1) The numerical evidence presented confirms the theoretical result (Dixon, 1971) that, with an accurate line search procedure for determining the step length, all Broyden's (1967) family form a subset of Huang's (1970) family that has identical behaviour on non-quadratic functions. On the two functions where this behaviour failed, due to rounding errors, the Broyden-Fletcher-Shanno algorithm was far more efficient.

(2) All the recent proposals [the Broyden-Fletcher-Shanno formula, total 1105 in Table 4; the Fletcher (1970) Switching policy, total 1403; the rank-one approach, total 1464; and the Biggs (1971) Dominant Degree method, total 1138], are confirmed to be more efficient than the original DFP implementation (total 2108F).

(3) Of these recent proposals, the combination given by Broyden and Fielding was the most efficient for this test series.

(4) The benefit of using the BFS formulae seems to be directly linked with not finding the minimum along the line. This was true for 9 of the 12 functions tested. As most of the theoretical studies of this formula have assumed that exact unidirectional minima were found, further investigation of this point seems necessary.

References

Bard, Y. (1968). On a numerical instability of Davidon-like Methods. *Math. Comput.* **22**, 665–666.

Bard, Y. (1970). Comparison of gradient methods for the solution of non-linear parameter estimation problems *SIAM Numer. Anal.* **7**, 157–186.

Biggs, M. C. (1970). 'Computational Experience with Variable Metric Methods for Function Minimisation'. The Hatfield Polytechnic, Numerical Optimisation Centre, T.R.7.

Biggs, M. C. (1971). 'A New Variable Metric Technique Taking Account of Non-quadratic Behaviour of the Objective Function.' The Hatfield Polytechnic, Numerical Optimisation Centre, T.R.17. To appear in *J. Inst. Maths Applics.*

Box, M. J. (1966). A comparison of several current optimisation methods, and the use of transformations in constrained problems. *Comput. J.* **9**, 67–77.

Broyden, C. G. (1967). Quasi-Newton methods and their application to function minimisation. *Math. Comput.* **21**, 368–381.

Broyden, C. G. (1970). The convergence of a class of double-rank minimisation algorithms. *J. Inst. Maths Applics* **6**, 76–90, and 222–231.

Colville, A. R. (1968). 'A Comparative Study of Non-linear Programming Codes'. I.B.M. New York Scientific Centre, T.R. 320–2925.

Davidon, W. C. (1959). 'Variable Metric Method for Minimisation'. A.E.C. R and D Report, ANL-5990.

Dixon, L. C. W. (1969). 'A Comparison of the Relative Efficiency of Several Methods of Finding the Minimum of a Function of one Variable'. The Hatfield Polytechnic, Numerical Optimisation Centre, T.R.2.

Dixon, L. C. W. (1971). 'Variable Metric Algorithms. Necessary and Sufficient Conditions for Identical Behaviour on Non-quadratic Functions'. The Hatfield Polytechnic, Numerical Optimisation Centre, T.R.26.

Fielding, K. (1970). Algorithm 387, Function minimisation and linear search (E4) *Comm. A.C.M.* **13**, No. 8, 509–510.

Fletcher, R., and Powell, M. J. D. (1963). A rapidly convergent descent method for minimisation. *Comput. J.* **6**, 163–168.

Fletcher, R. (1966). Certification of Algorithm 251, Function Minimisation. *Comm. A.C.M.* **9**, No. 9.

Fletcher, R. (1968). A review of methods for unconstrained optimisation. *In* 'Optimisation' (R. Fletcher, ed.), pp. 1–12. Academic Press, London.

Fletcher, R. (1970a). A new approach to variable metric algorithms. *Comput. J.* **13**, 317–322.

Fletcher, R. (1970b). 'An Efficient, Globally Convergent, Algorithm for Unconstrained and Linearly Constrained Optimisation Problems'. Paper presented at the 7th International Mathematical Programming Symposium, The Hague.

Goldfarb, D. (1970). A family of variable metric methods devised by varational means. *Math. Comput.* **24**, 23–26.

Goldstein, A. A., and Price, J. F. (1967). An efficient algorithm for minimisation. *Numer. Math.* **10**, 184–189.

Greenstadt, J. (1970). Variations on variable metric methods. *Math. Comput.* **24**, 1–22.

Huang, H. Y. (1970). Unified approach to quadratically convergent algorithms for function minimisation. *J. Optim. Theory Applns* **5**, 405–423.

Huang, H. Y., and Levy, A. V. (1970). Numerical experiments on quadratically convergent algorithms for function minimisation. *J. Optim. Theory Applns* **6**, 269–282.

Murtagh, B. A., and Sargent, R. W. H. (1970). Computational experience with quadratically convergent minimisation methods. *Comput. J.* **13**, 185–194.

Pearson, J. D. (1969). Variable metric methods of minimisation. *Comput. J.* **12**, 171–178.

Powell, M. J. D. (1962). An iterative method for finding stationary values of functions of several variables. *Comput. J.* **5**, 147–151.

Powell, M. J. D. (1970). Rank one methods for unconstrained optimisation. *In*: 'Integer and Nonlinear Programming' (J. Abadie, ed.), pp. 139–156. North Holland Publishing Co., Amsterdam.

Rosenbrock, H. H. (1960). An automatic method for finding the greatest or the least value of a function. *Comput. J.* **3**, 175–184.

Shanno, D. F. (1970). Conditioning of Quasi-Newton methods for function minimisation. *Math. Comput.* **24**, 647–657.

Shanno, D. F., and Kettler, P. C. (1970). Optimal conditioning of Quasi-Newton methods. *Math. Comput.* **24**, 657–665.

Wolfe, P. (1967). Another variable metric method (unpublished).

Wolfe, P. (1969). Convergence conditions for ascent methods. *SIAM Rev.* **11**, 226–235.

11. Some Aspects of Non-linear Least Squares Calculations

M. R. OSBORNE

Australian National University, Canberra, Australia

Summary

This paper considers a general method for minimizing a sum of squares which has the property that a linear least squares problem is solved at each stage, and which includes the methods of Gauss-Newton and Levenberg, Marquardt, and Morrison as particular special cases. First results relevant to linear least squares are summarized. Then the algorithm is defined and convergence and rate of convergence results derived. Finally, a computational scheme is suggested and test problems and numerical results given.

1. Introduction

The problem of minimizing a sum of squares arises naturally from the problem of determining parameters x_i, $i = 1, 2, \ldots, p$ in the model equation

$$y(t) = F(t, \mathbf{x}) \tag{1}$$

from observations

$$y_i = y(t_i) + \epsilon_i, \quad (i = 1, 2, \ldots, n), \tag{2}$$

where the ϵ_i (the experimental errors) are independent, normally distributed random variables with mean zero and standard deviation σ. In the case $n > p$ the appropriate maximum likelihood analysis indicates that $\mathbf{x}$ should be estimated by minimizing $\|\mathbf{f}(\mathbf{x})\|$, where

$$f_i(\mathbf{x}) = y_i - F(t_i, \mathbf{x}) \tag{3}$$

and

$$\|\mathbf{f}\|^2 = \sum_{i=1}^{n} f_i(\mathbf{x})^2.$$

This problem will be referred to as the *model problem*, and it is stressed that we have offered a statistical justification for minimizing a sum of squares. It is by no means clear that this is the best strategy if the aim is merely to make the components of a vector valued function small. A comparative study for this purpose of algorithms which use only function and first-derivative information is given in Osborne (1971). Here it is shown that a faster rate of convergence can frequently be obtained by using the maximum norm, and this norm also offers possibilities for economization of storage.

In this paper we treat algorithms for the model problem which have the property that a linear least squares problem is solved at each step of the iteration. This class of algorithms includes as special cases the Gauss-Newton algorithm and the method of Levenberg, Marquardt, and Morrison. We begin by summarizing results for the linear least squares problem and the related problem of calculating the generalized inverse of a rectangular matrix. Certain approximate methods are defined and algorithms for their implementation suggested. The key tool is the orthogonal reduction of a rectangular matrix to upper triangular form using elementary orthogonal (Householder) transformations given by Golub. The presentation is based on the report of Jennings and Osborne (1970), which also gives program details and numerical results. A general form for an algorithm for the non-linear problem is then presented, together with results on convergence and rate of convergence. Finally, a computational scheme for implementing the non-linear algorithm is suggested, and we give test problems and numerical results.

2. Linear Least Squares Problems

Consider the linear least squares problem

$$\mathbf{y} - A\mathbf{x} = \mathbf{r}, \tag{4}$$

where A is an $n \times p$ matrix, $n \geqslant p$. The vector of minimum norm minimizing $\|\mathbf{r}\|$ is given by

$$\mathbf{x} = A^+ \mathbf{y} \tag{5}$$

where the $+$ indicates the generalized inverse defined by

$$A^+ = \lim_{\nu \to 0} (A^T A + \nu^2 I)^{-1} A^T. \tag{6}$$

An alternative form for the generalized inverse can be given in terms of the singular value decomposition of A. Here we write

$$A = UDV^T \tag{7}$$

where D is diagonal and $D_i = [\lambda_i(A^T A)]^{1/2}$, $i = 1, 2, \ldots, p$; V is orthogonal and $A^T A \kappa_i(V) = \lambda_i \kappa_i(V)$, where $\kappa_i(V)$ denotes the ith column of V; U is an $n \times p$ matrix which is column wise orthogonal, i.e., $U^T U = I$, and $AA^T \kappa_i(U) = \lambda_i \kappa_i(U)$. The D_i are frequently called the singular values of A. In this notation we have

$$A^+ = VD^+ U^T \tag{8}$$

where

$$\begin{aligned} D_i^+ &= 1/D_i, & D_i \neq 0, \\ &= 0, & D_i = 0. \end{aligned} \tag{9}$$

Note that if any D_i is small but non-zero, then A^+ will have large elements. A consequence of this is that the calculation of the generalized inverse is likely to be numerically unstable.

This potential instability makes it important to consider approximate methods which can provide control over the size of the solution to the least squares problem. For example, if we are modelling a physical problem then we would expect continuous dependence of the solution on the data and some freedom from the effects of rounding error in our calculations. In this content the *truncation* estimate of A is important. Let

$$A_\nu = UD_\nu V^T \qquad (10)$$

where

$$(D_\nu)_i = D_i, \qquad D_i \geqslant \nu,$$
$$= 0 \text{ otherwise.} \qquad (11)$$

Let the rank of D_ν be $q \leqslant p$, then A_ν is the best approximation to A in the euclidean norm by matrices of rank q. This result is due to Eckart and Young (see Golub and Kahan, 1965).

More readily available computationally are the *damped least squares* estimates. In this case what is calculated is the solution to the linear least squares problem

$$\begin{bmatrix} A \\ B \end{bmatrix} \mathbf{x} = \begin{bmatrix} \mathbf{y} \\ \mathbf{0} \end{bmatrix} + \begin{bmatrix} \mathbf{r}^{(1)} \\ \mathbf{r}^{(2)} \end{bmatrix} \qquad (12)$$

so that $\mathbf{x}$ is given by

$$[A^T A + B^T B]\mathbf{x} = A^T \mathbf{y}. \qquad (13)$$

Two cases have been considered in the literature,

(a) $B^T B = \nu^2 I$, which corresponds to using a finite value of ν in equation (6), and
(b) $B^T B = \nu^2 I + \nu^2 (A^T A + \nu^2 I)^{-1}$.

This second form is due to Rutishauser (1969), who called it doubly relaxed least squares. Numerical experience with both cases is reported in Jennings and Osborne (1970).

Some idea of what can be gained by using these approximate methods (or perhaps smoothing methods is a better term) can be obtained by looking at explicit solutions. We have

(i) equation (6),

$$\mathbf{x} = \sum_{D_i > 0} \frac{\mathbf{y}^T \kappa_i(U)}{D_i} \kappa_i(V), \qquad (14)$$

(ii) truncation,
$$\mathbf{x}_T = \sum_{D_i \geq \nu} \frac{\mathbf{y}^T \kappa_i(U)}{D_i} \kappa_i(V), \tag{15}$$

(iii) damped least squares,

(a) $B^T B = \nu^2 I$,
$$\mathbf{x}_L = \sum_{i=1}^{P} \frac{D_i}{D_i^2 + \nu^2} [\mathbf{y}^T \kappa_i(U)] \kappa_i(V), \tag{16}$$

and

(b) $B^T B = \nu^2 I + \nu^2 (A^T A + \nu^2 I)^{-1}$,
$$\mathbf{x}_R = \sum_{i=1}^{P} \frac{D_i}{D_i^2 + \nu^2 + [\nu^2/(D_i^2 + \nu^2)]} [\mathbf{y}^T \kappa_i(U)] \kappa_i(V). \tag{17}$$

We note that the coefficients in $\mathbf{x}_L$ and $\mathbf{x}_R$ are in the ratio
$$\left(1 + \frac{\nu^2}{(D_i^2 + \nu^2)^2}\right) \bigg/ 1$$
so that, in particular, $\|\mathbf{x}_R\| < \|\mathbf{x}_L\|$. Also, the coefficients associated with D_i values small compared with ν are attenuated by a factor ν^2 in $\mathbf{x}_R$ as compared to $\mathbf{x}_L$. In this sense $\mathbf{x}_R$ is closer to $\mathbf{x}_T$ than is $\mathbf{x}_L$. Presumably, its best approximation property makes the truncation estimate attractive. However, its computation involves the calculation of singular values which could well be expensive in the inner loop of an iteration, and for this reason further investigation of doubly relaxed least squares would appear justified.

In computing the solution of the damped least squares problem it is now well appreciated that the formation of the normal matrix $M = A^T A + B^T B$ is a retrograde step, and that it is preferable to use a stable method for factorizing the rectangular matrix into the product of an orthogonal matrix times a rectangular matrix in upper triangular form (Golub and Wilkinson, 1966). In the case $B^T B = \nu^2 I$ it is convenient to build up the orthogonal factorization in two stages

$$(a) \begin{bmatrix} A \\ \nu I \end{bmatrix} \rightarrow \begin{bmatrix} Q_1 \\ \hline & I \end{bmatrix} \begin{bmatrix} R_1 \\ 0 \\ \nu I \end{bmatrix}, \quad (b) \begin{bmatrix} R_1 \\ 0 \\ \nu I \end{bmatrix} \rightarrow Q_2 \begin{bmatrix} R \\ 0 \\ 0 \end{bmatrix}$$

where Q_1 is orthogonal and R_1 upper triangular, and where Q_2 is orthogonal and R upper triangular. Note that the band of zeros introduced in the first step is unaffected by the subsequent calculation if Golub's algorithm is used.

11. NON-LINEAR LEAST SQUARES CALCULATION

This can lead to considerable savings if n is much larger than p and if calculations for several values of ν are required (this is often the case in the iterative methods to be considered subsequently). This suggestion is due to Golub.

If we now partition Q_2 in the form (where we indicate only the components of immediate interest)

$$Q_2 = \begin{bmatrix} & & \\ \hline S^T & & \end{bmatrix}$$

then we see that

$$S^T R = \nu I \tag{18}$$

whence

$$R^T R + \nu^2 (R^T R)^{-1} = R^T R + S^T S. \tag{19}$$

Thus the factorization in the case $B^T B = \nu^2 I + \nu^2 (A^T A + \nu^2 I)^{-1}$ can be obtained by a third step in which we compute the orthogonal factorization of

$$\begin{bmatrix} R \\ S \end{bmatrix}.$$

3. Non-linear Problems

We give now a general form for an algorithm for the non-linear problem in which a linear least squares problem is solved at each stage. We write

$$\mathbf{f}_i = \mathbf{f}(\mathbf{x}_i), \quad A_i = \nabla \mathbf{f}(\mathbf{x}_i).$$

Algorithm

(i) Solve the linear least squares problem

$$\begin{bmatrix} \mathbf{f}_i \\ 0 \end{bmatrix} + \begin{bmatrix} A_i \\ B_i \end{bmatrix} \mathbf{h}_i = \begin{bmatrix} r_i^{(1)} \\ r_i^{(2)} \end{bmatrix} = \mathbf{r}_i. \tag{20}$$

(ii) Determine a step length γ_i (for example by choosing γ_i to minimize $\|\mathbf{f}(\mathbf{x}_i + \gamma \mathbf{h}_i)\|$).

(iii) Set $\mathbf{x}_{i+1} = \mathbf{x}_i + \gamma_i \mathbf{h}_i$, test convergence, repeat if convergence test not satisfied.

Remark

(i) The case $B_i = 0$ corresponds to the Gauss-Newton algorithm, while $B_i = \nu_i^2 I$ gives the method of Levenberg, Marquardt, and Morrison.

(ii) The truncation estimate could be used in the first step of the algorithm. In this case equation (20) becomes

$$\mathbf{f}_i + A_{\nu_i}^{(i)} \mathbf{h}_i = \mathbf{r}_i. \tag{20a}$$

(iii) The number of function evaluations required to calculate γ_i is an important factor in determining the cost of the algorithm. For this reason it is of interest to determine when $\gamma = 1$ can be used. This case is referred to as the full step method.

We now give a lemma which gives optimality conditions in a form appropriate to the above algorithm.

Lemma 1

The following conditions are equivalent:

(i) $\begin{bmatrix} \mathbf{f}_i \\ 0 \end{bmatrix} = \mathbf{r}_i,$

(ii) $\|\mathbf{f}_i\| = \|\mathbf{r}_i\|$, and

(iii) $\mathbf{x}_i$ is a stationary point of $\|\mathbf{f}(\mathbf{x})\|$.

Proof

The solution of the linear least squares problem gives

$$[A_i^T \quad B_i^T] \mathbf{r}_i = 0 \tag{21}$$

whence

$$\mathbf{f}_i^T \mathbf{r}_i^{(1)} = \|\mathbf{r}_i\|^2. \tag{22}$$

Thus (ii) implies (i) as a consequence of the Cauchy inequality. If (i) holds then

$$0 = [\mathbf{f}_i^T, 0] \begin{bmatrix} A_i \\ B_i \end{bmatrix} = \tfrac{1}{2} \nabla(\|\mathbf{f}\|^2) \tag{23}$$

so that (i) implies (iii). Now assuming (iii) and taking the scalar product of equation (20) with

$$[\mathbf{f}_i^T, 0]$$

gives

$$\|\mathbf{f}_i\|^2 + \tfrac{1}{2} \nabla(\|\mathbf{f}\|^2) \mathbf{h}_i = \mathbf{r}_i^{(1)T} \mathbf{f}_i = \|\mathbf{r}_i\|^2 \tag{24}$$

so that (iii) implies (ii).

11. NON-LINEAR LEAST SQUARES CALCULATION

Remark

(i) The corresponding result for equation (20a) is that $\|\mathbf{f}_i\| = \|\mathbf{r}_i\|$ implies $\mathbf{f}_i = \mathbf{r}_i$ and either $\|\mathbf{f}_i\| = 0$ or $\|\nabla(\|\mathbf{f}\|)\| < \nu_i$. In this case the condition for a stationary point is satisfied to within the tolerance used in defining A_{ν_i}.

(ii) As $\|\mathbf{r}_i\| \leq \|\mathbf{f}_i\|$, we see that $\mathbf{h}_i$ is downhill for minimizing $\|\mathbf{f}\|$ at $\mathbf{x}_i$ unless $\mathbf{x}_i$ is a stationary point of $\|\mathbf{f}\|$.

We now analyse the convergence of the algorithm, assuming that γ_i is chosen to minimize $\|\mathbf{f}(\mathbf{x}_i + \gamma \mathbf{h}_i)\|$, that the iteration is bounded (so that R is bounded), that $\mathbf{f}(\mathbf{x})$ is sufficiently smooth for us to be able to write

$$\mathbf{f}(\mathbf{x} + \gamma \mathbf{t}) = \mathbf{f}(\mathbf{x}) + \gamma \nabla \mathbf{f}\, \mathbf{t} + \gamma^2 \|\mathbf{t}\|^2\, \mathbf{w}(\mathbf{x}, \gamma, \mathbf{t}) \tag{25}$$

and that $\|\mathbf{w}\| \leq W$ for points in R.

Lemma 2

Provided $0 \leq \gamma \leq 1$, then $\|\mathbf{f}(\mathbf{x}_i + \gamma \mathbf{h}_i)\| \leq Q_i(\gamma)$, where

$$Q_i(\gamma) = (1 - \gamma)\|\mathbf{f}_i\| + \gamma \|\mathbf{r}_i^{(1)}\| + \gamma^2 \|\mathbf{h}_i\|^2\, W. \tag{26}$$

Proof

Substituting $\mathbf{h}_i$ for $\mathbf{t}$ in equation (25), and using equation (20), gives

$$\mathbf{f}(\mathbf{x}_i + \gamma \mathbf{h}_i) = \mathbf{f}_i + \gamma(\mathbf{r}_i^{(1)} - \mathbf{f}_i) + \gamma^2 \|\mathbf{h}_i\|^2\, \mathbf{w}_i(\gamma).$$

The result now follows from the triangular inequality.

Lemma 3

Let $A_i^T A_i + B_i^T B_i = M_i$, and

$$\min_{\mathbf{t}, \|\mathbf{t}\|=1} (\mathbf{t}^T M_i \mathbf{t}) = \delta_i^2,$$

then

$$\|\mathbf{h}_i\| \leq \frac{1}{\delta_i} \{\|\mathbf{f}_i\|^2 - \|\mathbf{r}_i\|^2\}^{1/2}. \tag{27}$$

Proof

Taking the scalar product of equation (20) with itself gives

$$\mathbf{h}_i^T M_i \mathbf{h}_i = \left\| \begin{bmatrix} \mathbf{f}_i \\ 0 \end{bmatrix} - \mathbf{r}_i \right\|^2.$$

The result now follows from the definition of δ_i and equation (22).

Theorem 1

Let B_i be chosen so that the condition $\delta_i \geqslant \delta > 0$ is satisfied, then the sequence $\{\|\mathbf{f}_i\|\}$ is convergent, and the limit points of the sequence $\{\mathbf{x}_i\}$ are stationary points of $\|\mathbf{f}\|$.

Proof

We have necessarily that

$$\|\mathbf{f}_{i+1}\| \leqslant \min_{0 \leqslant \gamma \leqslant 1} Q_i(\gamma). \tag{28}$$

If the minimum of the right-hand side is attained for $\gamma < 1$ then $dQ_i/d\gamma$ must vanish at the minimum. We have

$$\frac{dQ_i}{d\gamma} = -(\|\mathbf{f}_i\| - \|\mathbf{r}_i^{(1)}\|) + 2\gamma \|\mathbf{h}_i\|^2 W$$

whence

$$\gamma = \frac{\|\mathbf{f}_i\| - \|\mathbf{r}_i^{(1)}\|}{2\|\mathbf{h}_i\|^2 W},$$

and the corresponding value of Q_i is

$$Q_i(\gamma) = \|\mathbf{f}_i\| - \frac{\frac{1}{2}(\|\mathbf{f}_i\| - \|\mathbf{r}_i^{(1)}\|)^2}{2\|\mathbf{h}_i\|^2 W} = \|\mathbf{f}_i\| - \frac{\gamma}{2}(\|\mathbf{f}_i\| - \|\mathbf{r}_i^{(1)}\|). \tag{29}$$

However, if $\gamma = 1$ gives the minimum, then

$$\frac{\|\mathbf{f}_i\| - \|\mathbf{r}_i^{(1)}\|}{2\|\mathbf{h}_i\|^2 W} \geqslant 1.$$

In this case we have

$$Q_i(1) = \|\mathbf{r}_i^{(1)}\| + \|\mathbf{h}_i\|^2 W \leqslant \|\mathbf{f}_i\| - \frac{1}{2}(\|\mathbf{f}_i\| - \|\mathbf{r}_i^{(1)}\|). \tag{30}$$

Let

$$\gamma_i^* = \min\left(1, \frac{\|\mathbf{f}_i\| - \|\mathbf{r}_i^{(1)}\|}{2\|\mathbf{h}_i\|^2 W}\right), \tag{31}$$

then

$$\|\mathbf{f}_{i+1}\| \leqslant \|\mathbf{f}_i\| - \frac{\gamma_i^*}{2}(\|\mathbf{f}_i\| - \|\mathbf{r}_i^{(1)}\|). \tag{32}$$

11. NON-LINEAR LEAST SQUARES CALCULATION

Using lemma 3 gives

$$\gamma_i^* \geq \min\left(1, \frac{\delta_i^2}{2W} \frac{1}{(\|\mathbf{f}_i\| + \|\mathbf{r}_i^{(1)}\|)}\right) \geq \min\left(1, \frac{\delta_i^2}{2W} \frac{1}{2\|\mathbf{f}_i\|}\right)$$

$$\geq \min\left(1, \frac{\delta_i^2}{2W} \frac{1}{2\|\mathbf{f}_0\|}\right) = \mu > 0. \tag{33}$$

Thus the sequence $\{\|\mathbf{f}_i\|\}$ is decreasing and bounded below, and therefore convergent. Further,

$$\|\mathbf{f}_i\| - \|\mathbf{r}_i\| \leq \|\mathbf{f}_i\| - \|\mathbf{r}_i^{(1)}\| \leq \frac{2}{\mu}(\|\mathbf{f}_i\| - \|\mathbf{f}_{i+1}\|)$$

so that the limit points of the sequence $\{\mathbf{x}_i\}$ are stationary points of $\|\mathbf{f}\|$ by lemma 1.

Corollary 1

If

$$\|\mathbf{f}_i\| \leq \frac{\delta_i^2}{4W},$$

then $\gamma^* = 1, j \geq i$. If, in addition, the stationary points of $\|\mathbf{f}\|$ are isolated, then the full step method $\mathbf{x}_{j+1} = \mathbf{x}_j + \mathbf{h}_j, j \geq i$ is convergent.

Remark

(i) The matrix B_i is available to ensure that the first condition of corollary 1 is satisfied.
(ii) A sufficient condition for an isolated minimum is

$$\|\mathbf{f}_i\| < \min_{1 \leq i \leq p} \frac{\lambda_i(A^T A)}{2W}.$$

Theorem 2

If the full step method is convergent and $\delta_i \geq \delta > 0$, then the rate of convergence is first order.

Proof

In this case the correction at $\mathbf{x}_{i+1}$ is given by

$$\begin{bmatrix} \mathbf{f}_{i+1} \\ 0 \end{bmatrix} + \begin{bmatrix} A_{i+1} \\ B_{i+1} \end{bmatrix} \mathbf{h}_{i+1} = \mathbf{r}_{i+1}$$

where

$$\mathbf{f}_{i+1} = \mathbf{f}_i + A_i \mathbf{h}_i + \|\mathbf{h}_i\|^2 \mathbf{w}_i(1)$$
$$= \mathbf{r}_i^{(1)} + \|\mathbf{h}_i\|^2 \mathbf{w}_i(1).$$

The normal equations become

$$-M_{i+1} \mathbf{h}_{i+1} = [A_{i+1}^T | B_{i+1}^T] \begin{bmatrix} \mathbf{r}_i^{(1)} + \|\mathbf{h}_i\|^2 \mathbf{w}_i(1) \\ \hline 0 \end{bmatrix}. \tag{34}$$

Equation (21) gives

$$A_i^T \mathbf{r}_i^{(1)} + B_i^T \mathbf{r}_i^{(2)} = A_i^T \mathbf{r}_i^{(1)} + B_i^T B_i \mathbf{h}_i = 0$$

so that equation (34) can be written

$$-M_{i+1} \mathbf{h}_{i+1} = (A_{i+1}^T - A_i^T) \mathbf{r}_i^{(1)} - B_i^T B_i \mathbf{h}_i + \|\mathbf{h}_i\|^2 A_{i+1}^T \mathbf{w}_i(1)$$
$$= \|\mathbf{h}_i\| \left\{ \overline{\frac{dA_i^T}{dt_i}} \mathbf{r}_i^{(1)} - B_i^T B_i \mathbf{t}_i \right\} + \|\mathbf{h}_i\|^2 A_{i+1}^T \mathbf{w}_i(1) \tag{35}$$

where $\mathbf{h}_i = \|\mathbf{h}_i\| \mathbf{t}_i$, d/dt_i denotes differentiation in the direction $\mathbf{t}_i$, and the bar denotes that mean values are appropriate. Equation (35) demonstrates the first-order convergence.

Corollary 2

The Gauss-Newton method is divergent from an arbitrarily good starting point if

$$\min_{\mathbf{t}} \left\| M^{-1} \frac{dA^T}{dt} \mathbf{f} \right\| > 1$$

for all directions **t** through the solution (Kowalik and Osborne, 1968).

Example

Let

$$\mathbf{f}(\mathbf{x}) = \begin{bmatrix} \eta_1 - x^2 \\ \eta_2 - x^3 \end{bmatrix}$$

then

$$A(x) = \begin{bmatrix} -2x \\ -3x^2 \end{bmatrix}.$$

11. NON-LINEAR LEAST SQUARES CALCULATION

Writing $\eta_1 = u^2 + \epsilon_1$, $\eta_2 = u^3 + \epsilon_2$ then $x = u$ is a solution provided

$$[2u \; 3u^2] \begin{bmatrix} \epsilon_1 \\ \epsilon_2 \end{bmatrix} = 0,$$

that is provided $\epsilon_1 = -(3u/2)\epsilon_2$. This solution is a minimum if $6u\epsilon_2 < 8u^2 + 18u^4$. We have $M = A^T A = 4x^2 + 9x^4$,

$$\frac{dA}{dx} = -\begin{bmatrix} 2 \\ 6x \end{bmatrix},$$

and

$$\left\| M^{-1} \frac{dA^T}{dx} \mathbf{f} \right\|_{x=u} = \frac{1}{4u^2 + 9u^4} |2\epsilon_1 + 6u\epsilon_2| = \frac{|3u\epsilon_2|}{4u^2 + 9u^4}.$$

This can be made arbitrarily large by an appropriate choice of ϵ_2.

Remark

It is of interest to ask if B_i can be chosen to improve the convergence rate. This would require that the term

$$\frac{dA_i^T}{dt_i} \mathbf{r}_i^{(1)} - B_i^T B_i \mathbf{t}_i$$

in equation (35) gets small. This requires that

$$\mathbf{t}_i^T \frac{dA_i^T}{dt_i} \mathbf{r}_i^{(1)} - \mathbf{t}_i^T B_i^T B_i \mathbf{t}_i$$

also gets small. As $B_i^T B_i$ is positive semidefinite, this requires that

$$\mathbf{t}_i^T \frac{dA_i^T}{dt_i} \mathbf{r}_i^{(1)} \geq 0$$

at least ultimately. In terms of the model problem we have at the minimum

$$\mathbf{t}^T \frac{dA^T}{dt} \mathbf{r}^{(1)} = -\sum_{i=1}^{n} \epsilon_i \sum_{p,q} \frac{\partial^2 F_i}{\partial x_p \partial x_q} t_p t_q$$

so that the expected value of

$$\mathbf{t}^T \frac{dA^T}{dt} \mathbf{r}^{(1)}$$

is

$$E\left\{\mathbf{t}^T \frac{\mathrm{d}A^T}{\mathrm{d}t}\mathbf{r}^{(1)}\right\} = -\sum_{i=1}^{n} E\{\epsilon_i\} \sum_{p,q} \frac{\partial^2 F_i}{\partial x_p \partial x_q} t_p t_q = 0.$$

We conclude that it is unlikely that the order of convergence can be improved by appropriate choice of B_i. However, this probable restriction is a shortcoming of our formalism. For example, an algorithm with super-linear convergence has been given by Goldstein and Price (1967).

Theorem 3

If the full step method is convergent, and if $\mathbf{f}(\mathbf{x}) = 0$ is compatible (so that there exists an $\mathbf{x}^*$ satisfying these equations) then the rate of convergence is second order provided

  (i) $\min_{1 \leq j \leq p} \lambda_j(A_i^T A_i) \geq \delta^2 > 0$, and

  (ii) $\|B_i^T B_i\| = \max_{\mathbf{t}} \mathbf{t}^T B_i^T B_i \mathbf{t} \leq \beta \|\mathbf{h}_i\|$.

Proof

The convergence rate equation (35) can be written

$$-M_{i+1} h_{i+1} = \|\mathbf{h}_i\| \left\{ \overline{\frac{\mathrm{d}A_i^T}{\mathrm{d}t_i}} \mathbf{r}_i^{(1)} - B_i^T B_i \mathbf{t}_i + \|\mathbf{h}_i\| L_i \right\}$$

so that

$$\|\mathbf{h}_{i+1}\| \leq \frac{\|\mathbf{h}_i\|}{\delta^2} \left\{ \left\|\overline{\frac{\mathrm{d}A_i^T}{\mathrm{d}t_i}}\right\| \|\mathbf{r}_i^{(1)}\| + \|B_i^T B_i\| + \|\mathbf{h}_i\| \|L_i\| \right\}. \quad (36)$$

Taylor expansion gives

$$\mathbf{f}_i + A_i(\mathbf{x}^* - \mathbf{x}_i) = -\|\mathbf{x}^* - \mathbf{x}_i\|^2 \mathbf{w} \quad (37)$$

and comparison of this equation with equation (20) gives

$$\|\mathbf{r}_i^{(1)}\| \leq \|\mathbf{r}_i\| \leq \{w^2 \|\mathbf{x}^* - \mathbf{x}_i\|^4 + \|B_i^T B_i\| \|\mathbf{x}^* - \mathbf{x}_i\|^2\}. \quad (38)$$

Thus, by condition (ii) of the Theorem,

$$\|\mathbf{r}_i^{(1)}\| = o(\|\mathbf{x}^* - \mathbf{x}_i\|). \quad (39)$$

Also, subtracting the first n equations of (20) from (37) gives

$$A_i(\mathbf{h}_i - (\mathbf{x}^* - \mathbf{x}_i)) = \mathbf{r}_i^{(1)} + \|\mathbf{x}^* - \mathbf{x}_i\|^2 \mathbf{w}$$

11. NON-LINEAR LEAST SQUARES CALCULATION

whence

$$\|\mathbf{h}_i - (\mathbf{x}^* - \mathbf{x}_i)\| \leq \frac{1}{\delta}\{\|\mathbf{r}_i^{(1)}\| + W\|\mathbf{x}^* - \mathbf{x}_i\|^2\} \tag{40}$$

which, by equation (39), implies that

$$\|\mathbf{x}^* - \mathbf{x}_i\| \leq K_i\|\mathbf{h}_i\|$$

where $K_i \to 1$ as $\|\mathbf{h}_i\| \to 0$. Thus $\|\mathbf{r}_i^{(1)}\| = o(\|\mathbf{h}_i\|)$, and the inequality (36) becomes

$$\|\mathbf{h}_{i+1}\| \leq \frac{\|\mathbf{h}_i\|^2}{\delta^2}\{\beta + \|\mathbf{L}_i\|\} + o(\|\mathbf{h}_i\|^2). \tag{41}$$

This inequality establishes second-order convergence.

Remark

Greenstadt (1967) notes in function minimization with positive definite Hessian H that (i) it is the eigenvectors of H corresponding to small eigenvalues that are likely to be profitable search directions (the others being likely to cause 'hemstitching'), and (ii) it is the components of the predicted search direction $\mathbf{h}$ which corresponds to these eigenvectors which are most sensitive to the effects of perturbation due to rounding error. This should be contrasted with our problem. Here

$$H_{pq} = \sum_{i=1}^{n}\left\{\frac{\partial f_i}{\partial x_p}\frac{\partial f_i}{\partial x_q} + f_i\frac{\partial^2 f_i}{\partial x_p \partial x_q}\right\}$$

so that in calculating $\mathbf{h}$ we are ignoring the contribution to the Hessian due to the terms

$$f_i\frac{\partial^2 f_i}{\partial x_p \partial x_q}.$$

Thus we are already introducing a significant perturbation unless $\|\mathbf{f}\|$ is small. In terms of our model problem this perturbation ultimately depends on the experimental errors, but the solution to the problem must be very largely independent of them providing the modelling is correct. This observation provides the justification for trying to control and smooth the $\mathbf{h}_i$ even though the cost may be some loss of efficiency, as reflected in the rate of convergence results. However, when the system is compatible (or nearly so) an efficient rate of convergence can be attained if $\|B_i^T B_i\|$ is allowed to get small as the minimum is approached.

4. Numerical Experience

Particular importance attaches to the special case $B_i = v_i I$ which gives the algorithm of Levenberg, Marquardt, and Morrison. In this case the following results are readily derived (see for example Kowalik and Osborne, 1968):

(i) $\|\mathbf{h}_i(v)\|$ is a monotonic decreasing function of v.
(ii) The angle between $\mathbf{h}_i(v)$ and $-\nabla(\|\mathbf{f}\|^2)(\mathbf{x}_i)$ decreases monotonically as v increases.
(iii) $\mathbf{x} = \mathbf{h}_i(v)$ minimizes $\|A_i \mathbf{x} - \mathbf{f}_i\|^2$ subject to the constraint $\|\mathbf{x}\| = \|\mathbf{h}_i(v)\|$.

The first result is of particular relevance for implementing the algorithm (note the way in which it supplements corollary 1). It is not difficult to show, using equation (17), that it holds also for doubly relaxed least squares.

A use of the result (i) is illustrated in the computational scheme outlined below. In this two constants $EXP > 1$ and $DECR < 1$ are kept. If the current value of v does not give an $\mathbf{h}_i$ such that $\|\mathbf{f}(\mathbf{x}_i + \mathbf{h}_i)\| < \|\mathbf{f}_i\|$, then the current stage is repeated with $v = EXP*v$. However, if the first attempt at the current stage gives an $\mathbf{h}_i$ reducing the sum of squares, then v is reduced by the factor $DECR$ before beginning the next stage. Hopefully, multiple passes through the current stage will not be too frequent, but when these do occur the amount of work resulting is significantly reduced by the device of the two stage orthogonal factorization described in Section 2.

4.1 Computational Scheme

We assume that $B_i = v_i B(v_i)$.

(i) Estimate v_1.
(ii) Calculate the orthogonal factorization of A_i and set $IC = 0$.
(iii) Complete the factorization of $\begin{bmatrix} A_i \\ B_i \end{bmatrix}$.

(iv) Calculate $\mathbf{h}_i$, $\|\mathbf{f}(\mathbf{x}_i + \mathbf{h}_i)\|$, and set $IC = IC + 1$.
(v) If $\|\mathbf{f}(\mathbf{x}_i + \mathbf{h}_i)\| < \|\mathbf{f}(\mathbf{x}_i)\|$ then

   begin $\mathbf{x}_i := \mathbf{x}_i + \mathbf{h}_i$;

   if $IC = 1$ then $v_i := DECR*v_i$;

   $i := i + 1$; go to (vi)

   end

else $v_i := EXP * v_i$;

go to (iii);

(vi) Test convergence. If no convergence, go to (ii).

A FORTRAN program implementing this scheme is given in Jennings and Osborne (1970). In the numerical experiments reported here we took

$$B_i = v_i I, \quad v_1 = \left\{ \sum_{i,j} \frac{A_{ij}^2}{np} \right\}^{1/2}, \quad EXP = 1 \cdot 5, \quad DECR = 0 \cdot 5.$$

The values of the parameters EXP and $DECR$ were chosen to favour decreasing v_i as a strategy.

Two problems are considered.

4.2 Exponential Fitting

Here the given data values are fitted by the model

$$F(t, \mathbf{x}) = x_1 + x_2 \exp(-x_4 t) + x_3 \exp(-x_5 t).$$

The data values are given in Table 1. They were supplied by Dr A. M. Sargeson of the Research School of Chemistry in the Australian National University. The progress of the algorithm is summarized in Table 2.

TABLE 1
Data for exponential fitting problem

i	t_i	y_i	i	t_i	y_i
1	0	0·844	18	170	0·558
2	10	0·908	19	280	0·538
3	20	0·932	20	190	0·522
4	30	0·936	21	200	0·506
5	40	0·925	22	210	0·490
6	50	0·908	23	220	0·478
7	60	0·881	24	230	0·467
8	70	0·850	25	240	0·457
9	80	0·818	26	250	0·448
10	90	0·784	27	260	0·438
11	100	0·751	28	270	0·431
12	110	0·718	29	280	0·424
13	120	0·685	30	290	0·420
14	130	0·658	31	300	0·414
15	140	0·628	32	310	0·411
16	150	0·603	33	320	0·406
17	160	0·580			

TABLE 2
Summary of numerical results in exponential fitting problem

i	ν	‖f‖²	x_1	x_2	x_3	x_4	x_5
0	18·586	0·879 E0	0·5	1·5	−1·0	0·01	0·02
1	18·586	0·161 E0	0·4984	1·4993	−1·0007	0·01443	0·02329
2	9·293	0·103 E0	0·4888	1·4967	−1·0034	0·01693	0·02879
3	4·646	0·523 E-1	0·4586	1·4896	−1·0106	0·01462	0·02606
4	2·323	0·941 E-2	0·4140	1·4796	−1·0209	0·01309	0·02634
5	1·116	0·771 E-3	0·3831	1·4799	−1·0211	0·01207	0·02573
6	0·5808	0·987 E-4	0·3715	1·4866	−1·0150	0·01174	0·02511
7	0·2904	0·770 E-4	0·3691	1·4893	−1·0134	0·01167	0·02489
8	0·1452	0·766 E-4	0·3690	1·4913	−1·0152	0·01167	0·02485
9	0·0726	0·754 E-4	0·3691	1·4987	−1·0227	0·01169	0·02479
10	0·0363	0·720 E-4	0·3697	1·5246	−1·0491	0·01179	0·02455
11	0·0182	0·782 E-4					
	0·0272	0·684 E-4	0·3705	1·5616	−1·0867	0·01191	0·02423
12	0·0272	0·650 E-4	0·3711	1·5914	−1·1169	0·01201	0·02401
13	0·0136	0·740 E-4					
	0·0204	0·629 E-4	0·3719	1·6322	−1·1158	0·01213	0·02371
14	0·0204	0·602 E-4	0·3723	1·6641	−1·1906	0·01222	0·02350
15	0·0102	0·679 E-4					
	0·0153	0·591 E-4	0·3729	1·7061	−1·2331	0·01234	0·02324
16	0·0153	0·573 E-4	0·3733	1·7380	−1·2652	0·01242	0·02306
17	0·0077	0·615 E-4					
	0·0115	0·567 E-4	0·3738	1·7779	−1·3056	0·01252	0·02284
18	0·0115	0·557 E-4	0·3742	1·8071	−1·3350	0·01259	0·02270
19	0·0057	0·570 E-4					
	0·0086	0·554 E-4	0·3745	1·8413	−1·3694	0·01267	0·02253
20	0·0086	0·549 E-4	0·3748	1·8648	−1·3932	0·01272	0·02242
21	0·0043	0·551 E-4					
	0·0065	0·548 E-4	0·3750	1·8898	−1·4183	0·01277	0·02231
22	0·0065	0·547 E-4	0·3751	1·9054	−1·4340	0·01281	0·02225
23	0·0032	0·547 E-4	0·3753	1·9250	−1·4538	0·01285	0·02217
24	0·0016	0·547 E-4	0·3754	1·9343	−1·4631	0·01286	0·02213
25	0·0008	0·546 E-4	0·3754	1·9358	−1·4646	0·01287	0·02212
26	0·0004	0·546 E-4	0·3754	1·9358	−1·4647	0·01287	0.02212
27	0·0002	0·546 E-4	0·3754	1·9358	−1·4647	0·01287	0·02212

4.3 *Fitting Gaussians plus an Exponential Background*

In this case the model has the form
$$F(t, \mathbf{x}) = x_1 \exp(-x_5 t) + x_2 \exp[-x_6(t - x_9)^2] + x_3 \exp[-x_7(t - x_{10})^2] + x_4 \exp[-x_8(t - x_{11})^2].$$

The data values are listed in Table 3, and the progress of the iteration is summarized in Table 4. The data was supplied by Dr W. J. Caelli of the

11. NON-LINEAR LEAST SQUARES CALCULATION

TABLE 3
Data for fitting gaussians plus exponential background

i	t_i	y_i	i	t_i	y_i
1	0·0	1·366	34	3·3	0·375
2	0·1	1·191	35	3·4	0·372
3	0·2	1·112	36	3·5	0·391
4	0·3	1·013	37	3·6	0·396
5	0·4	0·991	38	3·7	0·405
6	0·5	0·885	39	3·8	0·428
7	0·6	0·831	40	3·9	0·429
8	0·7	0·847	41	4·0	0·523
9	0·8	0·786	42	4·1	0·562
10	0·9	0·725	43	4·2	0·607
11	1·0	0·746	44	4·3	0·653
12	1·1	0·679	45	4·4	0·672
13	1·2	0·608	46	4·5	0·708
14	1·3	0·655	47	4·6	0·633
15	1·4	0·616	48	4·7	0·668
16	1·5	0·606	49	4·8	0·645
17	1·6	0·602	50	4·9	0·632
18	1·7	0·626	51	5·0	0·591
19	1·8	0·651	52	5·1	0·559
20	1·9	0·724	53	5·2	0·597
21	2·0	0·649	54	5·3	0·625
22	2·1	0·649	55	5·4	0·739
23	2·2	0·694	56	5·5	0·710
24	2·3	0·644	57	5·6	0·729
25	2·4	0·624	58	5·7	0·720
26	2·5	0·661	59	5·8	0·636
27	2·6	0·612	60	5·9	0·581
28	2·7	0·558	61	6·0	0·428
29	2·8	0·533	62	6·1	0·292
30	2·9	0·495	63	6·2	0·162
31	3·0	0·500	64	6·3	0·098
32	3·1	0·423	65	6·4	0·054
33	3·2	0·395			

Research School of Physical Sciences in the Australian National University.

It will be seen that the calculations are very satisfactory. The values chosen for EXP and $DECR$ have proved successful also in other cases, and the exponential fitting problem has proved to be one in which this comparatively simple strategy for adjusting ν was least successful. It is possible that the convergence analysis might give a useful strategy here. In constructing the dominating function $Q_i(\gamma)$ all that is actually required is the quantity

$$W_i = \max_{0 \leq \gamma \leq 1} \|\mathbf{w}_i(\gamma)\|.$$

TABLE 4
Summary of numerical results for fitting gaussians plus exponential background

i	ν	$\|\mathbf{f}\|^2$	x_1	x_2	x_3	x_4	x_5	x_6	x_7	x_8	x_9	x_{10}	x_{11}
0	0·3047	2·094	1·3	0·65	0·65	0·7	0·6	3·0	5·0	7·0	2·0	4·5	5·5
1	0·3407	0·7010	1·1140	0·08846	0·5711	0·7017	0·2778	3·1329	4·6853	6·7556	2·1401	4·5281	5·5507
2	0·1704	0·2984	1·1095	0·1124	0·4842	0·6124	0·3216	3·2142	3·9956	6·2774	2·6650	4·5478	5·5731
3	0·0852	0·1794	1·1410	0·1878	0·5028	0·6181	0·3725	2·6710	2·3188	5·3592	2·4151	4·5400	5·6054
4	0·0426	1·561											
	0·0639	0·3172											
	0·0958	0·1359	1·1949	0·2877	0·5538	0·5996	0·4654	2·0006	1·3213	5·0849	2·4858	4·5573	5·6565
5	0·0958	0·0874	1·2654	0·3686	0·6115	0·5623	0·5990	1·2003	1·1355	5·0755	2·4163	4·5754	5·6829
6	0·0479	0·0512	1·3046	0·4270	0·6382	0·5906	0·7220	0·9211	1·2792	4·9222	2·4048	4·5714	5·6819
7	0·0240	0·0402	1·3098	0·4317	0·6339	0·5993	0·7229	0·9062	1·3597	4·8302	2·3996	4·5694	5·6760
8	0·0120	0·0402	1·3100	0·4314	0·6336	0·5993	0·7539	0·9057	1·3650	4·8252	2·3988	4·5689	5·6754
9	0·0060	0·0402	1·3100	0·4315	0·6336	0·5993	0·7539	0·9056	1·3650	4·8248	2·3988	4·5689	5·6754
10	0·0030	0·0402	1·3100	0·4315	0·6336	0·5993	0·7539	0·9056	1·3651	4·8248	2·3988	4·5689	5·6754
11	0·0015	0·0402	1·3100	0·4315	0·6336	0·5993	0·7539	0·9056	1·3651	4·8248	2·3988	4·5689	5·6754

Let $\mathbf{f}^*_{i+1} = \mathbf{f}(\mathbf{x}_i + \mathbf{h}_i)$. Frequently, we can expect $\|\mathbf{w}_i(1)\|$ to be a reasonable estimate of W_i so that we can estimate the value of γ for which $dQ_i/d\gamma = 0$ by

$$\bar{\gamma} = \frac{\|\mathbf{f}_i\| - \|\mathbf{r}_i^{(1)}\|}{2\|\mathbf{f}^*_{i+1} - \mathbf{r}_i^{(1)}\|}.$$

Presumably, we would not want to reduce ν unless $\bar{\gamma} \geq 1$. Also, note that a strategy of using a full step provided $\|\mathbf{f}\|$ is decreasing is probably too simple a strategy to guarantee convergence to a point where $\|\mathbf{f}\| = \|\mathbf{r}\|$ (a related problem is discussed in Fletcher, 1970). However, a failure would be indicated by the sequence

$$\left\{ \frac{\|\mathbf{f}_i\| - \|\mathbf{r}_i\|}{\|\mathbf{f}_i\| - \|\mathbf{f}_{i+1}\|} \right\}$$

being unbounded.

References

Fletcher, R. (1970). An efficient, globally convergent algorithm for unconstrained and linearly constrained optimization problems. Paper presented at the 7th Mathematical Programming Symposium, The Hague, Netherlands.

Goldstein, A. A., and Price, J. F. (1967). An efficient algorithm for minimization. *Num. Math.* **10**, 184–189.

Golub, G., and Kahan, W. (1965). Calculating the singular values and pseudo inverses of matrices. *SIAM J. Numer. Anal.* **2**, 205–224.

Golub, G., and Wilkinson, J. H. (1966). Note on the iterative refinement of least squares solutions. *Numer. Math.* **9**, 139–148.

Greenstadt, J. (1967). On the relative efficiencies of gradient methods. *Math. Comput.* **21**, 360–367.

Jennings, L. S., and Osborne, M. R. (1970). 'Applications of Orthogonal Matrix Transformations to the Solution of Systems of Linear and Nonlinear Equations'. Technical Report 37, Computer Centre, Australian National University.

Kowalik, J., and Osborne, M. R. (1968). 'Methods for Unconstrained Optimization Problems'. Elsevier, New York.

Osborne, M. R. (1971). An algorithm for discrete, nonlinear, best approximation problems. *In*: 'Proceedings of the Conference on Numerical Methods in Approximation Theory'. Oberwolfach.

Rutishauser, H. (1968). The least square problem. *Linear Algeb. Appls* **1**, 479–488.

12. A Modified Form of Levenberg's Correction

M. DAVIES AND I. J. WHITTING

*Department of Mathematics, University of Surrey, Guildford,
Surrey, England and Operational Research Department,
The Gas Council, London, England*

Summary

Using a Taylor expansion of the gradient of a function correct to the first order in the displacements, a formula is derived for determining the length of one step of a linear search in any assigned direction to locate the minimum of the function along that direction. This includes the Newton-Raphson method as a special case. It is shown that Newton-Raphson is best possible in the sense that if by this procedure we estimate the location of the line-minimum in every direction and then find the minimum over the points thus generated, we obtain the Newton-Raphson solution.

Newton-Raphson displacements cannot in practice be guaranteed to have the order of smallness required to justify the foregoing approximations, however. The Levenberg-Marquardt technique of adding a parameter λ to the diagonal elements of the matrix of coefficients of the Newton-Raphson displacement equations is therefore applied. As λ varies, the end-point of the displacement vector describes a spiral. By analogy with linear search we estimate in terms of λ the point on this spiral where the function is minimum. Comparison is made with other methods of estimating this parameter for certain test problems.

1. Linearized Search Procedure using Gradient Methods

We are concerned with the problem of finding a local unconstrained minimum of a differentiable function

$$\Phi(\mathbf{x}) = \Phi(x_1, x_2, \ldots, x_n). \tag{1}$$

We consider firstly what step-length α we ought to take if, starting at some initial point $\mathbf{x}^{(0)}$ which is the current point of an iterative procedure, we proceed in a direction prescribed by a vector $\mathbf{v}$ in the space of the variables to a new point $\mathbf{x}^{(1)}$ of the iteration through a displacement $\boldsymbol{\delta}$, thus:

$$\boldsymbol{\delta} = \mathbf{x}^{(1)} - \mathbf{x}^{(0)} = \alpha \mathbf{v}. \tag{2}$$

The new value of the objective function is then
$$\Phi(\mathbf{x}^{(1)}) = \Phi(\mathbf{x}^{(0)} + \alpha \mathbf{v}) = F(\alpha), \tag{3}$$
say, where we have used a different symbol in order to particularize the objective function for a given $\mathbf{x}^{(0)}$ and $\mathbf{v}$ as a function of α only.

Ideally, we would like the displacement δ to terminate at a point of grazing contact with a contour of the objective function so as to reduce F as much as possible in the direction $\mathbf{v}$. So, expanding about $\mathbf{x}^{(0)}$, at the point of grazing contact we have

$$F = F_0 + \alpha \left(\frac{\mathrm{d}F}{\mathrm{d}\alpha}\right)_0 + \frac{\alpha^2}{2}\left(\frac{\mathrm{d}^2 F}{\mathrm{d}\alpha^2}\right)_0 + \cdots \tag{4}$$

and

$$\frac{\mathrm{d}F}{\mathrm{d}\alpha} = \left(\frac{\mathrm{d}F}{\mathrm{d}\alpha}\right)_0 + \alpha \left(\frac{\mathrm{d}^2 F}{\mathrm{d}\alpha^2}\right)_0 + \cdots = 0, \tag{5}$$

where a zero subscript signifies evaluation at $\mathbf{x}^{(0)}$. If we now differentiate (3) with respect to α we obtain for the derivatives in (4) and (5)

$$\frac{\mathrm{d}F}{\mathrm{d}\alpha} = \sum_{i=1}^{n} \frac{\partial \Phi}{\partial x_i} v_i = \mathbf{v}^T \mathbf{g} \tag{6}$$

and

$$\frac{\mathrm{d}^2 F}{\mathrm{d}\alpha^2} = \sum_{i,j=1}^{n} \frac{\partial^2 \Phi}{\partial x_i \partial x_j} v_i v_j = \mathbf{v}^T \mathbf{G} \mathbf{v} \tag{7}$$

where $\mathbf{g}$ is the gradient vector of the function Φ and $\mathbf{G}$ is its Hessian matrix of second derivatives with respect to the x_i.

In general, the only practicable method of accurately locating the point of grazing contact would be via some iterative process along the direction $\mathbf{v}$. We can, however, use the results (6) and (7) to estimate the point of grazing contact correct to the first order in the displacements by truncating equation (5). This gives as an estimate of the step-length in any direction $\mathbf{v}$:

$$\alpha = -\left(\frac{\mathrm{d}F}{\mathrm{d}\alpha}\right)_0 \bigg/ \left(\frac{\mathrm{d}^2 F}{\mathrm{d}\alpha^2}\right)_0 = -\mathbf{v}^T \mathbf{g}/\mathbf{v}^T \mathbf{G} \mathbf{v} \tag{8}$$

where $\mathbf{g}$ and $\mathbf{G}$ are evaluated at $\mathbf{x}^{(0)}$.

Regarding the choice of $\mathbf{v}$, in the wide class of so-called quasi-Newton techniques $\mathbf{v}$ is taken to be of the form

$$\mathbf{v} = -\mathbf{H}\mathbf{g}$$

where the matrix $\mathbf{H}$ is specific to the particular technique used. For such techniques, (8) has the form of a Rayleigh quotient

$$\alpha = \mathbf{g}^T \mathbf{H} \mathbf{g} / \mathbf{g}^T \mathbf{H}^T \mathbf{G} \mathbf{H} \mathbf{g}.$$

Special cases of this general method are:

(i) *steepest descent*: $\mathbf{H} = \mathbf{I}$, the unit matrix, and hence $\alpha = \mathbf{g}^T \mathbf{g}/\mathbf{g}^T \mathbf{G} \mathbf{g}$.
(ii) *Newton-Raphson*: $\mathbf{H} = \mathbf{G}^{-1}$, and hence $\alpha = 1$.

The fixed value obtained for α in the Newton-Raphson method arises, of course, because the method yields both the direction and the magnitude of the displacement.

2. The 'Best' Direction and its Geometric Interpretation

Having obtained in Section 1 an explicit formula for estimating the step-length α, we proceed next to examine the choice of direction for $\mathbf{v}$. Were it possible to determine the point of grazing contact accurately, this would immediately

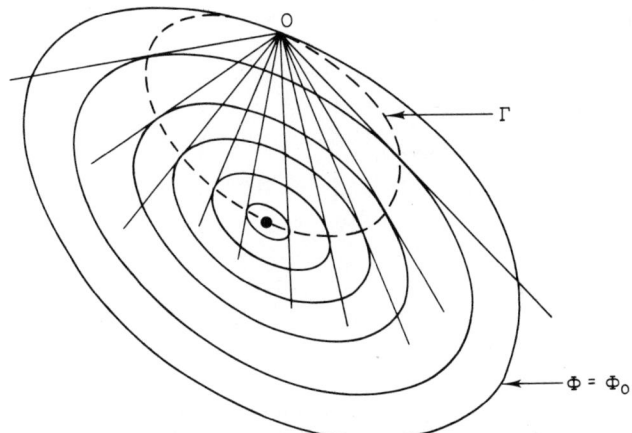

FIG. 1. Linearized search: Locus Γ of points of grazing contact with the contours of the objective function Φ.

result in a reduction of the dimensionality of the problem because the points of grazing contact define a hypersurface which must necessarily contain the minimum we seek. (See Fig. 1.) It would then suffice to minimize on this hypersurface.

Using equations (2) and (8), this hypersurface is estimated, to the order of accuracy defined by the truncated expansions of Section 1, as the hyperquadric Γ whose equation relative to $\mathbf{x}^{(0)}$ as origin is

$$\boldsymbol{\delta}^T \mathbf{G} \boldsymbol{\delta} + \boldsymbol{\delta}^T \mathbf{g} = 0, \tag{9}$$

and which touches the contour $\Phi = \Phi_0$ at $\mathbf{x}^{(0)}$.

Thus all the first-order displacements defined in Section 1 take us to points on Γ and, in particular, by writing (9) in the form

$$\delta^T(\mathbf{G}\delta + \mathbf{g}) = 0$$

we immediately verify that the Newton-Raphson displacement $\delta = -\mathbf{G}^{-1}\mathbf{g}$ itself belongs to this class. Since the centre C of Γ is the point $\delta = -\tfrac{1}{2}\mathbf{G}^{-1}\mathbf{g}$, it follows that the Newton-Raphson displacement takes us along a diameter to the point on Γ where the tangent is parallel to the tangent to the contour at $\mathbf{x}^{(0)}$. This is shown, together with the steepest-descent displacement, in Fig. 2.

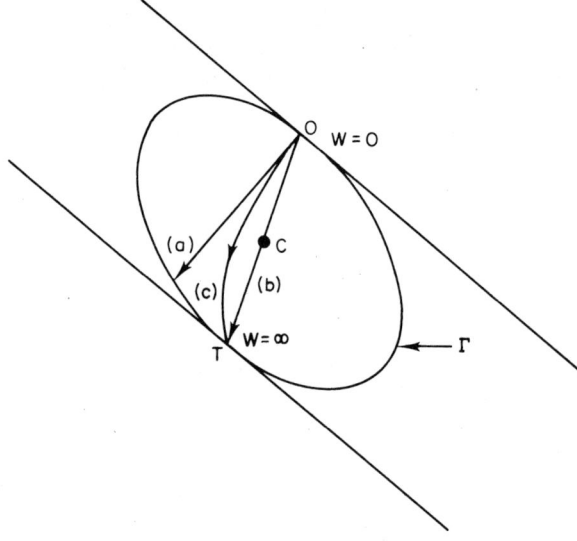

FIG. 2. The fundamental quadric Γ for gradient methods correct to the first order in the displacements. (a) Steepest descent displacement; (b) Newton-Raphson displacement; (c) SPIRAL.

Our next step is to estimate the point on Γ where Φ is a minimum. To the order of accuracy justified by the foregoing displacement estimates, we have, on substituting for $d^2 F/d\alpha^2$ from (8) in (4):

$$F - F_0 = \tfrac{1}{2}\alpha \left(\frac{dF}{d\alpha}\right)_0,$$

and this is stationary when

$$-\alpha \left(\frac{dF}{d\alpha}\right)_0 = (\mathbf{v}^T \mathbf{g})^2/\mathbf{v}^T \mathbf{G}\mathbf{v}$$

is stationary.

Making unrestricted variations with respect to **v**:

$$\frac{\partial}{\partial \mathbf{v}} \{(\mathbf{v}^T \mathbf{g})^2 / \mathbf{v}^T \mathbf{G} \mathbf{v}\} = \mathbf{0},$$

and performing the differentiation we find that the appropriate solution is

$$\mathbf{g} = (\mathbf{v}^T \mathbf{g} / \mathbf{v}^T \mathbf{G} \mathbf{v}) \, \mathbf{G} \mathbf{v}$$
$$= -\alpha \mathbf{G} \mathbf{v}$$
$$= -\mathbf{G} \boldsymbol{\delta},$$

i.e. is the system of Newton-Raphson equations. Thus to the superior convergence properties of the Newton-Raphson method must be added the further property that it is best possible in the sense that if for every direction we determine to the first order of small quantities the displacement to a point of grazing contact with a contour and find the minimum on the hypersurface thus generated, we obtain the Newton-Raphson solution.

3. Levenberg's Method of Damping

Since the displacements computed on the basis of the foregoing procedure cannot be guaranteed to have the order of smallness assumed in their theoretical derivation, the Newton-Raphson method is not always the best method to use in practice. In fact a Newton-Raphson displacement is initially in a direction of (absolutely) decreasing gradient which may not even be in a direction of descent unless **G** is positive definite. For the commonly occurring class of functions which are sums of squares:

$$\Phi = \frac{1}{2} \sum_{j=1}^{m} f_j^2(x_1, \ldots, x_n),$$

this shortcoming may be obviated by replacing the individual functions f_j by their local linear approximations, and minimizing the resulting quadratic approximation to Φ which always has a positive definite Hessian matrix. This is known as the Gauss-Newton method, and on account of the guaranteed initial descent this formulation is frequently used in least squares problems.

However, even when **G** is positive definite there is a danger of 'overshooting' due to the finite size of the calculated displacements. This can cause the computation to diverge (see Barnes, 1968). Levenberg's (1944) proposal for overcoming this difficulty was to modify the objective function by adding to it a positive definite quadratic form in the displacements, and locate instead a local minimum of the function

$$w\Phi(x_1, \ldots, x_n) + \tfrac{1}{2}(a_1 \delta_1^2 + \cdots + a_n \delta_n^2) \tag{10}$$

where w and the a_i are certain suitably chosen positive numbers. This has the effect of increasing the convexity of the function and bringing the minimum closer to the initial point $\mathbf{x}^{(0)}$. Since if the method converges we must have $\delta_i \to 0$, it follows that at a minimum of the modified function (10) the function Φ will also be minimized.

The Newton-Raphson equations corresponding to the first-order Taylor expansions of the space-derivatives of the function (10) are, in matrix-vector notation,

$$(\mathbf{G} + \mathbf{\Lambda}) \boldsymbol{\delta} = -\mathbf{g} \tag{11}$$

in which the diagonal elements of the coefficients-matrix are modified by adding the diagonal matrix

$$\mathbf{\Lambda} = w^{-1} \begin{bmatrix} a_1 & & 0 \\ & a_2 & \\ & & \ddots \\ 0 & & & a_n \end{bmatrix}. \tag{12}$$

It will be appreciated that, for a suitable normalization of the a_i, it is the parameter w which basically governs the extent of modification of the original function and thus the degree of damping which is applied, whereas the a_i are simply scaling factors which determine the local metric of the variable-space.

As w increases from zero, the end-point of the solution vector traces out a spiral path starting at $\mathbf{x}^{(0)}$ when $w = 0$ and terminating at the unmodified Newton-Raphson displacement as $w \to \infty$, as illustrated in Fig. 2. When $\mathbf{G}$ is positive-definite, so that Γ is a closed hyper-ellipsoid, this path is always inside Γ, as may be seen by pre-multiplying (11) by $\boldsymbol{\delta}^T$ to give

$$\boldsymbol{\delta}^T \mathbf{G} \boldsymbol{\delta} < -\boldsymbol{\delta}^T \mathbf{g} \quad \text{for } \boldsymbol{\delta} \neq \mathbf{0}.$$

This strict inequality shows that the spiral path cannot get outside Γ, and justifies our previous statement that the effect of the procedure is to pull back the displacement towards the starting point.

3.1. *Levenberg's Parameter*

In order to assign a value to w, Levenberg (1944) uses the fact that if Φ is expanded in powers of w about $w = 0$, then in a sufficiently close neighbourhood of $\mathbf{x}^{(0)}$:

$$\Phi = \Phi_0 + w \left(\frac{d\Phi}{dw} \right)_0 \tag{13}$$

approximately, and on the assumption that Φ is small compared with Φ_0, he takes

$$w = -\Phi_0 \Big/ \left(\frac{d\Phi}{dw} \right)_0. \tag{14}$$

Using

$$\frac{d\Phi}{dw} = \sum_{i=1}^{n} \frac{\partial \Phi}{\partial \delta_i} \frac{d\delta_i}{dw} \tag{15}$$

and substituting the solution $\delta_1, \ldots, \delta_n$ of (11) obtained using Cramer's rules, we find

$$\delta_i = -g_i a_i^{-1} w + O(w^2). \tag{16}$$

Therefore

$$\left(\frac{d\delta_i}{dw}\right)_{w=0} = -g_i a_i^{-1} \tag{17}$$

and

$$\left(\frac{d\Phi}{dw}\right)_0 = -\sum_{i=1}^{n} g_i^2 a_i^{-1}. \tag{18}$$

Thus Levenberg's own estimate of his parameter is

$$\left.\begin{array}{c} w = \Phi_0/\mathbf{g}^T \mathbf{v} \\[2mm] \mathbf{v} = \{g_1 a_1^{-1}, \ldots, g_n a_n^{-1}\}. \end{array}\right\} \tag{19}$$

where

3.2. *Scaling of the Variables*

Although the unmodified Newton-Raphson and Newton-Gauss methods are invariant under linear transformation of the variables, this no longer holds once the diagonal elements of the coefficients matrix have been modified.

The simplest form of scaling is to take $a_i = 1$ ($i = 1, \ldots, n$), which causes the spiral of Fig. 2 to start off in the direction of steepest descent. Equation (11) can then be written in the form

$$(\mathbf{G} + \lambda \mathbf{I}) \boldsymbol{\delta} = -\mathbf{g} \tag{20}$$

and Levenberg's estimate of the parameter λ in this case is

$$\lambda = \mathbf{g}^T \mathbf{g}/\Phi_0. \tag{21}$$

The parameter λ occurring in (20) is sometimes referred to as Marquardt's parameter, after Marquardt (1963) who put forward equation (20) independently, although he used an empirical trial and error procedure to estimate the value of λ at each iteration. Marquardt recommended a scaling of the variables which would make the diagonal elements of $\mathbf{G}$ unity in (20). Levenberg achieves an exactly equivalent result without explicitly scaling the variables by taking $a_i = G_{ii}$ ($i = 1, \ldots, n$), i.e. by replacing the diagonal elements of $\mathbf{G}$ by $G_{ii}(1 + \lambda)$, so that the equation system to be solved becomes

$$(\mathbf{G} + \lambda \operatorname{diag} \mathbf{G}) \boldsymbol{\delta} = -\mathbf{g} \tag{22}$$

where diag **G** denotes the diagonal matrix of diagonal elements of **G**. Levenberg's method of estimating the parameter λ in this equation then gives

where
$$\left. \begin{array}{l} \lambda = \mathbf{g}^T \mathbf{v}/\Phi_0 \\ \\ \mathbf{v} = \{g_1 G_{11}^{-1}, \ldots, g_n G_{nn}^{-1}\}. \end{array} \right\} \quad (23)$$

Both forms of scaling referred to above are used in practice (Matz, 1964; Barnes, 1968). In necessarily very general terms, the formulation (22) would seem most appropriate when there are large contrasts between the diagonal elements of the matrix of unmodified coefficients, and (20) otherwise.

4. SPIRAL and the Modified Levenberg Parameter

The problem of the optimum choice of the Levenberg parameter $\lambda = w^{-1}$ is, of course, the same as trying to predict the point on the spiral path where Φ is least. In his computer program SPIRAL, Jones (1970) estimates this path by determining the constants in an equation representing a suitable spiral through the points O and T in Fig. 2 which is also tangential to the steepest descent direction at O (thus corresponding to the scaling $a_i = 1$), and then searches along the fitted spiral to reduce Φ as much as possible.

SPIRAL involves a rather elaborate procedure at each iteration. It would therefore be of value to develop a single predictive formula for the parameter w, improving if possible on Levenberg's suggestion of estimating w from a local linear approximation to Φ. Since the need for damping implies that an interior minimum exists on the spiral path, in contrast with a monotonic decrease of Φ, we proceed by constructing the analogue of the linearized search procedure of Section 1, and estimate the point on the spiral in the neighbourhood of $w = 0$ for which Φ is stationary with respect to w, thus:

$$\frac{d\Phi}{dw} = \left(\frac{d\Phi}{dw}\right)_0 + w\left(\frac{d^2\Phi}{dw^2}\right)_0 + \cdots = 0. \quad (24)$$

Hence to the first order of small quantities

$$w = -\left(\frac{d\Phi}{dw}\right)_0 \bigg/ \left(\frac{d^2\Phi}{dw^2}\right)_0. \quad (25)$$

The first-order derivative in (25) has already been evaluated in (18). To determine the second-order derivative we need to extend the solution $\boldsymbol{\delta}$ of (11) as given by (16) to the next highest order of small quantities in w. Again using Cramer's rules, we find after some reduction that

$$\delta_i = -g_i a_i^{-1} w + a_i^{-1} \sum_j g_j a_j^{-1} G_{ij} w^2 + O(w^3). \quad (26)$$

Differentiating (15),

$$\frac{d^2\Phi}{dw^2} = \sum_{i,j} \frac{\partial^2 \Phi}{\partial \delta_i \partial \delta_j} \frac{d\delta_i}{dw} \frac{d\delta_j}{dw} + \sum_i \frac{\partial \Phi}{\partial \delta_i} \frac{d^2 \delta_i}{dw^2},$$

and using (26) we then find that

$$\left(\frac{d^2\Phi}{dw^2}\right)_0 = 3\mathbf{v}^T \mathbf{G} \mathbf{v}$$

where $\mathbf{v}$ is defined in (19). Hence, substituting in (25), we obtain the formula

$$w = \tfrac{1}{3}\mathbf{v}^T \mathbf{g}/\mathbf{v}^T \mathbf{G} \mathbf{v}. \qquad (27)$$

As particular cases of this result, for the scaling $a_i = 1$ ($i = 1, \ldots, n$) our modified form of the Levenberg parameter is therefore

$$\lambda = 3\mathbf{g}^T \mathbf{G} \mathbf{g}/\mathbf{g}^T \mathbf{g}, \qquad (28)$$

and for the scaling $a_i = G_{ii}$, corresponding to a Marquardt scaling of the variables,

$$\left.\begin{array}{c} \lambda = 3\mathbf{v}^T \mathbf{G} \mathbf{v}/\mathbf{g}^T \mathbf{v} \\ \\ \mathbf{v} = \{g_1 G_{11}^{-1}, \ldots, g_n G_{nn}^{-1}\}. \end{array}\right\} \qquad (29)$$

where

5. Results and Discussion

Table 1 shows the results obtained from applying the modified Levenberg procedure to the first seven test problems for which data is given by Jones (1970). As these are all least-squares problems ($\Phi = \tfrac{1}{2} \sum f_j^2$) the Newton-Gauss matrix $\mathbf{G} = \mathbf{J}^T \mathbf{J}$, where $\mathbf{J}$ is the Jacobian matrix of partial derivatives $\partial f_j/\partial x_i$, was used. This ensures that $\mathbf{G}$ is at least positive semi-definite, and hence that λ is never negative. In the table, comparison is made with Levenberg's original method and also with Marquardt's procedure and with the results of SPIRAL as quoted by Jones (1970). The results of the Newton-Gauss method are also given where possible, i.e. in those cases for which $\mathbf{G}$ is strictly positive definite initially.

Both forms of scaling described above were used in the unmodified [equations (21) and (23)] and the modified [equations (28) and (29)] Levenberg procedures, but it was found that the change of scaling made negligible difference to the rates of convergence. The problems chosen had zero sums of squares for their exact solutions, and the criterion used to terminate the current computations (those corresponding to the last three columns of the table) was that $\Phi < 10^{-14}$. This enables a representative comparison to be made with the results quoted by Jones, which are shown in the second and third columns of Table 1.

The significance of the asterisked entries is that convergence had not been completed after 100 iterations and the computation was therefore stopped. It would appear from these results that the rate of convergence of the modified Levenberg method is too slow to be of practical value. However, it will be seen that with one exception (problem 5) the Newton-Gauss method is so greatly superior to the other methods that any form of damping by increasing the diagonal elements of the coefficients-matrix has an adverse effect on the rate of convergence. (The exception arises because the coefficients-matrix is initially singular in problem 5.) What the table clearly demonstrates is that the damping produced by the modified Levenberg method is large. Whilst this is a disadvantage for those cases where no or little damping is required, investigations

TABLE 1

Number of iterations required to reduce sum of squares to approximately 10^{-14} in various test problems

Problem number	Marquardt	SPIRAL	Levenberg	Modified Levenberg	Newton-Gauss
1	92	17	71	100*	2
2	72	27	69	100*	2
3	49	39	58	94	9
4	98	66	45	100*	13
5	103	76	46	100*	—
6	61	9	42	60	1
7	21	13	54	78	1

by gradient methods of large-scale multi-parameter problems, arising in civil engineering (Buchholdt and McMillan, 1971) and in the physical sciences (Ahmad et al., 1971) for example, have shown the need for the application of strong damping in these problems.

Some idea of the comparative rates of convergence of the unmodified and modified Levenberg processes for well-behaved functions may be obtained by considering the function $\Phi = \frac{1}{2}(x^2 + y^2)$ whose contours are concentric circles, and for which **G** is the unit matrix. For this function the unmodified and modified Levenberg parameters are $\lambda = 2, 3$ and so reduce the distance from the minimum at (0,0) by factors of $\frac{2}{3}$ and $\frac{3}{4}$ respectively at each iteration. Hence if ν is the number of iterations required to reduce the sum of squares from 1 to 10^{-14}, we have

$$(\tfrac{2}{3})^\nu = 10^{-7}, \qquad (\tfrac{3}{4})^\nu = 10^{-7}$$

giving $\nu = 40$ and 56 iterations respectively.

For badly distorted contours there might well be merit in the heavier damping of the modified Levenberg process as a method of safeguarding convergence even if it is at the expense of an increase in the number of iterations.

It is worthy of note that the parameter in (28), without the numerical factor 3, has been introduced independently by Smith and Shanno (1971) from semi-empirical reasoning, and it would certainly seem that a damping parameter of this form is a natural one for non-linear optimization problems. It is clear that further numerical experiments are required before definite conclusions can be established, and in particular it would be interesting to study the effect on the rate of convergence of varying the numerical factor in (28), even though it might be difficult to justify a departure from 3 on purely theoretical grounds.

References

Ahmad, M. S., Barrow, D. E., Little, E. A., and Szkopiak, Z. C. (1971). Computer analysis of complex relaxation spectra. *J. Phys. D: appl. Phys.* **4**, 1460–1469.

Barnes, V. (June 1968). GASPAN—an advanced computer code for the analysis of high resolution gamma-ray spectra. *IEEE Trans. Nucl. Sci. NS* **15**, No. 3, 437–454.

Buchholdt, H. A., and McMillan, B. R. (1971). Iterative methods for the solution of pretensioned cable structures and pinjointed assemblies having significant geometrical displacements. Int. Ass. Shell Struct. Pacific Symposium on Tension Structures and Space Frames. Part 2. Tokyo, Japan (to be published).

Jones, A. (1970). Spiral—a new algorithm for non-linear parameter estimation using least squares. *Comput. J.* **13**, 301–308.

Levenberg, K. (1944). A method for the solution of certain non-linear problems in least squares. *Q. appl. Math.* **2**, 164–168.

Marquardt, D. W. (1963). An algorithm for least-squares estimation of non-linear parameters. *J. Soc. ind. appl. Math.* **11**, 431–441.

Matz, A. W. (1964). Automating damped least squares to solve the equations determining refractive index of crystals. *Appl. Stat., J. R. Stat. Soc.* (Ser. C) **13**, 118–127.

Smith, F. B., and Shanno, D. F. (1971). An improved Marquardt procedure for non-linear regressions. *Technometrics* **13**, 63–74.

13. A Dynamic Programming Algorithm for Load Duration Curve Fitting

S. T. LONEY
*South of Scotland Electricity Board,
Glasgow, Scotland*

Summary

The power demand on a network is measured each half-hour for a week. These 336 readings, arranged in descending order, define a profile characteristic of the week as a whole. Conventionally, this profile is expressed in condensed form as an approximating function of six steps. Calculation of the optimum approximation is numerically intractable, and is best performed by a dynamic programming algorithm accelerated by use of a recurrence relation which restricts the region to be explored.

1. Introduction

A curve-fitting problem of unusual type arises in power system operation. Let the demand on the network be measured each half-hour for a week, and the 336 readings sorted into descending order of magnitude. Plotting these points as in Fig. 1, a cumulative distribution curve $y(x)$ is defined, signifying 'Demand is at or above y Megawatts on n occasions, that is for $x\%$ of the time, where $x = n/3\cdot 36$'.

For engineering use in generation scheduling it is customary to condense this profile to a six-step approximation, as illustrated in Fig. 1. Each step specifies a requirement for $(y_2 - y_1)$ MW at $x\%$ load factor, and the various load factors may be associated with different types of generating plant. Thus the top step (low load factor) represents plant used in short bursts interspersed with long idle periods, for which gas turbine plant is appropriate. The lowest step (high load factor) represents plant used in steady-state fashion appropriate to nuclear plant, and the intermediate steps may be identified with coal-fired, oil-fired, hydroelectric and other types of equipment.

Hitherto, these approximations have been produced by manual sketching. A self-contained algorithm was required as part of an integrated data-processing routine.

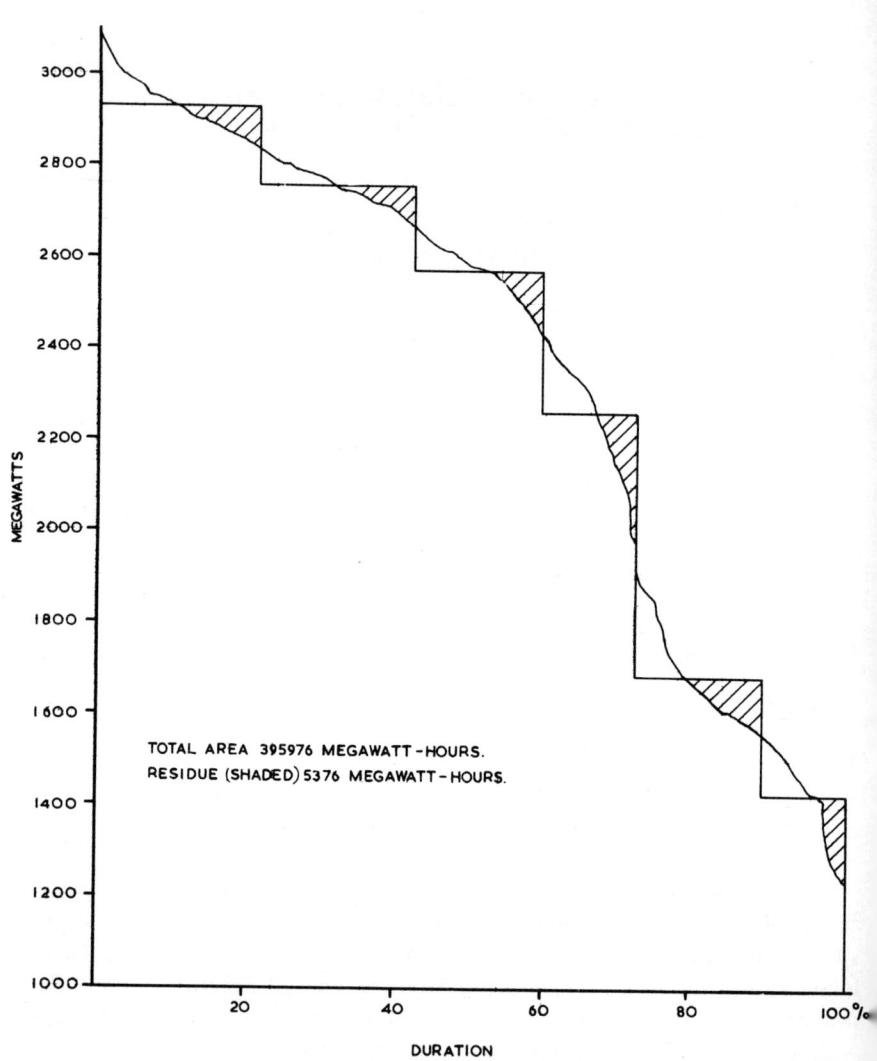

MINIMUM-RESIDUE SIX-STEP APPROXIMATION

Fig. 1.

2. Definition of Problem

2.1. If the lowest step is required to end at 100% time in all cases, any 6-step approximation is specified by the six step heights and the five inter-step break-points, eleven variables in all. The general problem is to derive the best approximation by minimizing an objective function of these eleven variables. The computation is reduced to manageable length by two simplifications:

(a) Introduce the constraint that, between each pair of adjacent break-points, the integral of the approximation shall equal the integral of the original curve. This implies that the height of the step is uniquely determined by the two break-points. The step-heights are no longer free variables, and the problem reduces to five dimensions corresponding to the break-points x_1, x_2, x_3, x_4, x_5.

(b) For practical purposes, a mesh of 336 data points is unnecessarily fine. Acceptable accuracy remains if the data, after sorting, are averaged in successive groups to three to supply a problem in 112 data points.

2.2. The objective function to be minimized may be defined as $A(x_1, x_2, x_3, x_4, x_5)$, the total residual *non-overlapping* area between the two curves. Since the integrals of the curves are equal [from 2.1(a)], it is necessary to evaluate only one side of this residue (Fig. 1).

2.3. For the condensed problem in 112 points, there are $^{112}C_5 \approx 10^8$ possible approximations, any one of which may minimize the objective function A. Examination of the structure of A for a typical set of data reveals an extremely convoluted function, commonly having 30 or 40 local minima within a small region of the feasible space. Gradient methods will not consistently produce the global minimum, and some search procedure short of complete enumeration is required.

Section 3 describes a successful solution by Dynamic Programming (White, 1969) which may be accelerated by the addition of constraints as in Section 4.

3. Dynamic Programming Formulation

3.1. This approach is based on the Principle of Optimality (White, 1969). Computation starts by determining the optimum 2-step approximation to the 112 points. A single break-point defines two single steps with an associated total residue, which may be minimized with respect to all possible positions of the break-point. Two-step optimal approximations to segments of the data may then be combined with single steps to provide optimal 3-step approximations, and the process extended until the desired 6-step solution is reached; see Section 3.3. Define:

$\phi(L, M)$ = residual area corresponding to the unique one-step approximation to the data for points L to M inclusive.

$A(K,J)$ = minimum total residual area attainable in K steps for points J to 112 inclusive.
$N(K,J)$ = end-point of top step of solution corresponding to $A(K,J)$.
$H(K,J)$ = height of top step of solution corresponding to $A(K,J)$.

3.2. If $J+K = 113$, the number of steps is equal to the number of data points. Fit is exact, with zero residue. Then
$$A(K,J) = 0,$$
$$N(K,J) = J,$$
$$H(K,J) = D(J),$$
the value of datum point J.

If $J+K > 113$, the quantities are undefined.

If $J+K < 113$,
$$A(1,J) = \phi(J, 112), \text{ and}$$
$$A(K+1,J) = \min_M \{\phi(J, M) + A(K, M+1)\}$$
$$= \phi(J, N(K+1,J)) + A(K, N(K+1,J)+1).$$

3.3. The computation proceeds thus:

(a) Evaluate $A(1,J)$ for all J, 1 to 112.
(b) Evaluate $\{\phi(1, M) + A(1, M+1)\}$ for all M. The minimal residue specifies $N(2,1)$, $H(2,1)$ and $A(2,1)$. Multiple equal minima may be encountered. In this event, an arbitrary selection is made and the calculation proceeds.
(c) Repeat (b) to produce $A(2,J)$, etc. for all J.
(d) Repeat (b) and (c) to produce $A(3,J)$ for all J, and continue the process until the desired $A(6,1)$ is achieved.

4. Range Limitation

4.1. By considering the geometry of the triangular residual areas (Fig. 1), two recurrence relations may be established:

(a) $N(K+1,J) \leqslant N(K,J)$. That is, when the solution in K steps to a given set of data is known, and the solution in $(K+1)$ steps is to be found, the top step will in no case be longer.
(b) $N(K,J+1) \geqslant N(K,J)$. That is, when the solution in K steps to a given set of data is known, and the solution in K steps, omitting the first datum point, is to be found, the end of the top step will in no case move to an earlier datum point.

13. LOAD DURATION CURVE FITTING

4.2. Thus, when seeking the minimum residue as in 3.3, it is not necessary to explore the entire feasible region, but only that portion which lies between the limits (a) and (b) above. This accelerates the solution of the condensed problem by a factor of 5 or 6 in a typical case.

5. Specimen Results

5.1. For the typical curve of Fig. 1, the solutions in six and fewer steps are summarized in Table 1.

5.2. Using ICL System 4-70, Fortran IV half-word integer arithmetic, computation times were:

Complete data (336 points), Dynamic Programming, $\frac{1}{2}$ hr.
Condensed data (112 points), DP, 1 min.
Condensed data, DP with range limitation, 10 sec.

TABLE 1
Solutions of specimen problem

	1-step	2-step	3-step	4-step	5-step	6-step
Residual area (Megawatt-hours)	39816	14552	9760	7631	6380	5376
Height of 1st step (MW)	2357	2692	2831	2887	2930	2930
End of 1st step	100%	68·8%	43·8%	30·4%	20·5%	20·5%
Height of 2nd step		1619	2426	2639	2757	2757
End of 2nd step		100%	70·5%	56·3%	41·1%	41·1%
Height of 3rd step			1590	2300	2570	2570
End of 3rd step			100%	70·5%	58·0%	58·0%
Height of 4th step				1590	2258	2258
End of 4th step				100%	71·4%	71·4%
Height of 5th step					1578	1682
End of 5th step					100%	88·4%
Height of 6th step						1426
End of 6th step						100%

Reference

White, D. J. (1969). 'Dynamic Programming'. Oliver and Boyd, Edinburgh.

14. Attempts to Calculate Global Solutions of Problems that may have Local Minima

GARTH P. MCCORMICK

George Washington University, Washington, D.C., U.S.A.

Summary

Proposals for obtaining global solutions to not necessarily convex programming problems are examined, with emphasis on the associated pitfalls. Included are penalty function methods, Lagrangian methods, grid methods, heuristic methods, random methods, and a branch and bound method for separable programming problems.

1. Introduction

A particularly vexing problem in the solution of non-linear optimization problems is the possibility that algorithms for solving the problem converge to local as opposed to global solutions. In this paper is contained an all too brief survey of proposals for obtaining global solutions to not necessarily convex optimization problems. In all cases an attempt will be made to point out the difficulties associated with the ideas.

2. Convex Programming, Convex Envelopes and the Global Solution

It is well known that a local solution to the problem

$$\min f(x) \text{ subject to } g_i(x) \geqslant 0, \quad i = 1, \ldots, m,$$

where $x \in E^n$, $f(x)$, $\{-g_i(x)\}$, are convex functions is also a global solution. An interesting proposal for solving the global solution when the convexity assumptions do not hold has been proposed by Kleibohm (1967). The problem he addressed was slightly more general, given as

$$\min f(x) \text{ subject to } x \in B$$

where f is continuous, B a compact set.

Define $\text{CON}(B)$ (the convex hull of B) as

$$\text{CON}(B) = \left\{ x \mid x = \sum_{j=1}^{n} \lambda_j x_j, \ x_j \in B \right\}$$

where
$$\lambda_j \geqslant 0, \quad \sum_{j=1}^{n} \lambda_j = 1.$$

Define the largest convex subfunction $u(\cdot)$ as

(1) $u(\cdot)$ is convex on $\text{CON}(B)$;
(2) $u(x) \leqslant f(x)$ for all $x \in B$;
(3) $u(x) \geqslant q(x)$ for all $x \in B$ for all functions $q(\cdot)$ having the properties that
(4) $q(\cdot)$ is convex on $\text{CON}(B)$;
(5) $q(x) \leqslant f(x)$ for all $x \in B$.

He proves the following result. The solution set of

$$\min u(x) \text{ subject to } x \in \text{CON}(B)$$

contains the solution set of

$$\min f(x) \text{ subject to } x \in B.$$

This interesting proposal has much theoretical interest and points out, at least in an abstract geometrical manner, how to solve the problem. The difficulties with it are threefold. The major problem is that it is impossible, except in certain special cases, to describe algebraically $\text{CON}(B)$ and $u(\cdot)$. In more concrete terms, one cannot implement the algorithm on a digital computer. Second, it seems clear from some small examples that the function $u(\cdot)$ does not maintain the differentiability properties of the original function. Hence existing algorithms for solving a convex program could not be used. Finally, the solution to the convex programming problem need not be feasible to the original problem, although the solution values are the same.

3. The Use of Grids

One of the most obvious ways to solve a programming problem, and one which is often suggested by those new to the field, is the grid approach. The algorithm assumes lower and upper bounds on each variable, divides the resulting interval into I_j equally spaced intervals and finds the point over the grid with minimum objective function value which is feasible. The difficulty with this approach is the extraordinarily high number of function evaluations required for a problem with many variables. The number of possibilities, or combinations of points, involved is

$$\prod_{j=1}^{n} I_j.$$

Many methods for obtaining global solutions essentially are of the grid type. The branch and bound technique for separable programming described later has elements of a grid search but is intended, like all branch and bound methods, to reduce the number of points considered by showing that subsets of the combinations cannot possibly contain the solution to the problem. Recent work by Hartley et al. (1970) falls into the category of a grid search method.

4. Random Methods

An idea akin to the grid approach for obtaining global solutions is the random method, and variations on this method. Basically, the idea is to assume lower and upper bounds on the range of the variables and generate points within that range by a randomized procedure. Methods vary in the way the points are generated and how much use they make of previous information. A limited amount of experience by the author's student with the Russian approaches (see Mátyás, 1965, and Motkus, 1962), showed evidence that the methods had a very slow rate of convergence and took an enormous number of function evaluations. Theoretically, of course, it is possible to show that the methods converge in the limit to the global solution with probability one.

5. The Lagrangian Approach

One often heralded approach to the programming problem is the use of the Lagrangian function $L(x,u) = f(x) - \sum u_i g_i(x)$ to obtain global solutions. Roughly stated, one generates a sequence of non-negative multiplier vectors $\{u^k\}$ and an associated sequence $\{x^k\}$ of points which are the *global* unconstrained minimizers of the Lagrangian. If the values of $\{g_i(x^k)\}$ tend to zero for non-zero u_i^k limits, then limit points of $\{x^k\}$ are global solutions to the programming problem. The alleged ease of implementing this method for separable problems has made it attractive. Several points should be noted. First, in the case when there are no constraints the algorithm reduces to the tautology—'to find the global unconstrained minimizer of $f(x)$, find the global unconstrained minimizer of $f(x)$', i.e., there is no algorithm in the sense of a prescribed iterative procedure for doing this. Second, in general for the non-convex optimization problem, the Lagrange multipliers associated with the global solution and the solution point do not constitute a pair for which the Lagrangian has a global unconstrained minimizer at x^*. In a non-convex problem the Lagrangian is usually non-convex and has no unconstrained minimizer at all. In the convex problem the difficulty with the Lagrangian approach is just to find the proper multipliers. Falk (1967) has shown the conditions under which such an algorithm would work for the convex case.

In another paper, Falk (1969a) characterized the solution obtained by the Lagrangian approach to the problem: $\min f(x)$ subject to $x \geqslant b$, $x \in G$, where G is a compact convex set as the solution to the minimization of $c(x)$ subject to $Ax \geq b$, $x \in G$, where $c(x)$ is the convex envelope of $f(x)$ over G. (Recall from the theorem of Kleibohm the minimization should be of the convex envelope of $f(x)$ over $\{G \cap Ax \geq b\}$ to obtain the same solution.) For more complicated problems the solution obtained is not easy to characterize but it is certainly not the correct one.

Example 1

To make the point more explicitly, consider the problem

Minimize $-x_1^2 - x_2^2$

Subject to $-x_1 - 4x_2 + 5 \geq 0$, $-x_1 + 1 \geq 0$, $x_1 \geq 0$, $x_2 \geq 0$.

This problem has two local minimizers. Form the Lagrangian function

$$L(x, u) = -x_1^2 - x_2^2 - u_1(-x_1 - 4x_2 + 5) - u_2(-x_1 + 1) - u_3(x_1) - u_4(x_2).$$

Now, for no choice of multipliers does this Lagrangian function have a finite unconstrained minimizer.

The second-order necessary conditions characterizing a local minimizer to a non-convex programming problem indicate very clearly that the Lagrangian is, in some sense, minimized (again, speaking roughly) at x^* only in directions orthogonal to the gradients of the binding constraints. Hence, basing a method on finding the unconstrained minimizer of the Lagrangian is doomed to fail since it does not take into account the fundamental characteristics which apply to a local (and hence global) minimizer to a non-linear programming problem.

6. The Penalty Function Approach

An easy theorem to prove is that if one applies an interior point unconstrained penalty function to a non-convex programming problem, such as that given by (1) and (2), e.g., if one obtains the global minimizer x^k of $f(x) - r_k \sum \ln g_i(x)$ for a decreasing sequence of values $\{r_k\}$ which tend to zero, then the global unconstrained minimizers approach the global unconstrained solution ($x^k \to x^*$) in the limit as $r_k \to 0$. Unlike the Lagrangian approach, the existence of the global unconstrained minimizer in the interior of the feasible region usually obtains in the non-convex situation. For example, if $\{x | f(x) \leq M, g_i(x) \geq 0, i = 1, \ldots, m\}$ is bounded for all M (which it usually is in practice), then a global minimizer of the penalty function exists. The difficulty here is that there is no guarantee that one can obtain a global—as opposed to local—unconstrained minimizer. In fact, one can show that a sequence of unconstrained local minimizers exists which approach every isolated compact set of local constrained minimizers. In the problem of Example 1, using a penalty function approach would yield either of the local minimizers depending upon the initial starting point. The penalty function approach then is not primarily intended to be a global method in the nonconvex case.

(It should be noted that the existence of a global unconstrained minimizer for *exterior* point penalty functions does not follow under the same circumstances and may generate unbounded sequences of points.)

7. Methods of Successive Feasibility

Many investigators have produced algorithms for obtaining global solutions which are variations of the general idea of the elimination of local minimizers once they are obtained by the addition of constraints which eliminate the local minimizer. Such an idea where spheres around the local minimizer are introduced is contained in Hesse (1971a). As a heuristic such ideas might have merit, but in general there is a great deal of parameter selection necessary (such articles are full of experimentation, with $\epsilon = 0.1, 0.001, \ldots$), and as a theoretical device are fraught with difficulties. Two objections are immediate. In some cases the new constraints introduced generate programming problems where the local minimizers are not local minimizers of the original problem. In others—such as min $f(x)$ subject to $g_i(x) \geq 0$, $i = 1, \ldots, m$, $f(x) \leq f(x^k) - \epsilon$, where x^k is a local minimizer previously found—the question is begged, since in a neighbourhood of x^k there is no point feasible to the problem. The problem of finding a feasible point is itself an optimization problem, and is subject to the same difficulties as the original one.

8. Non-Convex Quadratic Programming

The area of finding global solutions to not necessarily convex quadratic programming problems has an extensive literature and will not be discussed here. The first major paper was published by Ritter (1966). Recent work of Cottle and Mylander (1969) explains and expands this work. When the quadratic form (to be minimized) is negative semi-definite, the global (and any local solution) is at a vertex. Using this fact several authors have made suggestions on how best to find it—Tui (1964a), Hu (1969), Cabot and Francis (1970).

Other work on the quadratic indefinite form has been done by Mueller and Cooper (1971).

9. Geometric Programming and Differentiably Unimodal Functions

The class of functions for which local minimization implies global minimization is not restricted to those where the functions are convex. A very general class for which this is true is (see Mangasarian, 1969) minimize $f(x)$ (a pseudo-convex function) subject to $g_i(x) \geq 0$, $i = 1, \ldots, m$, where each g_i is a quasi-concave function. It is well known also that geometric programming problems have the local global property even though the functions involved are not pseudo-convex or quasi-concave. In a recent paper, Zwart (1970) showed for a class of functions with certain properties, local solutions are global solutions. His classification covered the case of geometric programming. The main development is repeated here. A function f is a *differentiably unimodal function* on an open set R if f is differentiable on R and if $\nabla f(\bar{x}) = 0$ for some point $\bar{x} \in R$ implies that $\bar{x}$ is a global minimizer for f restricted to R (such a point need not exist).

Theorem (Zwart, 1970, p. 157)

Suppose that F is a family of functions for which (i) $f \in F \Rightarrow f$ is differentiably unimodal on R, (ii) $f \in F \Rightarrow \alpha f \in F$ for any positive real number α, and (iii) $f_1 \in F, f_2 \in F \Rightarrow f_1 + f_2 \in F$. Then any problem of the form:

Minimize $f(x)$

Subject to $g_i(x) \geq 0$, $i = 1, \ldots, m$,

where $f, \{-g_i\} \in F, i = 1, \ldots, m$, must have the property that any local minimizer is a global minimizer.

Analyses such as that above may serve to bring the special characteristics of geometric programming into a synthesis with the properties of convex programming. Work on obtaining global solutions to geometric programming problems when the signs on the posynomial coefficients are improper is currently under way—see Duffin and Peterson (1970), and the references therein.

10. Miscellaneous Methods

Some interesting results on a theoretical level which may lead to procedures for obtaining global minimizers have been suggested in several places. There is no way to categorize these, except possibly to say that they all involve using the integrals of the objective function rather than the derivatives. Abbreviated summaries of these results follow.

Theorem (Falk, 1969b)

Suppose the programming problem is

$\max f(x)$ (continuous)

subject to $x \in S$, the closure of a bounded domain where without loss of generality it can be assumed that $f(x) \geq \alpha > 0$ for all $x \in S$. Then a point $x^* \in S$ is *not* a global minimizing point for this problem if, for some $n > 0$,

$$\int_S [f(x)/f(x^*)]^{n+1} > \int_S [f(x)/f(x^*)]^n.$$

Theorem (Falk, 1969b)

A necessary and sufficient condition that a point x^* solve the problem above is that

$$\limsup_{t \to \infty} \int_S [f(x)/f(x^*)]^t < \infty.$$

An explicit representation of the point which is the global solution of the above problem was given by Pincus (1968).

Assume that for the above problem, f attains its global minimum at exactly one point x^*. Then the co-ordinates of the minimizing point are given as

$$x_j^* = \lim_{\lambda \to \infty} \frac{\int_S x_j \exp[\lambda f(x)]}{\int_S \exp[\lambda f(x)]}.$$

11. The Branch and Bound Approach

Almost all optimization problems which can be implemented on a digital computer are capable of being converted into equivalent separable programming problems by the addition of variables and equality constraints. A method for solving a non-linear optimization problem with convex constraints using the branch and bound approach was suggested by Falk and Soland (1969) when the objective function was separable in the non-convex portion. Later the algorithm was extended by Soland (1971b) to handle separable non-convex constraints.

Computer implementation of the idea, which relies heavily on the concept of convex envelopes, has been highly successful in certain cases, particularly when the subproblems generated by the branch and bound algorithm could be handled by linear programming subroutines. A brief description of the algorithm is as follows.

The problem addressed is

$$\text{Minimize } f(x) = \sum_{j=1}^{n} f_j(x_j)$$

subject to

$$x \in G \text{ (a closed set)},$$

$$x \in C = \{x | l \leq x \leq L\},$$

$$g_i(x) = \sum_{j=1}^{n} g_{ij}(x_j) \leq 0, \qquad i = 1, \ldots, m.$$

For each j, f_j and all g_{ij} must be lower semi-continuous on the finite interval $[l_j, L_j]$. It is further assumed that the set of points $G \cap H$ is non-empty where H is defined as

$$H = \{x | x \in C; g_i(x) \leq 0, \qquad i = 1, \ldots, m\}.$$

These assumptions are enough to ensure that f attains its minimum over the set $G \cap H$.

The algorithm produces a sequence of (not necessarily feasible) points $\{x^k\}$. Each x^k is a solution of problem P^{kv_k} which involves the minimization of a convex function over the intersection of G with a convex set contained in C. Branching is the partitioning of C into smaller and smaller rectangles, and the lower bounds are lower bounds of f over the intersection of each of these rectangles with $G \cap H$.

Crucial to the algorithm is the use of convex envelopes to provide underestimating convex functions of the original problem functions. Let

$$C^{kv} = \{x | l^{kv} \leq x \leq L^{kv}\}.$$

In problem P^{kv}, f_j is replaced by its convex envelope ψ_j^{kv} over $[l_j^{kv}, L_j^{kv}]$, and each g_{ij} by its convex envelope θ_{ij}^{kv} over $[l_j^{kv}, L_j^{kv}]$. Let

$$\psi^{kv}(x) \equiv \sum_{j=1}^{n} \psi_j^{kv}(x_j),$$

$$\theta_i^{kv}(x) \equiv \sum_{j=1}^{n} \theta_{ij}^{kv}(x), \qquad i = 1, \ldots, m$$

so that ψ^{kv} is the convex envelope of f over C^{kv} and θ_i^{kv} is the convex envelope of g_i over C^{kv}.

As a computational aside, it is very simple to compute the convex envelope of a simple function of a *single* variable over an *interval* in most practical cases. For example, if the function is concave, its convex envelope is a straight line. The process can be implemented on a computer so that this is an automatic procedure for standard functions as $\sin(x)$, e^x, $x^{0.1}$.

The programming problem P^{kv} associated with any rectangle C^{kv} is then

$$\text{Minimize } \psi^{kv}$$

subject to

$$x \in G,$$

$$x \in C^{kv} = \{x | l^{kv} \leq x \leq L^{kv}\},$$

$$\theta_i^{kv}(x) \leq 0, \quad i = 1, \ldots, m.$$

By construction it is easy to show that x^{kv}, any solution point to P^{kv}, is a lower bound to f over $G \cap H \cap C^{kv}$.

At any stage k the original rectangle C has been subdivided into p_k rectangles which together constitute a *partition* $P^k = \{C^{k1}, \ldots, C^{kp_k}\}$. Associated with each rectangle C^{kv} are convex underestimating functions $\{\psi^{kv}\}$, $\{\theta_i^{kv}\}$, and a programming problem P^{kv} with solution point x^{kv}. Attention is focused on the rectangle whose objective function value is smallest. That is, let v_k denote an integer where

$$\psi^{kv_k}(x^{kv_k}) = \min_v \psi^{kv}(x^{kv}), \quad v = 1, \ldots, p_k.$$

For simplicity, let x^k denote x^{kv_k}. The termination rule for the algorithm is that if $f(x^k) = \psi^{kv_k}(x^k)$, and if $x^k \in H$, then the problem is solved by x^k. This is true because

$$\psi^{kv_k}(x^k) \leq \psi^{kv}(x) \leq f(x)$$

for all $x \in G \cap H \cap C^{kv}$ and $v = 1, \ldots, p_k$. If $f(x^k) > \psi^{kv_k}(x^k)$ and/or $x^k \notin H$ the algorithm proceeds to stage $k + 1$ by dividing the rectangle C^{kv_k} into two or more rectangular subsets.

The branching part of the algorithm takes two forms depending on whether or not the problem functions are continuous or merely lower semi-continuous

Weak Branching Rule

Choose any j that maximizes the difference

$$f_j(x_j^k) - \psi_j^{kv_k}(x_j^k)$$

or

$$g_{ij}(x_j^k) - \theta_{ij}^{kv_k}(x_j^k)$$

where i is restricted to the infeasible constraints, i.e., those for which $g_i(x^k) > 0$. Then $p_{k+1} = p_k + 1$, i.e., two new rectangles are formed by splitting C^{kv_k} into two parts. The bounds for both new rectangles are the same for all components except the jth which in one case has $[l_j^{kv_k}, x_j^k]$ as its bounds, and in the other, $[x_j^k, L_j^{kv_k}]$.

Strong Branching Rule

For *every j* such that

$$f_j(x_j^k) - \psi_j^{kv_k}(x_j^k) > 0$$

or

$$g_{ij}(x_j^k) - \theta_{ij}^{kv_k}(x_j^k) > 0$$

for those i such that $g_i(x^k) > 0$, divide the corresponding interval $[l_j^{kv_k}, L_j^{kv_k}]$ into the two intervals $[l_j^{kv_k}, x_j^k]$ and $[x_j^k, L_j^{kv_k}]$, creating a new rectangle for stage $(k+1)$ for every such j.

Note that the strong branching rule in general generates many more new rectangles (and hence programming problems to be solved) than the weak branching rule. Its use is to be avoided if possible. However, it may be needed to guarantee convergence of the algorithm. Two statements about convergence are stated in the following theorems.

Theorem

If the strong branching rule is used to generate new rectangles, then any limit point of $\{x^k\}$ is a solution of problem P.

Theorem

If the functions $f, \{g_i\}$ are continuous, and if the weak branching rule is used to generate the new rectangles, then every limit point of $\{x^k\}$ is a solution of problem P.

More details, and illustrative examples are contained in Falk and Soland (1969), and Soland (1971b). A similar approach to the problem using concepts of *special ordered sets* has been proposed by Beale and Tomlin (1970), and Tomlin (1970). Their piece-wise-linear approximation approach allows the use of linear programming codes to solve the sequences of subproblems generated by this branch and bound method.

A general approach which uses the integral of the function over the feasible domain has been proposed by Graves and Whinston (1969). This proposal has elements of the grid approach but with a refinement procedure for creating smaller rectangles similar to that of Falk and Soland discussed in the previous section. Instead of choosing the rectangle with the smallest lower bound for branching, they compute the average value of the function over the region and subdivide the rectangle with smallest average value. As the area of the nested rectangles goes to zero, the average value approaches the value of the limiting point. (This is an abbreviated description of their more general approach.)

They have successfully solved some small problems. Difficulties in using this method stem computationally from the problem of computing the average value (they give several approximation schemes for this) and the lack of a valid convergence criterion other than an exhaustive subdivision of all the rectangles.

Conclusion

Almost all of the algorithms suggested for obtaining global solutions to non-convex programming problems contain some aspect which make their implementation impossible, or elements which require a combinatorially unacceptable amount of computer work. The branch and bound approach, relying on use of underestimating convex functions, seems the most reasonable approach at this time. The efficiency it offers depends upon how quickly the regions which do not contain the global solution are eliminated.

In many instances the solution to a non-convex problem obtained by an algorithm which obtains local minimizers can be seen to be the solution. The branch and bound method takes no advantage of the fact that a good guess at the solution is available. Rather than as an algorithm for solving the problem, one should probably regard the branch and bound algorithm as a *verification* procedure. Then its use, which is invariably longer in computer time than an algorithm which directly tries to obtain local solutions, can be made greater or lesser by those who formulate the problem to be solved.

References

Beale, E. M. L., and Tomlin, J. A. (1970). Special facilities in a general mathematical programming system for nonconvex problems using ordered sets of variables. *In*: 'Proceedings of the Fifth International Conference on Operational Research' (J. Laurence, ed.), pp. 447–454. Tavistock Publications, London.

Cabot, A. V., and Francis, R. L. (1970). Solving nonconvex quadratic minimization problems by ranking the extreme points. *Operat. Res.* **18**, 82–86.

Cottle, R. W., and Mylander, W. C. (1969). 'Ritter's Cutting Plane Method for Nonconvex Programming'. Technical Report No. 69-11, Operations Research House, Stanford University, Stanford, California.

Duffin, R. J., and Peterson, E. L. (1970). 'Geometric Programming with Signomials.' Report 70-38, Department of Mathematics, Carnegie-Mellon University, Pittsburgh, Pennsylvania.

Falk, J. E. (1967). Lagrange multipliers and nonlinear programming. *J. Math. Anal. Appls* **19**, 141–159.

Falk, J. E. (1969a). Lagrange multipliers and nonconvex programs. *SIAM J. Control*, **7**, 534–545.

Falk, J. E. (1969b). 'Conditions for Global Optimality in Nonlinear Programming.' Report No. 69-2, Series in Applied Mathematics, Northwestern University, Evanston, Illinois.

Falk, J. E., and Soland, R. M. (1969). An algorithm for separable nonconvex programming problems. *Managmt Sci.* **15**, 550–569.

Fletcher, R. (1970). 'An Efficient, Globally Convergent, Algorithm for Unconstrained and Linearly Constrained Optimization Problems.' T. P. 431, Atomic Energy Research Establishment, Harwell, England.

Fricks, R. E. (1970). Nonconvex Linear Programming. Doctoral Thesis, Case Western Reserve University, Department of Operations Research, Cleveland, Ohio.

Gould, F. J. (1969). Extensions of Lagrange multipliers in nonlinear programming. *SIAM J. appl. Math.* **17**, 1280–1297.

Gran, R. (1970). 'On the Convergence of Random Search Algorithms in Continuous Time with Applications to Adaptive Control'. Grumman Aircraft Engineering Corporation, Report No. RE-369J, Grumman Aircraft Engineering Corporation, Bethpage, New York.

Graves, G. W., and Whinston, A. B. (1969). 'An Algorithm for Nonconvex Programming'. Report of Krannert Graduate School of Industrial Administration, Purdue University, Lafayette, Indiana.

Hartley, H. O., George, M. D., and LaMotte, L. R. (1970). 'Mixed Convex and Nonconvex Programming.' Project Themis Report No. 21, Texas A and M University, College Station, Texas.

Hesse, R. (1971a). 'Some Systematic Methods for Leaving a Local Optimum'. Draft Memorandum, University of Southern California.

Hesse, R. (1971b). 'A Suboptimal Method for the Global Solution of the Nonlinear Programming Problem'. Report No. COO-1493-16, Washington University, St. Louis, Missouri.

Hu, T. C. (1969). 'Minimizing a Concave Function in a Convex Polytope'. MRC Report No. 1011, Mathematics Research Center, University of Wisconsin, Madison, Wisconsin.

ten Kate, A. (1971). 'Conditions for Global Optimality in Non-Convex Decomposable Mathematical Programs'. Discussion Paper No. 8, Centre for Development Planning, Netherlands School of Economics, Rotterdam.

Kleibohm, K. (1967). Remarks on the nonconvex programming problem (Bemerkungen zum Problem der Nichtkonvexen Programmierung). *Unternehmensf.* **11**, 49–60.

Krolak, P. D. (1968). Further extensions of fibonaccian search to nonlinear programming problems. *SIAM J. Control*, **6**.

Mangasarian, O. L. (1969). 'Nonlinear Programming'. McGraw-Hill, New York.

Mátyás, J. (1965). Random optimization. *Automat. Telemechan.* **26**, 246–253.

McCormick, G. P. (1968). 'Global Solutions to Optimization Problems'. Unpublished Working Papers, George Washington University, Washington, D.C.

Meyer, R. (1970). The validity of a family of optimization problems. *SIAM J. Control*, **8**, 41–54.

Motkus, I. B. (1962). On a method of distributing random tests in solving many-extremum problems. *J. Higher Math. math. Phys.* **12**, 380–385.

Mueller, R., and Cooper, L. (1971). 'The Indefinite Quadratic Programming Problem'. Department of Applied Mathematics and Computer Sciences, School of Engineering and Applied Science, Washington University, Report No. COO-1493-12.

Pascual, L. D., and Ben-Israel, A. (1969). 'Constrained Maximization of Posynomials by Geometric Programming'. Report No. 69-3, Series in Applied Mathematics, Northwestern University, Evanston, Illinois.

Pincus, M. (1968). A closed form solution of certain programming problems. Letter to the Editor, *Operat. Res.* **16**, 690–694.

Rech, P., and Barton, L. G. (1970). A non-convex transportation algorithm. *In*: 'Applications of Mathematical Programming Techniques' (E. M. L. Beale, ed.). American Elsevier Publishing Company, New York.

Ritter, K. (1966). A method for solving maximum-problems with a non-concave quadratic objective function. *Z. Wahrscheinlichk. Verw. Geb.* **4**, 340–351.

Szegö, G. P. (1968). The theorem of Rolle's type in E^n for functions of the class C^1. *Pacif. J. Math.* **27**, 193–195.

Soland, R. M. (1971a). 'Optimal Plant Location with Concave Costs'. Paper presented at 39th National Meeting of the Operations Research Society of America in Dallas, Texas, May 5–7, 1971.

Soland, R. M. (1971b). An algorithm for separable nonconvex programming problems II: nonconvex constraints. *Managmt. Sci.* **17**, 759–773.

Tomlin, J. A. (1970). Branch and bound methods for integer and non-convex programming. *In*: Integer and Nonlinear Programming' (J. Abadie, ed.), pp. 437–450. North-Holland Publishing Company, Amsterdam.

Tui, H. (1964a). Concave programming under linear constraints. *Sov. Math.* **5**, No. 6 (translation published by American Mathematical Society).

Tui, H. (1964b). Concave programming under linear constraints. *Sov. Math.* Doklady **5**, 1437–1440.

Zidov, N. P., and Scedrin, B. M. (1968). A certain method of search for the minimum of a function of several variables. *In*: 'Computing Methods and Programming' (Russian) IZAT. Moscow University, Moscow, 203-210.

Zwart, P. B. (1969). 'Global Maximization of a Convex Function with Linear Inequality Constraints.' Atomic Energy Commission, Report No. COO-1493-24.

Zwart, P. B. (1970). Nonlinear programming: global use of the Lagrangian. *J. Optim. Theory Applns* **6**, 150–160.

15. A Combinatorial Method to Compute a Global Solution of Certain Non-Convex Optimization Problems

U. UEING*

University of Bonn, Institut für Gesellschafts-und Wirtschaftswissenschaften, Bonn, Germany

Summary

Except for Ritter's method for maximizing an indefinite quadratic function over a feasible set given by linear constraints (Ritter, 1965), the mathematical methods available for solving non-convex programming problems generally yield local solutions. In the following, a method is presented for reaching the global solution of the non-convex problem of maximizing a strictly convex function over a non-convex, possibly not connected, feasible set. The problem is decomposed into a finite number of subproblems generated by a certain combination of the constraints and the objective function. The decomposition is such that the solution properties of the original problem are not changed. The set of solution points of the subproblems contains the global solution of the non-convex problem, which can finally be found by a selection routine.

1. Introduction

We consider the question of how a non-convex maximization problem can be decomposed into a finite number of convex subproblems. It turns out to be possible to represent the solution properties of a certain type of non-convex problem by the solutions of convex subproblems.

The maximization problem is defined as

$$\max \{f(x) | g_i(x) \geq 0, i = 1, \ldots, m\} \qquad (1)$$

with the properties:

(a) the function $f(x)$ is a strictly convex and differentiable function of $x \in R^n$;

* During 1972 at Stanford University, Department of Chemical Engineering, Stanford, California, U.S.A.

(b) the functions $g_i(x)$, $i = 1, \ldots, m$, are convex and differentiable functions of $x \in R^n$;
(c) the problem (1) has N local maxima $x_1^0, \ldots, x_N^0$ on the boundary of the feasible set $B = \{x | g_i(x) \geq 0, i = 1, \ldots, m\}$.

The feasible set B is non-convex and may be connected or disconnected. The aim is to determine the global maximum among these N local maxima. The solution procedure uses mainly the behaviour of the local solutions in order to compute the global solution of the non-convex problem.

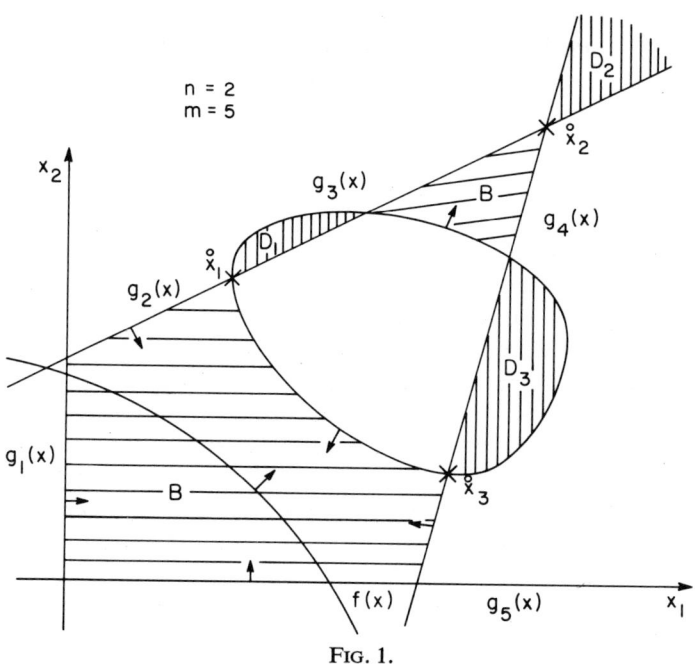

FIG. 1.

Figure 1 shows the geometric properties of a maximization problem of this type. The feasible set B is disconnected into two regions, and there are three local maxima x_k^0 ($k = 1, 2, 3$), each associated with a set D_k obtained by reversing the signs of the n constraints active at the kth local maximum. Property 1, proven in Section 2.1, is that the local maxima must be in one of the corners of B where n constraints are active. Property 2, proven in Section 2.2, is that only those corners where the objective function is globally minimum over a convex set D_k correspond to local maxima [equation (4)]. Section 3 gives two branch and bound procedures for finding the global maximum. The first requires tests on all theoretically possible sets D_k, although not all minimization subproblems need be solved, whereas the second can reject some of the sets D_k without testing them.

2. Properties of the Local Solutions

2.1. First Property

As $f(x)$ is strictly convex, and the $g_i(x)$, $i = 1, \ldots, m$, are convex the following statement holds:

Proposition 1

In every local solution x_k of the problem (1) at least n constraints are active. It is assumed that a constraint qualification holds in x_k, implying that all $\nabla g_i(x_k)$, $i \in B_k = \{i \mid g_i(x_k) = 0\}$, are linearly independent.

Proof

Suppose that x_k is a local solution of (1); furthermore, assume that for the number of active constraints n', $n' < n$ holds. Then x_k is in the intersection

$$S = \{x \mid g_i^l(x) = 0, i \in B_k\},$$

and there exists a neighbourhood V_k of x_k such that $V_k \cap S \subset B$. The functions $g_i^l(x)$ are linear approximations to $g_i(x)$ in x_k. Now consider a point $x_k' = x_k + y$, $y \neq 0$. The point y is chosen in such a way that x_k' lies in $V_k \cap S$. As $f(x)$ is strictly convex it follows that x_k is the solution of

$$\min\{f(x) \mid x \in V_k \cap S\}. \tag{2}$$

Hence, if one moves from x_k to x_k' in $V_k \cap S$, it follows that

$$f(x_k) < f(x_k'). \tag{3}$$

This is a contradiction to the assumption that x_k is a local maximum for $n' < n$.

2.2. Second Property

The second property of the local solutions of problem (1) can be expressed in

Proposition 2

In every local solution x_k

(a) $\max\{f(x) \mid x \in B \cap V_k\} = \min\{f(x) \mid x \in D_k\}, \tag{4}$

where the set V_k is a neighbourhood of x_k, the set J_k is a set of n different elements j which is contained in $\{1, \ldots, m\}$, and the number $\bar{k}$ is the maximal number of different sets J_k which can be formed by the elements of $\{1, \ldots, m\}$. Finally,

$$d_j(x) = -g_j(x), j \in J_k, \qquad 1 \leq k \leq \bar{k},$$
$$D_k = \{x \mid d_j(x) \geq 0, \qquad j \in J_k\};$$

(b) the set D_k is a convex set.

Proof

(a) If x_k is a local maximum of

$$\max\{f(x)|x \in B\} \tag{5}$$

in which n constraints given by the elements of one set of indices J_k are active, then the Kuhn-Tucker conditions are (we assume again the validity of the constraint qualification):

$$\nabla f(x_k) = -\sum_{j \in J_k} u_{jk} \nabla g_j(x_k), \tag{6}$$

$$u_{jk} g_j(x_k) = 0, \qquad j \in J_k,$$

$$u_{jk} \geq 0, \qquad j \in J_k.$$

Equation (6) can be written as

$$\nabla f(x_k) = \sum_{j \in J_k} u_{jk}(-1) \nabla g_j(x_k), \tag{7}$$

$$\nabla f(x_k) = \sum_{j \in J_k} u_{jk} \nabla d_j(x_k), \tag{8}$$

$$u_{jk} d_j(x_k) = 0, \qquad j \in J_k,$$

$$u_{jk} \geq 0, \qquad j \in J_k.$$

Equation (8) is the optimality condition for the minimum of $f(x)$ over the set

$$D_k = \{x | d_j(x) \geq 0, j \in J_k\}.$$

(b) As all $g_j, j \in J_k$, are convex functions, all $d_j, j \in J_k$, are concave functions. Therefore, D_k is given by concave functions and this always results in a convex set.

The consequence of proposition 2 is obvious: one only has to consider solutions of

$$\min\{f(x)|x \in D_k\}, \qquad 1 \leq k \leq \bar{k} \tag{9}$$

that have n active constraints at the minimum solution x_k with the additional property $x_k \in B$. Then all local maxima of $f(x)$ over the non-convex set B are determined. All solutions of (9) in which $n' < n$ constraints are active can be neglected, because they are not feasible in B.

As D_k is a convex set only one point fulfills statement 2(a), and if this point is not feasible in B, no point in D_k is a candidate for a local maximum of $f(x)$ over B even if $B \cap D_k$ is non-empty.

For practical purposes the main interest has to be concentrated on constructing the sets D_k and deciding, (i) whether the solution x_k of

$$\min\{f(x)|x \in D_k\}$$

is feasible in B or not, and (ii) whether n constraints are active in x_k or not. In order to regard all candidates for a local solution one has to construct

$$\bar{k} = \binom{m}{n}$$

sets D_k. Now the question is how to find a suitable form of constructing and solving the possibly numerous convex subproblems.

3. The Convex Subproblems

Two kinds of procedures are proposed for handling the subproblems

$$\min\{f(x)|x \in D_k\}, \quad k = 1, \ldots, \bar{k}.$$

The first is to make some test on all D_k; the second is to exclude some of the theoretically possible D_k during the computation.

3.1. *First Procedure*

The aim is to know which one of the solutions of $\min\{f(x)|x \in D_k\}$, $k = 1, \ldots, \bar{k}$, is the global maximum of $f(x)$ over B.

First, one has to decide whether feasible points y_k exist in the sets D_k. This can be done as proposed by Fiacco and McCormick (1968). Take a certain set D_k given by the constraints $d_j(x) \geq 0$, $j \in J_k$. Assume that a point y_k satisfies only some of the constraints. Accordingly, there are two sets of indices:

$$T_k = \{t|d_t(y_k) > 0, t \in J_k\} \tag{10}$$

and

$$S_k = \{s|d_s(y_k) \leq 0, s \in J_k\}. \tag{11}$$

Now a sequence of points has to be computed which increases the value of

$$\sum_{s \in S_k} d_s(x) \tag{12}$$

without violating any of the constraints already satisfied.

This is one procedure to fulfill the above goal; one could construct others not needing differentiability of the objective function and the constraints. In any case the applied procedure must provide a set of indices $\{k\} = L$ and a set of points y_k such that y_k is feasible in the corresponding set D_k.

Now the problems

$$\min\{f(x)|x \in D_k\}, \quad k \in L \tag{13}$$

can be solved, starting the iterations at the point y_k, $k \in L$.

It might not be necessary to solve all the problems (13) entirely. For example, if a solution $\bar{x}$ of (13) is feasible in B and therefore a local solution of

$$\max \{f(x)|x \in B\}, \tag{14}$$

then all sets D_k can be rejected for which the corresponding feasible point y_k satisfies

$$f(y_k) \leqslant f(\bar{x}), \qquad k \in L. \tag{15}$$

This can be shown by considering the subproblems (13). When a point y_k has been chosen which is rejected by (15), a higher function value than $f(\bar{x})$ can never be reached, because (13) is a convex minimization problem.

The procedure has four steps:

Step 0: Choose $i = 1$, $\bar{x} = y_1 \in B$.
Step 1: Choose a point y_{l_i}, $l_i \in L$ for which $f(\bar{x}) \leqslant f(y_{l_i})$.
Step 2: Take the corresponding set D_{l_i} and solve $\min\{f(x)|x \in D_{l_i}\}$, starting the iteration at point y_{l_i}.
Step 3: (a) If the solution $\hat{x}_{l_i}$ of step 2 is feasible in B, then replace L by $L - \{l_i\}$ and i by $i+1$; if $f(\hat{x}_{l_i}) > f(\bar{x})$ replace $\bar{x}$ by $\hat{x}_{l_i}$; go to step 1.
(b) If $\hat{x}_{l_i} \notin B$, then replace L by $L - \{l_i\}$, and i by $i+1$, and go to step 1.

If step 3(a) is reached, $\bar{x}$ is replaced by a local solution $\hat{x}_{l_i}$ of

$$\min \{f(x)|x \in B\}.$$

Now the set L is not only reduced by $\{l_i\}$, but also by the elements $l \in L$, for which

$$f(y_l) \leqslant f(\hat{x}_{l_i}). \tag{16}$$

This can be done because the aim of the procedure is not to compute all local solutions of $\max\{f(x)|x \in B\}$. Only a global solution is of interest, and it cannot be rejected by (16), because all starting points of the iteration process can only yield a local solution with a lower function value—the latter holds because $f(x)$ is minimized over the set D_l, $l \in L$.

This implies that $\hat{x}_{l_i}$ in step 3(a) is the global solution, if $L - \{l_i\}$ is empty either in step 3(a) or in 3(b). For to stop the computing process, the set $L - \{l_i\}$ has to be tested in steps 3(a) and 3(b), whether it is empty or not, or whether no point y_{l_i} exists which satisfies $f(\bar{x}) \leqslant f(y_{l_i})$.

The procedure computes the global solution in a finite number of steps, because in each loop the set L is reduced by at least one element.

3.2. Second Procedure

In Section 3.1 some tests on all theoretically possible D_k had to be made to compute the global solution. Now an approach will be proposed which rejects part of the possible sets D_k without testing them.

For this purpose all the sets D_k have to be built up in a special structure.

Let $M = \{1, \ldots, m\}$ be the set of indices of the constraints $g_i(x)$, $i = 1, \ldots, m$, and $P_{n'}$ a set of $n' < n$ different elements p with $p \in M$. Then there exist corresponding sets

$$D_{n'} = \{x \mid d_p(x) \geq 0, \quad p \in P_{n'}\}$$

with

$$d_p(x) = -g_p(x), \quad p \in P_{n'}.$$

By the procedure in Section 3.1 one can test whether $D_{n'}$ has a feasible point or not.

(a) If the set $D_{n'}$ has a feasible point, no information about rejecting sets D_k is given.
(b) If there is no feasible point in $D_{n'}$, all sets D_k can be rejected in which $d_p(x) \geq 0$, $p \in P_{n'}$ are side conditions.

For test problems of the above kind this was proposed by Ueing (1971). In general the number of sets D_k,

$$1 \leq k \leq \bar{k} = \binom{m}{n},$$

which can be rejected is

$$k' = \frac{(m-n')!}{(m-n)!(n-n')!}, \quad m > n, \quad 2 \leq n' \leq n.$$

For a given value of m and n the number k' depends to a great extent on the value of n'. For small values of n', k' is a great number compared with $\bar{k}$ so that the amount of computations can be reduced effectively.

On the other hand, when it turns out that by testing the sets $D_{n'}$ the possibility (a) is fulfilled, one has to test possibly all sets $D_{n'}$ for all $P_{n'}$ and all n' with $2 \leq n' \leq n$. In this case, this procedure would be an inefficient instrument for computing feasible points of the sets D_k.

In general it would be useful to apply the proposal in this section only when additional information about the constraints of the non-convex problem is available such as: for certain sets $D_{n'}$ there exists no feasible point.

Acknowledgement

I want to thank Professor K. Ritter for very helpful discussions.

References

Fiacco, A. V., and McCormick, G. P. (1968). 'Nonlinear Programming: Sequential Unconstrained Minimization Techniques'. Wiley, New York.

Ritter, K. (1965). 'Über das Maximumproblem für nichtkonkave quadratische Funktionen.' Report, Deutsche Versuchsanstalt für Luft- und Raumfahrt e.V., München.

Ueing, U. (1971). Zwei Lösungsverfahren für nichtkonvexe Programmierungsprobleme. *In*: 'Lecture Notes in Operations Research and Mathematical Systems'. Springer, Berlin.

16. A Method for Finding Multiple Extrema of a Function of n Variables

FRANKLIN H. BRANIN, JR. AND STANLEY K. HOO
Systems Development Division Laboratory,
IBM Corporation, Kingston, New York

1. Introduction

The problem of determining the global maximum and/or global minimum of a scalar function of n variables may be approached by a method which is able to locate many—and sometimes all—of the stationary points of the function. This is accomplished by computing the zeroes of the gradient of the function using a procedure for obtaining multiple solutions of a system of non-linear equations.

By examining the Hessian matrix of the function at each stationary point, a determination can be made as to whether the point is a maximum, a minimum, or a saddle point. The 'global' maximum or minimum—at least among those stationary points actually found—can then be identified. Although this approach cannot guarantee that the true global extrema will be found, it does represent a useful advance in the field of unconstrained optimization of non-linear functions.

2. The Method

Given a scalar function $f(x)$ of an n-dimensional vector variable x, our objective is to find all the zeroes of the gradient vector by solving the equation

$$g(x) = \frac{\partial f}{\partial x} = 0. \tag{1}$$

We assume that $f(x)$ has continuous second partial derivatives so that the Hessian matrix exists. As explained in detail by Branin (1971b), multiple solutions of equation (1) can be computed by solving the related differential equation

$$\frac{dg}{dt} \pm g(x) = 0. \tag{2}$$

This can be written in the form

$$\left(\frac{\partial g}{\partial x}\right)\frac{dx}{dt} \pm g(x) = 0 \qquad (3)$$

or, alternatively,

$$\frac{dx}{dt} = \mp \left(\frac{\partial g}{\partial x}\right)^{-1} g(x), \qquad (4)$$

where $(\partial g/\partial x)$ is the Jacobian matrix of the gradient, or equivalently, the Hessian matrix, $H(x)$ of the function $f(x)$ itself.

The exact solution of equation (2) in terms of the gradient vector is

$$g[x(t)] = g[x(0)]\exp(\mp t), \qquad (5)$$

which requires the gradient at every point along the trajectory $x(t)$ to be proportional to—hence codirectional with—the initial gradient vector. This constraint provides a valuable side condition which enables very accurate solutions of equation (4) to be obtained not only with the crudest of integration formulas but even without resorting to numerical integration at all (Branin, 1971b).

Since we are interested primarily in multiple solutions of equation (1), we must necessarily encounter singularities in the Hessian matrix. If we happen to be integrating equation (4) taken with, say, the negative sign, the inverse Hessian matrix will cease to exist when we arrive at a singularity. But if we jump through this singular point and change the sign in equation (4), we can continue along our trajectory by integrating the conjugate differential equation.

By applying this rule of sign reversal not only each time the Hessian determinant $|\partial g/\partial x|$ changes sign but also whenever we arrive at a solution point of equation (1), the trajectory $x(t)$ will proceed from one solution to the next (Branin, 1971b).

An effective numerical procedure for integrating equation (4) is a predictor-corrector scheme using Euler's predictor formula

$$x_{n+1} = x_n + h_n \dot{x}_n \qquad (6)$$

where $\dot{x} = dx/dt$. When this formula is applied to equation (4), it yields the expression

$$x_{n+1} = x_n \mp h_n \left(\frac{\partial g}{\partial x}\right)^{-1} g(x), \qquad (7)$$

which is equivalent to Newton's method with variable step size h_n and alterable sign.

The corrector formula is based on the idea of 'steering' the Newton vector, $(\partial g/\partial x)^{-1} g(x)$, by adding a slight increment to it so that the resultant gradient vector will satisfy equation (5). First, the gradient corresponding to the predicted x_{n+1} is resolved into a component parallel to g_0, the initial gradient vector, and a component orthogonal thereto; this orthogonal component is computed using the equation

$$v = g_{n+1} - kg_0, \qquad (8)$$

where kg_0 is the projection of g_{n+1} along g_0 and $k = g_{n+1}^T g_0 / g_0^T g_0$. A correction vector which will, to a first approximation, annihilate the v vector is then computed from the relation

$$\delta x = -\left(\frac{\partial g}{\partial x}\right)^{-1} v, \qquad (9)$$

where the inverse Hessian (or better still, its LU factors) already obtained for use in equation (7) may be used unchanged. Finally, the corrected x vector is calculated according to the expression

$$\bar{x}_{n+1} = x_{n+1} + \delta x. \qquad (10)$$

By repeated application of this corrector scheme, equation (5) can be satisfied to any desired degree of accuracy.

Now the principal theoretical difficulty that the method encounters is not due to the inevitable singularities in the Hessian matrix *per se*, but rather to a secondary effect, namely the existence of singular points of the differential equation, equation (4). As described by Branin (1971b), the solution points of $g(x) = 0$ are *essential* singularities of equation (4) and correspond to degenerate nodes; this is most desirable. But *extraneous* singularities may also occur when $g(x) \neq 0$ but $|H(x)| = 0$, and these can cause serious trouble. Only saddle points and vortex points have been observed so far, the latter always giving rise to regions of non-convergence. Saddle points may also occasionally generate regions of non-convergence. But more frequently they have the adverse effect of deflecting some trajectories away from certain solution points of $g(x) = 0$ which neighbouring trajectories will find. The net effect of these extraneous singular points, then, is often to prevent the method from being globally convergent and/or to preclude finding all solutions of $g(x) = 0$ with a single trajectory. Even so, cases have been observed where global convergence has been obtained and where every trajectory passes through all solutions both in the presence and in the absence of extraneous singularities. The behaviour of these singularities in any particular case determines the success of the method, of course, but a definitive theory is not yet available to predict this behaviour.

3. Computational Considerations

Given a starting point x_0, we compute the initial gradient $g_0 = g(x_0)$, the Hessian matrix $H_0 = H(x_0)$ and its LU. Then the initial sign of equation (7) is made to agree with the sign of the determinant $|H_0|$. This will, of course, establish a certain sense of direction along the corresponding trajectory.

Using an appropriate step size h, the predictor-corrector scheme described above can be applied for as many steps as may be desired, the sign of equation (7) being reversed whenever $|H|$ reverses sign.

For exploratory studies of the method, it has been thought desirable to compute accurate trajectories. Accordingly, the Newton vector, which changes radically in magnitude from step to step, is first normalized and then a fixed length step along the trajectory is taken. Finally, the corrector formulas are applied for as many iterations as needed to meet the desired accuracy criterion in satisfying equation (5).

A more efficient computational scheme would, of course, attempt to take steps as large as possible, consistent with being able to satisfy equation (5) and find all the solutions of $g(x) = 0$ actually traversed by the trajectory being computed. Towards this end, some exploration of a more efficient algorithm has been made, the principal result of which is that the Euler integration formula can be supplanted with a polynomial (or spline) curve-fitting process for predicting x_{n+1}. At present, however, no conclusive procedure has been developed for step size control.

Whenever a solution point of $g(x) = 0$ is found, the trajectory is projected through this point, and the sign of equation (7) reversed. This effectively converts the solution point from a stable node, which had been attracting the trajectory, to an unstable node which now repels it. Thus, the trajectory continues moving with the same sense along the direction it followed in arriving at the solution point.

When all the solution points that lie on this limb of the trajectory have been found, the trajectory will pass off to infinity. Accordingly, a test for divergence must be included. Next, the other limb of the trajectory must be explored by returning to the starting point x_0 and using the opposite of the initial sign in equation (7). This reverses the sense of direction along the trajectory so that all solution points lying on the path in that direction can be picked up.

As each solution point of $g(x) = 0$ is found, the corresponding value of $f(x)$ is, of course, preserved. Also, the point must be identified as a maximum, a minimum, or a saddle point. If the Hessian matrix is positive definite, the point is a local maximum. If the Hessian is negative definite, the point is a minimum. Finally, if the Hessian is indefinite, the point is a saddle point.

The necessary and sufficient condition for the Hessian to be positive definite is that its determinant and the determinants of all of its principal minors must be positive. These determinants may be computed by multiplying the appropriate diagonal terms of the U factor (upper triangular matrix) obtained from the LU decomposition provided that proper account has been taken of the effects of row-column interchanges on the signs of these determinants (Wilkinson, 1965). Accordingly, the 'adjusted' signs of all the diagonal elements of U (adjusted with respect to row-column interchanges) must be positive if $f(x)$ is to be a local maximum.

Correspondingly, the necessary and sufficient condition for the Hessian to be negative definite is that the adjusted signs of the diagonal terms of U be negative; this will indicate that $f(x)$ is a local minimum. Finally, if the adjusted signs of these diagonal terms are mixed, a saddle point has been found. Thus, for negligible extra cost beyond that required for the LU decomposition of H—which is needed anyway in equation (7)—the type of each stationary point can be identified.

4. Use of quasi-Newton Methods

In order to avoid the computational cost of evaluating the Hessian matrix and carrying out the LU decomposition at each iteration for equations (7) and (9), it is possible to use Broyden's approximation method for updating the inverse Hessian (Broyden, 1965). This may be particularly advantageous when the Hessian involves very complicated functions. Using the definitions

$$H_n = \frac{\partial g(x_n)}{\partial x}, \tag{11}$$

$$B_n \to H_n^{-1}, \tag{12}$$

$$p_n = -B_n g_n, \tag{13}$$

and

$$y_n = g_{n+1} - g_n, \tag{14}$$

then Broyden's 'full step' formula for updating the approximation to the inverse Hessian is

$$B_{n+1} = B_n + \frac{(\pm h_n p_n - B_n y_n) p_n^T B_n}{p_n^T B_n y_n}. \tag{15}$$

By factoring out B_n, the right-hand side of equation (15) can be written in a more convenient form so that

$$B_{n+1} = (I + a_n b_n^T) B_n \tag{16}$$

where I is the identity matrix,

$$a_n = \pm h_n p_n - B_n y_n \tag{17}$$

and

$$b_n = \frac{p_n}{p_n^T B_n y_n}.\tag{18}$$

Now equation (16) can be used to compute the determinant of B_{n+1} as follows:

$$|B_{n+1}| = |I + a_n b_n^T| \cdot |B_n| = (1 + a_n^T b_n) \cdot |B_n|.\tag{19}$$

If the initial Hessian, H_0, has been calculated explicitly and its determinant found, this defines the determinant of B_0 and permits the determinant of B_n to be computed by the expression

$$|B_n| = |B_0| \prod_{k=1}^{n} (1 + a_k^T b_k).\tag{20}$$

Thus, the *changes* in sign of $|B_n|$ can be monitored simply by observing the changes in sign of the successive expressions $(1 + a_n^T b_n)$.

If B_n is a good enough approximation to H_n^{-1}, then the sign of $|B_n|$ will change at the same time as that of $|H_n|$, thus enabling equation (20) to be used for deciding when to change signs in equation (7). In practice, however, this procedure proves to be unreliable, since $|B_n|$ often changes sign a few steps before or after $|H_n|$ does. It is possible to recover from this discrepancy at the cost of actually evaluating H_n whenever $|B_n|$ changes sign and retracing any erroneous steps. But this is not recommended.

If a polynomial or spline curve-fitting technique is used in place of equation (7) to predict x_{n+1}, the B_n matrix can still be used in equation (9) during application of the steering correction. Thus, the computational advantages of Broyden's updating algorithm can be realized. This approach has not yet been fully developed, but preliminary experiments indicate that it has considerable promise.

5. Summary and Conclusions

A new method for finding multiple extrema of a scalar function of n variables has been described. This method is based on a procedure for finding successively multiple zeroes of the gradient function by means of a differential equation directly related to the gradient equations. Since a detailed description of this procedure has been given elsewhere, only a brief explanation of it has been included here.

A discussion of computational considerations and the possible use of Broyden's method for updating the inverse Hessian matrix has been given. Broyden's method is not recommended for use in integrating the basic differential equation; but it can be used in conjunction with the 'steering' technique, provided that the integration process is supplanted with a curve-fitting method for predicting the next point on a trajectory.

The method is not globally convergent in general, but may be in certain instances. Moreover, not every trajectory passes through all solution points, although some may. Both of these limitations of the method are due to the existence of extraneous singularities of the basic differential equation. Even so, the method has a much wider region of convergence than other methods and is able to find multiple extrema, one after the other.

References

Branin, F. H., Jr. (1971a). 'Solution of Nonlinear DC Network Problems via Differential Equations', Memoirs Mexico 1971 IEEE Conference on Systems, Networks, and Computers. Oaxtepec, Mexico.

Branin, F. H., Jr. (1971b). 'A Widely Convergent Method for Finding Multiple Solutions of Simultaneous Nonlinear Equations', Tech. Report 21.466, IBM Systems Development Division Laboratory, Kingston, New York. (To appear in *IBM J. Res. Dev.* Sept. 1972).

Wilkinson, J. H. (1965). 'The Algebraic Eigenvalue Problem', pp. 237–239. Clarendon Press, Oxford.

Broyden, C. G. (1965). A class of methods for solving nonlinear simultaneous equations. *Math. Comput.* **19**, 577–593.

17. Extensions of Newton's Method and Simplex Methods for Solving Quadratic Programs

D. GOLDFARB

*The City College of the City University of New York,
New York, U.S.A.*

Summary

Two closely related methods, that may be viewed either as extensions of Newton's method to handle linear equalities and inequalities or as quadratic analogues of the gradient projection method, are presented for solving strictly convex quadratic programs. One of these methods and the Simplex method for quadratic programming are shown to follow the same solution path if started at a vertex. A useful relationship between the Lagrange multiplier variables for a nested set of constraint bases is also given.

1. Introduction

Two methods are described in this paper for solving strictly convex quadratic programming problems. These methods may both be viewed as extensions of Newton's method for minimizing an unconstrained quadratic function and were originally suggested by the author elsewhere (Goldfarb, 1966; 1968). Subsequently, Fletcher also discovered one of these methods and has published an effective algorithm for solving general quadratic programming problems (Fletcher, 1970a; 1971).

Both methods are based upon the following observation. Let

$$f(x) = f_0 + a'x + (1/2)x'Gx, \qquad (1)$$

where x is the n-vector of variables, a is a constant n-vector (the prime denotes transposition), and G is a constant and symmetric positive definite $(n \times n)$ matrix. The minimum of the quadratic function $f(x)$ in a linear manifold M_q, can then be found in one step from the formula $x^{\min} = x^0 - P_q G^{-1} g^0$, where $P_q = I - G^{-1} N_q (N_q' G^{-1} N_q)^{-1} N_q'$ is a weighted projection operator that takes any vector into the linear subspace L_q parallel to M_q, N_q is an $(n \times q)$ matrix, the columns of which are the normals to q linearly independent hyperplanes that intersect in M_q, and x^0 is some point in M_q at which $f(x^0)$ has the gradient $g^0 = Gx^0 + a$.

Both techniques require a feasible starting point and remain feasible throughout the subsequent computations. From a point x^0 in manifold M_q, a step is taken in the direction $-P_q G^{-1} g^0$ to the point $x^{\min}$ or, if some inactive constraint is violated by this move, to the point of intersection of this path with the first such constraint. In one of the methods as many constraints as possible are dropped from the current constraint basis before a step is taken, while in the other method—the one corresponding to Fletcher's algorithm—no constraints are dropped unless a stationary point is reached in the manifold and then one, and only one, constraint is dropped.

After a presentation of some preliminaries in the next section, including properties of non-orthogonal projections and some useful recursion formulas, the two algorithms are given in Section 3. In Section 4, theorems are given that ensure that the algorithms will only terminate at a global minimum and that they will choose step directions that are feasible. Of particular interest is the proof of lemma 3, for it provides a useful relationship between the Lagrange multiplier variables for a nested set of constraint bases. Finite convergence is proved in Section 5. The relationship to other published methods is discussed in Section 6 where it is shown that in certain cases one of the algorithms and Dantzig's simplex algorithm for quadratic programming follow the same solution path. In the last section, the merits of the algorithms are discussed.

2. Preliminaries

In this paper we shall consider the problem: minimize the strictly convex quadratic function (1) subject to the linear inequalities

$$A' x \geq b, \qquad (2)$$

where A is an $(n \times m)$ matrix and b is an m-vector. A point x satisfying (2) is called feasible and the set of all such points is called the feasible region R. The columns of A will be denoted by $n_1, \ldots, n_m$ and may also be viewed as the inward normals to the set of defining (affine) hyperplanes whose associated half spaces intersect in R.

Let us first consider the problem of locating the minimum of the quadratic function (1) over some linear manifold $M_q = \{x | n_i' x = b_i, i = 1, 2, \ldots, q\}$. A necessary and sufficient condition for $x^* \in M_q$ to be such a minimum is that the gradient of $f(x)$ at x^* be orthogonal to M_q, that is $g^* = g(x^*) = N_q \alpha$, where N_q is the $(n \times q)$ matrix $(n_1, \ldots, n_q)$ of rank q and α is some q-vector of constants (i.e. Lagrange multipliers). If x is any other point in M_q, then $g^* - g = G(x^* - x)$ or $x^* - x = G^{-1}(N_q \alpha - g)$. The point $(x^* - x)$ is in L_q. Therefore, $N_q'(x^* - x) = N_q' G^{-1} N_q \alpha - N_q' G^{-1} g = 0$, and hence $\alpha = (N_q' G^{-1} N_q)^{-1} N_q' G^{-1} g$, since $N_q' G^{-1} N_q$ can be shown to be non-singular. Thus,

$$x^* = x - P_q G^{-1} g, \qquad (3)$$

where $P_q = I - G^{-1} N_q (N_q' G^{-1} N_q)^{-1} N_q'$ is, in general, a non-orthogonal projection operator that projects any vector x into a vector in L_q, and in particular, all and only those vectors of the form $G^{-1} N_q v$ into 0 (v is a q-vector). The projection operator $\tilde{P}_q = I - P_q$ projects any vector x into a vector of the form $G^{-1} N_q v$, and in particular, all and only those vectors in L_q into 0.

In the unconstrained case (i.e. $q = 0$), $P_q = I$, and formula (3) reduces to the well-known Newton method, which motivates the methods to be described.

Properties of $P_q G^{-1}$ that will be important to the subsequent discussion can be stated as

Lemma 1

  (i) $P_q G^{-1} u = 0$ holds if, and only if, $u = N_q v$.
  (ii) $P_q G^{-1}$ is positive semidefinite, and $u' P_q G^{-1} u = 0$ if, and only if, $u = N_q v$.

Proof

(i) follows directly from the property of P_q just discussed, i.e. $P_q w = 0$ if, and only if, $w = G^{-1} N_q v$. Since $P_q P_q = P_q$ and $P_q G^{-1}$ is symmetric, $P_q G^{-1} = P_q G^{-1} G P_q G^{-1} = G^{-1} P_q' G P_q G^{-1}$ and hence (ii) follows from (i).

In the algorithms to be described in the next section the vector of Lagrange multipliers α and the vector $P_q G^{-1} g$ are needed. Therefore, it is important to provide efficient recursion relations for computing $P_{q+1} G^{-1}$ from $P_q G^{-1}$ and N_{q+1}^* from N_q^* and vice versa when linearly independent constraints are added to and dropped from the constraint basis, where

$$N_q^* = (N_q' G^{-1} N_q)^{-1} N_q' G^{-1}.$$

The recursion relations which follow have been given by Fletcher (1971).

$$P_{q+1} G^{-1} = P_q G^{-1} - \frac{P_q G^{-1} n_{q+1} (P_q G^{-1} n_{q+1})'}{n_{q+1}' P_q G^{-1} n_{q+1}}, \tag{4}$$

$$N_{q+1}^* = \begin{pmatrix} N_q^* \\ 0 \end{pmatrix} + \begin{pmatrix} -N_q^* n_{q+1} \\ 1 \end{pmatrix} \frac{(P_q G^{-1} n_{q+1})'}{n_{q+1}' P_q G^{-1} n_{q+1}}, \tag{5}$$

$$P_q G^{-1} = P_{q+1} G^{-1} + \frac{n^* n^{*'}}{n^{*'} G n^*}, \tag{6}$$

and

$$\begin{pmatrix} N_q^* \\ 0 \end{pmatrix} = N_{q+1}^* - \frac{N_{q+1}^* G n^* n^{*'}}{n^{*'} G n^*}, \tag{7}$$

where $n^{*\prime}$ is the $q+1$st row of N^*_{q+1}. If the pth rather than the $q+1$st hyperplane is dropped from the constraint basis, formulas (6) and (7) may still be used provided the pth and $(q+1)$st columns of N^*_{q+1} are interchanged before they are applied. Formulas (4) and (5) require a total of $2n(n+q+1)$ multiplications and divisions as do formulas (6) and (7). If symmetry is taken into account, these operational counts are reduced by $n(n-1)/2$.

3. Algorithms

The two algorithms presented here are essentially extensions of Newton's method to handle linear inequalities. They are also very similar in approach to the author's (Goldfarb, 1966; 1969) extension of the Davidon-Fletcher-Powell method and are in a sense quadratic analogues of the gradient projection method (Rosen, 1960). To simplify matters we shall assume at each point all active constraints are linearly independent. A technique for handling linear dependence has been described by Rosen (1960) for use with his gradient projection method. It can be applied with some obvious changes to the algorithms below.

Algorithm 1

   Step 0. Find some $x^0 \in R$ and compute $g^0 = Gx^0 + a$. Compute $P_q G^{-1}$ and N^*_q. Note that $n_1, \ldots, n_q$ are the normals to the q linearly independent defining hyperplanes on which x^0 lies. Set $i = 0$.
   Step 1. Compute $s^i = P_q G^{-1} g^i$. If $s^i = 0$, go to step 3.
   Step 2. If $P_q G^{-1} g^i \neq 0$, compute $x^{i+1} = x^i - \tau s^i$ and $g^{i+1} = Gx^{i+1} + a$, where $\tau = \min\{1, \hat{\tau}\}$, and

$$\hat{\tau} = \min_{\substack{n'_j s^i > 0 \\ q+1 \leq j \leq m}} \left(\frac{n'_j x^i - b_j}{n'_j s^i} \right). \tag{8}$$

If $\tau < 1$ uses formulas (4) and (5) to update $P_q G^{-1}$ and N^*_q. (We have assumed for simplicity that the minimum in formula (8) is achieved for $j = q+1$). Set $i = i+1$ and $q = q+1$ and go to step 1. If $\tau = 1$ set $i = i+1$ and go to step 3.
   Step 3. Compute $\alpha^q = N^*_q g^i$. Determine $\min \alpha^q_j = \alpha^q_r$. (Assume, for simplicity that $r = q$). If $\alpha^q_q \geq 0$, terminate; x^i is the global minimum of $f(x)$ over R. Otherwise, $\alpha^q_q < 0$. Use formulas (6) and (7) with q replaced by $q-1$ to update $P_q G^{-1}$ and N^*_q. Set $q = q-1$ and go to step 1.

Algorithm 2

   Step 0. Same as in Algorithm 1.
   Step 1. Compute $\alpha^q = N^*_q g^i$. Set $p = 0$. Determine $\min \alpha^q_j = \alpha^q_r$. If $\alpha^q_r \geq 0$, go to step 3. Otherwise, $\alpha^q_r < 0$. Go to step 2.

Step 2. Use formulas (6) and (7) with q replaced by $q-p-1$ to update $P_{q-p}G^{-1}$ and N^*_{q-p}. (For simplicity we have assumed that $r = q-p$.) Set $p = p+1$, compute $\alpha^{q-p} = N^*_{q-p}g^i$, and determine

$$J = \{j \mid \alpha_j^{q-p} < 0,\ \alpha_j^i \leqslant \alpha_j^{i+1},\quad \text{for } i = q-p,\ldots,q-1\}.$$

If $J = \emptyset$, set $q = q-p$ and go to step 3. Otherwise, determine $\min_{j \in J} \alpha_j^{q-p} = \alpha_r^{q-p}$ and repeat this step.

Step 3. Compute $s^i = P_q G^{-1} g^i$. If $s^i = 0$, terminate; x^i is a global minimum of $f(x)$ over R. If $s^i \neq 0$, compute $x^{i+1} = x^i - \tau s^i$ and $g^{i+1} = g(x^{i+1})$, where $\tau = \min\{1, \hat{\tau}\}$ and $\hat{\tau}$ is given by (8) with q replaced by $q+p$. If $\tau = 1$ and $p = 0$, terminate; x^{i+1} is a global minimum of $f(x)$ over R. If $\tau = 1$ and $p \neq 0$ set $i = i+1$ and go to step 1. If $\tau < 1$ use formulas (4) and (5) to update $P_{q-p}G^{-1}$ and N^*_{q-p}. Set $i = i+1$ and $q = q-p+1$ and go to step 1.

Linear equalities were not specifically included in the formulation of the quadratic programming problem given by (1) and (2). However, such constraints can be conveniently handled by slightly modifying Algorithms 1 and 2 so as to include them in the initial constraint basis and prohibit their removal thereafter. In step 0, phase I of the Simplex method or any other routine finding a feasible point for linear or quadratic programs can be used.

For Algorithm 1, a procedure developed by Fletcher (1970b), which computes a feasible vertex if one exists and $N_n^* = N_n^{-1}$, the inverse of the basis matrix, is recommended. (Note that $P_n G^{-1} = 0$.) For Algorithm 2, however, a variant of Rosen's (1960) procedure for finding a feasible point is probably more suitable. If the matrices $N_q^+ = (N_q' N_q)^{-1} N'$ and $P_q = I - N_q N_q^+$ are replaced by N_q^* and $P_q G^{-1}$ wherever they appear in Rosen's procedure and formulas (4) through (7) are used for updating them, this results in a method that yields a feasible point, (if one exists), that is not necessarily a vertex and provides the required initial matrices N_q^* and $P_q G^{-1}$ as well. It should be noted that Fletcher's feasible point routine is similar in many ways to Rosen's original method if one starts from a point that lies on the intersection of n linearly independent hyperplanes.

4. Properties of the Algorithms

The following well-known theorem ensures that both algorithms terminate only when a global minimum is reached.

Theorem 1

Let x^0 be a point in R lying on exactly q linearly independent hyperplanes with normals $n_1, \ldots, n_q$. Then x^0 is a global minimum of the quadratic programming problem (1), (2) if, and only if,

$$\text{(i)} \quad P_q G^{-1} g^0 = 0$$

and

(ii) $N_q^* g^0 > 0.$

A proof of this theorem can be given that is almost identical to the proof of Theorem 3 in Goldfarb (1969). In the constructive proof of the necessity of conditions (i) and (ii) essential use is made of Algorithm 1 to locate points with lower function values when these conditions are not satisfied. It is also easy to show that conditions (i) and (ii) are equivalent to those of the Kuhn-Tucker theory.

We now proceed to show that both algorithms always choose directions that are feasible. In the following lemmas $v_0 = G^{-1} g^0$, and it is assumed that normals $n_1, n_2, \ldots, n_n$ are linearly independent. The proofs will be by induction.

Lemma 2

$$n_k' P_q v_0 = \sum_{j=q+1}^{k} \alpha_j^j n_k' P_{j-1} G^{-1} n_j, \qquad 0 \leqslant q < k \leqslant n.$$

Proof

$$v_0 = \tilde{P}_{q+1} v_0 + P_{q+1} v_0 = \sum_{j=1}^{q+1} \alpha_j^{q+1} G^{-1} n_j + P_{q+1} v_0.$$

From properties of the operator P_{q+1} we have that

$$P_q v_0 = \alpha_{q+1}^{q+1} P_q G^{-1} n_{q+1} + P_{q+1} v_0$$

and

$$n_k' P_q v_0 = \alpha_{q+1}^{q+1} n_k' P_q G^{-1} n_{q+1} + n_k' P_{q+1} v_0$$

$$= \alpha_{q+1}^{q+1} n_k' P_q G^{-1} n_{q+1} + \sum_{j=q+2}^{k} \alpha_j^j n_k' P_{j-1} G^{-1} n_j,$$

for $0 \leqslant q < k-1 \leqslant n-1$, where the last term follows from the induction hypothesis. Since $n_{q+1}' P_{q+1} v_0 = 0$ we also have that

$$n_{q+1}' P_q v_0 = \alpha_{q+1}^{q+1} n_{q+1}' P_q G^{-1} n_{q+1}, \qquad 0 \leqslant q < n, \tag{9}$$

which completes the induction.

Lemma 3

$$n_k' P_q G^{-1} n_{q+1} = C_{k-q-1}^k n_{q+1}' P_q G^{-1} n_{q+1}, \qquad 1 \leqslant q+1 < k \leqslant n \tag{10}$$

where C_{k-q-1}^k is given recursively by

$$C_{k-q-1}^k = \begin{cases} \sum_{j=q+1}^{k-1} \dfrac{\alpha_j^{k-1} - \alpha_j^k}{\alpha_k^k} C_{j-q-1}^j, & q+1 < k \\ 1, & q+1 = k. \end{cases} \tag{11}$$

17. SOLVING QUADRATIC PROGRAMS

Proof

The lemma is obviously true for $k = q+1$, since in that case $C_{k-q-1}^{k} = 1$ by definition. The vector v_0 can be expressed as

$$v_0 = \tilde{P}_k v_0 + P_k v_0 = \sum_{j=1}^{k} \alpha_j^k G^{-1} n_j + P_k v_0.$$

The relation

$$P_q G^{-1} n_k = \frac{1}{\alpha_k^k} \left\{ P_q v_0 - \sum_{j=q+1}^{k-1} \alpha_j^k P_q G^{-1} n_j - P_k v_0 \right\}$$

then follows from the properties of the projection P_q assuming that $q+1 < k$. Also from these properties and equation (9) we have that

$$n_{q+1}' P_q G^{-1} n_k = \frac{1}{\alpha_k^k} \left\{ n_{q+1}' P_q v_0 - \sum_{j=q+1}^{k-1} \alpha_j^k n_{q+1}' P_q G^{-1} n_j \right\}$$

$$= \frac{1}{\alpha_k^k} \left\{ \alpha_{q+1}^{q+1} - \sum_{j=q+1}^{k-1} \alpha_j^k C_{j-q-1}^{j} \right\} n_{q+1}' P_q G^{-1} n_{q+1}. \quad (12)$$

Addition and subtraction of

$$\sum_{j=q+1}^{k-1} \alpha_j^{k-1} C_{j-q-1}^{j}$$

to and from the term in brackets on the right-hand side of the above equation yields

$$n_k' P_q G^{-1} n_{q+1} = \left\{ \sum_{j=q+1}^{k-1} \frac{\alpha_j^{k-1} - \alpha_j^k}{\alpha_k^k} C_{j-q-1}^{j} \right.$$

$$\left. + \frac{1}{\alpha_k^k} \left[\alpha_{q+1}^{q+1} - \sum_{j=q+1}^{k-1} \alpha_j^{k-1} C_{j-q-1}^{j} \right] \right\} n_{q+1}' P_q G^{-1} n_{q+1}.$$

The symmetry of $P_q G^{-1}$ has been used to interchange the order of n_k and n_{q+1} on the left hand side of the above equation. It is obvious that the lemma will be proved if we can show that for $q+2 \leq k$,

$$\alpha_{q+1}^{q+1} - \sum_{j=q+1}^{k-1} \alpha_j^{k-1} C_{j-q-1}^{j} = 0. \quad (13)$$

But by induction and equating the coefficients of $n_{q+1}' P_q G^{-1} n_{q+1}$ in equations (10) and (12) with k replaced by $k-1$, we must have that

$$C_{k-q-2}^{k-1} = \begin{cases} \dfrac{1}{\alpha_{k-1}^{k-1}} \left[\alpha_{q+1}^{q+1} - \displaystyle\sum_{j=q+1}^{k-2} \alpha_j^{k-1} C_{j-q-1}^{j} \right], & q+2 < k, \\ 1, & q+2 = k. \end{cases}$$

Subtracting $C_{k-1}^{k-1}{}_{q-2}$ from both sides of this equation yields

$$0 = \frac{1}{\alpha_{k-1}^{k-1}} \left\{ \alpha_{q+1}^{q+1} - \sum_{j=q+1}^{k-1} \alpha_j^{k-1} C_{j-q-1}^j \right\}, \quad q+2 \leq k.$$

Hence, equation (13) is valid and the lemma has been proved.

Theorem 2

If

(i) $\alpha_i^{j-1} \leq \alpha_i^j$, $\quad i = q+1, \ldots, j-1,$
$\quad j = q+2, \ldots, k,$

and

(ii) $\alpha_j^j \leq 0$, $\quad j = q+1, \ldots, k,$

then

(iii) $n_j' P_q v_0 \leq 0$, $\quad j = q+1, \ldots, k.$

Moreover, if the inequality is strict in (ii) it is also strict in (iii).

Proof

From Lemmas 2 and 3 we have that

$$n_j' P_q v_0 = \sum_{i=q+1}^{j} \alpha_i^i n_j' P_{i-1} G^{-1} n_i = \sum_{i=q+1}^{j} \alpha_i^i C_{j-i}^j n_i' P_{i-1} G^{-1} n_i,$$

where C_{j-i}^j is given recursively by (11). The proof then follows straightforwardly from the fact that $n_i' P_{i-1} G^{-1} n_i > 0$, which is itself a consequence of lemma 1 and the assumed linear independence of the set of normals $n_1, \ldots, n_i$.

Lemmas 2 and 3 and Theorem 2 apply equally well to gradient projection methods if G^{-1} is replaced by the identity matrix wherever it appears.

The conditions of Theorem 2 are clearly sufficient, but not necessary for the direction $-P_{q-p} G^{-1} g^i$ to be feasible in the non-degenerate case. The following simple example, however, shows that the conditions $\alpha_{q-j}^{q-j} < 0, j = 0, 1, \ldots, p$, by themselves are not sufficient to ensure feasibility after dropping the corresponding $p+1$ hyperplanes from the constraint basis.

Consider the quadratic programming problem:

$$\text{Minimize } \tfrac{1}{2}(x_1^2 + x_2^2 + x_3^2) - 20(x_1 + x_2)$$
$$\text{Subject to } 7x_1 + 7x_2 + \sqrt{2} \cdot x_3 \leq 20$$
$$\text{and } x_1, x_2, x_3 \geq 0.$$

Starting at the vertex $x^0 = (0, 0, 10\sqrt{2})$, the inward normals corresponding to the constraints active at that point are

$$n_1' = (10, 0, 0), \quad n_2' = (0, 10, 0) \quad \text{and} \quad n_3' = (-7, -7, -\sqrt{2}).$$

We have normalized the lengths of all normals to remove the effect of their relative lengths from consideration. At x^0, $g^0 = (-20, -20, 10\sqrt{2})'$,

$$N_3^* = N_3^{-1} = \begin{bmatrix} \frac{1}{10} & 0 & -\frac{7}{20}\sqrt{2} \\ 0 & \frac{1}{10} & -\frac{7}{20}\sqrt{2} \\ 0 & 0 & -\frac{1}{2}\sqrt{2} \end{bmatrix}, \quad \alpha^3 = \begin{bmatrix} -9 \\ -9 \\ -10 \end{bmatrix}, \quad -P_3 G^{-1} g^0 = (0, 0, 0)'$$

$$N_2^* = \begin{bmatrix} \frac{1}{10} & 0 & 0 \\ 0 & \frac{1}{10} & 0 \end{bmatrix}, \quad \alpha^2 = \begin{bmatrix} -2 \\ -2 \end{bmatrix}, \quad -P_2 G^{-1} g^0 = (0, 0, -10\sqrt{2})'$$

$$N_1^* = [\tfrac{1}{10} \ 0 \ 0], \quad \alpha^1 = (-2) \quad \text{and} \quad -P_1 G^{-1} g^0 = (0, 20, -10\sqrt{2})'$$

where

$$N_3 = [n_1, n_2, n_3], \ N_2 = [n_1, n_2] \text{ and } N_1 = [n_1].$$

The direction $-P_2 G^{-1} g^0$ is feasible $-P_1 G^{-1} g^0$ is not since $-n_3' P_1 G^{-1} g^0 = -120$.

5. Convergence

As we have done previously, we shall assume that all active constraints are linearly independent. In order to prove that Algorithm 2 converges in a finite number of iterations, we need the following anti-zigzag rule employed by Zoutendijk (1960) with several of his algorithms.

Rule 1

If a constraint that has previously been left is returned to, it is kept in the constraint basis until a minimal solution is obtained for this revised quadratic programming problem.

Theorem 3

Algorithm 1 and Algorithm 2 (with Rule 1) terminate in a finite number of iterations.

Proof

Under the assumption of linear independence the value of the objective function strictly decreases whenever a step is taken. (This follows from the fact that $g' P_q G^{-1} g > 0$ at each step.) Therefore, whenever a constraint is dropped from the constraint basis in Algorithm 1, that constraint basis (or face) can never be returned to, since in Algorithm 1 constraints are only dropped if the current point is the minimum of $f(x)$ on that face. Since there can only be a finite number of steps between such points (i.e. at most n constraining hyperplanes can be added successively to any basis) and since there are only a finite number of combinations of m constraints possible, the algorithm must terminate in a finite number of steps.

In Algorithm 2, the anti-zigzag rule and the strict decrease in $f(x)$ for steps of nonzero length ensure that some revised quadratic programming problem (i.e. the original problem with some inequalities replaced by equalities), will be solved in a finite number of steps. This solution corresponds to the minimum of $f(x)$ on some face of the feasible polyhedral region. If it is not also the solution to the original quadratic programming problem, that face will be left on the next iteration and will never be returned to. Finiteness then follows by the same reasoning as already given for Algorithm 1.

6. Relationship to Other Methods

Algorithm 1 was independently developed by Fletcher (1971). His algorithm, however, is more general than the one presented here and can be applied to quadratic functions with positive semidefinite G, and even to those with indefinite G. In the former case, Fletcher's method will locate a global constrained minimum or indicate that an unbounded solution exists. In the latter case, it will terminate at a constrained local minimum. Using a Lagrangian approach Fletcher has shown that G need not be invertible for the basic procedure of Algorithm 1 to work, for the required operators $P_q G^{-1}$ and N_q^* must exist if $f(x)$ has a unique minimum in M_q. When this is not the case—something that can occur when dropping a constraining hyperplane from the constraint basis—the required step direction is shown to be given by that row of N_q^* corresponding to the constraint being dropped.

One further, and very important, point, is that Fletcher provides alternate strategies and updating formulae, and criteria for choosing among them, that not only allow his method to be used for general G, but also avoid ill-conditioned intermediate stages as much as possible. This, of course, can be important in problems with a positive definite G, where the ratio of the smallest to the largest eigenvalue is very close to zero. Such considerations are important when implementing an algorithm on a computer. In Fletcher (1970b) one can find a FORTRAN program of Fletcher's algorithm.

Several quadratic programming methods are often described as 'simplex' methods. These include Dantzig's (1963) method [also discovered by van de Panne and Whinston (1964)], Wolfe's (1959) method, and Beale's (1955, 1959) method. These three different methods all lay claim to the name 'simplex' by virtue of their having one or more of the following properties: using Simplex tableaux, performing Simplex pivots (i.e. Gauss-Jordan pivots), or reducing to the Simplex method for linear programming for the degenerate case of $G = 0$. The Dantzig algorithm has all three properties, while Wolfe's algorithm has only the first two. Beale's method may also be considered to have all three properties if one allows for variable-sized simplex tableaux in the course of the algorithm. On first consideration Algorithms 1 and 2 seem to be very different from the linear programming-based simplex methods. In fact, both algorithms seem to be very close in spirit to a method suggested by Theil and van de Panne (1960). Therefore, it is somewhat surprising that Algorithm 1 and Dantzig's method follow the same path to the optimal solution if the initial feasible point is the minimum of some face of R, as we shall now demonstrate. This has also been observed by Fletcher (1971).

Dantzig's simplex method for the quadratic programming problem (1), (2) is based upon the Kuhn-Tucker conditions.

Expressed in tableau form, these are

	x	y^1	y^2	u^1	u^2	
	A_1'	$-I_q$	0	0	0	b^1
T_0	A_2'	0	$-I_{m-q}$	0	0	b^2
	G	0	0	$-A_1$	$-A_2$	$-a$

Here, y and u are non-negative m-vectors of surplus variables and Lagrange multipliers (or dual variables), respectively, and must satisfy the complementarity condition $y'u = 0$. In the tableau, A', y, u and b have each been partitioned into two parts, the first containing the first q rows and the second containing the latter $m - q$ rows of the respective matrix or vector. Consider the complementary tableau. T_1—the term standard is used by van de Panne and Whinston (1964)—with x, y^2, and u^1 basic, and y^1 and u^2 non-basic, obtained from tableau T_0 by a sequence of simplex pivots or, equivalently by premultiplying T_0 by the inverse of the basis matrix. It is easy to show that this inverse is given by

$$B_1^{-1} = \begin{pmatrix} C' & 0 & H \\ A_2'C' & -I_{m-q} & A_2'H \\ (A_1'G^{-1}A_1)^{-1} & 0 & -C \end{pmatrix},$$

where $C = (A_1'G^{-1}A_1)^{-1}A_1'G^{-1}$, and $H = G^{-1} - G^{-1}A_1(A_1'G^{-1}A_1)^{-1}A_1'G^{-1}$. Hence, $T_1 = B_1^{-1}T_0$.

	x	y^1	y^2	u^1	u^2	
	I_n	$-C'$	0	0	$-HA_2$	$C'b^1 - Ha$
T_1	0	$-A'_2 C'$	I_{m-q}	0	$-A'_2 HA_2$	$A'_2(C'b^1 - Ha) - b^2$
	0	$-(A'_1 G^{-1} A_1)^{-1}$	0	I_q	CA_2	$(A'_1 G^{-1} A_1)^{-1} b^1 + Ca$

In the basic solution corresponding to tableau T_1, $y^1 = 0$, i.e. $A'_1 x = b^1$. Therefore, $A_1 = N_q$ in the terminology of Algorithm 1 and since $x = C'b^1 - Ha = G^{-1}(g - a)$,

$$H = P_q G^{-1}$$
$$C = N_q^*$$

and

$$u^1 = (A'_1 G^{-1} A_1)^{-1} A'_1 G^{-1} g = N_q^* g = \alpha^q.$$

Starting from a complementary tableau, say T_1, with $y^2 \geq 0$, i.e. the minimum on some face of R, $(P_q G^{-1} g = 0)$, if $u^1 \geq 0$, ($\alpha^q \geq 0$), Dantzig's method terminates: the optimal solution has been found. Expressions in parentheses refer to equivalent statements in the terminology of Algorithm 1. Otherwise, both Dantzig's algorithm and Algorithm 1 leave the constraining hyperplane corresponding to the most negative dual variable, (Lagrange multiplier), say u_q^1 (α_q^q). In the Dantzig algorithm, this corresponds to introducing y_q^1 into the basis by a positive amount θ and changing x by θc, where c is the qth column of C'. The step direction in Algorithm 1 is given by $-P_{q-1} G^{-1} g$ which, by formula (6), equals

$$-\left(\frac{n^{*\prime} g}{n^{*\prime} G n^*}\right) n^* = -(\alpha_q^q / n^{*\prime} G n^*) n^*$$

where n^* is the qth column of N_q^*, since $P_q G^{-1} g = 0$. Since $C = N_q^*$ and $\alpha_q^q / n^{*\prime} G n^* < 0$, this direction is just equal to c multiplied by a positive scalar. Hence both algorithms take a step in the same direction.

Dantzig's algorithm increases y_q^1 until some component of y^2, say y_t^2, becomes zero, ($\tau < 1$) or u_q^1 becomes zero ($\tau = 1$). If the latter occurs first u_q^1 is made non-basic and complementary tableau results after a 'simplex' pivot. Otherwise y_t^2 is made non-basic and a non-complementary tableau results.

At some step, therefore, suppose that we have a non-complementary tableau with x, y^2, u_1^1, ..., u_{q-1}^1, and u_1^2 basic. In this situation Dantzig's algorithm introduces u_q^1, the dual variable of the non-basic pair, into the basis. Increasing its value from zero to θ changes x by θp where $-p$ is a vector of the first n elements of the u_q^1 column of T_2. Since T_2 can be obtained by pivoting T_1, taking the last element in the u_1^2 column as the pivot, it is easy to see that $p = -Ha_2/e'_q Ca_2$ where a_2 is the first column of A_2 and e_q is the q-vector $(0, 0, \ldots, 0, 1')$. Likewise, the value of x in the basic solution corresponding

to T_2 is given by
$$x = C'b^1 - Ha - \beta p$$
where
$$\beta = e_q'((A_1'G^{-1}A_1)^{-1}b_1 + Ca).$$
Since in T_2, $y^1 = 0$, $b^1 = A_1'x$ and $x = G^{-1}(g - a)$,
$$p = \frac{1}{\beta}[(C'A_1' - I)x - Ha] = \frac{1}{\beta}[-H(g-a) - Ha] = \frac{-1}{\beta}Hg$$

One can prove that β must be positive. Thus, in this case also, both algorithms take steps in the same direction. The remainder of our demonstration is just a restatement of the paragraph before the last with y_q^1 and u_q^1 replaced by u_q^1 and u_1^2, respectively. Since there can be only complementary or non-complementary tableaux, our demonstration is complete.

Even computationally there are instances in which these methods are very closely related. For example, in the Dantzig algorithm if one step takes us from one complementary tableau to another, this corresponds to choosing some diagonal element, say the qth, of the matrix $-(A_1'G^{-1}A_1)^{-1}$ in tableau T_1 as the pivot. The effect of this pivot step on the top row of matrix partitions of the inverse basis matrix B_1^{-1} can be expressed as

$$\left(I_n \mid 0 \mid \frac{C'e_q e_q'}{e_q'(A_1'G^{-1}A_1)^{-1}e_q}\right) B_1^{-1} =$$

$$\left(C' - \frac{C'e_q e_q'(A_1'G^{-1}A_1)^{-1}}{e_q'(A_1'G^{-1}A_1)^{-1}e_q} \mid 0 \mid H + \frac{C'e_q e_q'C}{e_q'(A_1'G^{-1}A_1)^{-1}e_q}\right)$$

where e_q is the q-dimensional vector $(0,\ldots,0,1)'$. Since $(A_1'G^{-1}A_1)^{-1} = CGC'$, it is apparent that these formulas are equivalent to formulas (6) and (7) used by Algorithm 1 and by Fletcher's algorithm when moving from a minimum in a face of R to a minimum in a face of higher dimension by dropping one constraint.

Van de Panne and Whinston (1966) have shown that Beale's method and Dantzig's method will generate the same sequence of points until and including the first point at which they both encounter a constraint before a minimum is reached along the current step direction, if they both start from the same vertex of R. That is, all points except the last are the minima of faces of R of increasing dimension. The same statement can be made with regard to Algorithm 1 and Beale's method, since we have already shown that it generates the same set of points as the Dantzig method in this case. If one analyses Beale's method from the point of view of conjugate directions one can also show that it and Algorithm 1, and hence Dantzig's algorithm, continue to generate the same set of points even if a constraint is added to the basis as long as that constraint was the last one that was dropped. In general, however, when a constraining hyperplane is added to the constraint basis, Algorithm 1 and Beale's method will thereafter follow different paths. On the next iteration, Algorithm 1 will proceed in the direction of the minimum in the new constraint manifold, whereas Beale's method will take as many as p steps, where p is the dimension of the manifold, before making a step towards that minimum. During the process of building up the necessary conjugacy information in Beale's method one may intersect some other constraint. This will result in constraint bases that do not occur when using Algorithm 1. If the same point is reached by both algorithms and this point is the minimum of some face of R, the two methods will again follow the same path until, in general, a new constraining hyperplane is added to the constraint basis.

Algorithm 1 is also related to the principal pivoting method of quadratic programming developed by Cottle and Dantzig (1968). (See also Cottle, 1968.) The connection follows from the observation that Dantzig's simplex method is a special case of the principal pivoting method.

Algorithm 2 does not, in general, follow the same solution path as any 'simplex' type method because it allows more than one constraint to be dropped at a time. It will, of course, follow the same solution path as Dantzig's method whenever it behaves identically to Algorithm 1.

It is interesting to view Algorithm 2 as a method of 'feasible directions' and compare how it determines directions to one of Zoutendijk's (1970) methods. Zoutendijk's direction generator 2·4 determines a feasible direction s, by solving the problem

$$\min\{(g^i)'s | N_q's \leq 0, s'Gs \leq 1\}$$

by principal pivoting. This yields $s = -P_l G^{-1} g^i$ where P_l is the projection matrix incorporating the minimum number of hyperplanes necessary for s to be a feasible direction. Algorithm 2 does not, in general, compute the projection matrix corresponding to this minimal set of hyperplanes since it relies only on sufficient conditions to drop hyperplanes from the constraint set. More important, however, is that the direction finding approach of Algorithm 2 is more efficient than is Zoutendijk's method.

7. Discussion

In the last section we showed that the Dantzig algorithm and Algorithm 1 followed the same solution path. These methods are not, however, computationally equivalent. In fact the operational counts (i.e. multiplications and divisions) per iteration are $(m+n)(m+1)+p+1$ for the Dantzig algorithm and $m(4n+m+q+p+1+d)$ for Algorithm 1, where n, m, q, and p are, respectively, the number of variables, constraints (i.e. all equalities and inequalities, including non-negativity restrictions), active constraints, and non-active constraints whose inward directed normals make an obtuse angle with the step direction. The quantity d equals 1 or q depending upon whether the iteration of Algorithm 1 starts by adding a constraint to, or dropping one from, the constraint basis. If symmetry is accounted for, the above figures can be reduced by $(n+q)$ $(n+q-1)/2$ and $n(n-1)/2$, respectively. If we compare these figures under the most unfavourable circumstances with respect to Algorithm 1, i.e. $q = n$, $p = m - n$, and $d = q = n$, we find that Algorithm 1 requires less multiplications and divisions than does the Dantzig algorithm for m slightly greater than $3n$.

Techniques based upon matrix factorizations are known that can be adopted to both the Dantzig method and Algorithm 1 to make them more numerically stable. (See Bartels et al., 1970.) Therefore, on this score, we cannot judge either method to be the better one.

Algorithm 2 can be modified to solve semidefinite quadratic programs. The same techniques as those used by Fletcher (1970a) can be employed—although their incorporation in Algorithm 2 would be somewhat more involved than it is in Algorithm 1. However, Algorithm 2 is best suited to the positive definite case. In the semidefinite case where G is of low rank, Beale's method is probably the most suitable choice. As G increases in rank, the quadratic programming problem becomes less and less like a linear programming problem and Beale's method becomes less attractive. In fact when G is of full rank there are instances in which Beale's method will require $n - q$ steps to reach the same point that Algorithm 1 or Dantzig's algorithm reach in one step, where q is the number of constraints in the basis.

Although only preliminary computational tests have been performed to compare Algorithms 1 and 2, they support the author's belief that, in general, Algorithm 2 will take less steps to find the global minimum of a strictly convex quadratic program. This conjecture is also supported indirectly by Fletcher's (1971) numerical results. Starting from a vertex with n constraints in the basis, Fletcher found that his algorithm required approximately $n \pm n$ basis changes to terminate. Starting from a feasible point $(1, 1, \ldots, 1)$ with no constraints in the basis required roughly $k \pm k$ basis changes where k was the number of active constraints in the final basis. In Fletcher's randomly generated quadratic programming problems k was less than $\frac{1}{2}n$ on the average. All of this implies that it is beneficial to drop as many constraints as possible from the basis at each stage of the solution process. This is just what Algorithm 2 attempts to do.

References

Bartels, R. H., Golub, G. H., and Saunders, M. A. (1970). Numerical techniques in mathematical programming. *In*: 'Nonlinear Programming' (J. B. Rosen, O. L. Mangasarian, and K. Ritter, eds.), pp. 123–176. Academic Press, London.

Beale, E. M. L. (1955). On minimizing a convex function subject to linear inequalities. *J. R. Stat. Soc.* (b), **17**, 173–184.

Beale, E. M. L. (1959). On quadratic programming. *Naval Res. Log. Quart.* **6**, 227–243.

Cottle, R. W. (1968). The principal pivoting method of quadratic programming. *In*: 'Lectures in Applied Mathematics', II, Mathematics of the Decision Sciences, Part 1 (G. B. Dantzig and A. F. Veinott, eds.), pp. 144–161. American Mathematical Society.

Cottle, R. W., and Dantzig, G. B. (1968). Complementary pivot theory of mathematical programming. *In*: 'Lectures in Applied Mathematics', II, Mathematics of the Decision Sciences, Part 1 (G. B. Dantzig and A. F. Veinott, eds.), pp. 115–136. American Mathematical Society.

Dantzig, G. B. (1963). 'Linear Programming and Extensions'. Princeton University Press, Princeton. (See especially Ch. 24, Section 4.)

Fletcher, R. (1970a). 'The Calculation of Feasible Points for Linearly Constrained Optimization Problems'. UKAEA Research Group Report, AERE R.6354.

Fletcher, R. (1970b). 'A FORTRAN Subroutine for Quadratic Programming'. UKAEA Research Group Report, AERE R.6370.

Fletcher, R. (1971). A general quadratic programming algorithm. *J. Inst. Maths Applics* **7**, 76–91.

Goldfarb, D. (1966). A conjugate gradient method for nonlinear programming. Doctoral thesis, Department of Chemical Engineering, Princeton University.

Goldfarb, D. (1968). Analogs of Newton's method for quadratic programming. *Notices Am. math. Soc.* **15**, No. 2, p. 400.

Goldfarb, D. (1969). Extension of Davidon's variable metric method to maximization under linear inequality and equality constraints. *SIAM J. appl. Math.* **17**, 739–764.

Panne, C., van de, and Whinston, A. (1964). The simplex and the dual method for quadratic programming. *Operat. Res. Quart.* **15**, 355–389.

Panne, C., van de, and Whinston, A. (1966). A comparison of two methods for quadratic programming. *Operat. Res.* **14**, 422–441.

Rosen, J. B. (1960). The gradient projection method for nonlinear programming, Part I, Linear constraints. *SIAM J. appl. Math.* **8**, 181–217.

Theil, H., and van de Panne, C. (1960). Quadratic programming as an extension of conventional quadratic maximization. *Managmt Sci.* **7**, 1–20.

Wolfe, P. (1959). The simplex method for quadratic programming. *Econometrica* **27**, 382–398.

Zoutendijk, G. (1960). 'Methods of Feasible Directions'. Elsevier, Amsterdam.

Zoutendijk, G. (1970). Some algorithms based upon the principle of feasible directions. *In*: 'Nonlinear Programming' (J. B. Rosen, O. L. Mangasarian, and K. Ritter, eds.), pp. 93–121. Academic Press, London.

18. A Primal-Dual Method for Quadratic Programming with Bounded Variables

AMILCAR S. GONÇALVES

Universidade de Coimbra, Instituto de Matemática,
Coimbra, Portugal

Summary

A primal-dual algorithm for quadratic programming with bounded variables is established here, which has the same characteristics as the method by Eisemann (1963) for the linear programming case, and which may be considered as an extension of that method. However, distinct from the linear programming case, the restricted problems constructed here have no bounds in their variables. The computations may be performed by pivotal operations on a tableau. Programming of the algorithm is facilitated by its efficient use of the product form of the inverse mechanism available in most commercial linear programming systems.

1. Introduction

Primal-dual algorithms in mathematical programming are sometimes preferred since they are easier to start with and cut short the number of iterations for solving some problems. The method presented here may be considered as an extension of the primal-dual method for linear programming with bounded variables developed by Eisemann (1963). It follows as a special case from a generalized primal-dual technique for quadratic programming which is an extension of a unified theory of primal-dual methods for linear programming (Gonçalves, 1969). However, distinct from the linear case, the method constructed here has at each iteration restricted problems with no bounds in their variables.

The algorithm may be carried out in tableau form (see Section 4), and it is fitted for use of the product form of the inverse.

Consider the following quadratic programming problem with bounded variables:

$$P_1 \begin{cases} \text{Minimize } \sum_{j \in J} c_j x_j + \tfrac{1}{2} \sum_{k, j \in J} r_{kj} x_k x_j \text{ subject to} \\ \sum_{j \in J} a_{ij} x_j = b_i \quad \text{for } i \in I, \\ m_j \leqslant x_j \leqslant M_j \quad \text{for } j \in J, \end{cases}$$

where $I = \{1, \ldots, m\}$, $J = \{1, \ldots, n\}$, the square matrix $(r_{kj}; k,j \in J)$ is symmetric and positive (semi)-definite, and $m_j < M_j$ are the bounds for the x_j variables.

In order to find a bounded optimal solution for P_1—or to discover that none exists—we replace problem P_1 by the following parametric problem P whose dual is D:

$$P \begin{cases} \text{Minimize } \tfrac{1}{2} \sum_{k,j \in J} r_{kj} x_k x_j + \sum_{j \in J} c_j x_j + \lambda \sum_{i \in I} z_i \text{ subject to} \\ \sum_{j \in J} a_{ij} x_j + z_i = b_i & \text{for } i \in I, \\ -x_j \geqslant -M_j & \text{for } j \in J, \\ x_j \geqslant m_j & \text{for } j \in J, \\ z_i \geqslant 0 & \text{for } i \in I. \end{cases}$$

$$D \begin{cases} \text{Maximize } -\tfrac{1}{2} \sum_{k,j \in J} r_{kj} x_k x_j + \sum_{i \in I} b_i y_i - \sum_{j \in J} M_j v_j + \sum_{j \in J} m_j s_j \text{ subject to} \\ y_i \text{ unrestricted} & \text{for } i \in I, \\ v_j \geqslant 0 \text{ and } s_j \geqslant 0 & \text{for } j \in J, \\ \sum_{i \in I} a_{ij} y_i - v_j + s_j - \sum_{k \in J} r_{kj} x_k = c_j & \text{for } j \in J, \\ y_i \leqslant \lambda & \text{for } i \in I, \end{cases}$$

where λ is a parameter and z_i are artificial variables.

If one of the two dual problems has an optimal solution, then both do, and moreover there exists a joint solution. Joint solutions are the only ones we shall consider. They are characterized by feasibility for both problems and the complementarity conditions:

$$\sum_{j \in J} v_j(M_j - x_j) = 0, \qquad (1)$$

$$\sum_{j \in J} s_j(x_j - m_j) = 0, \qquad (2)$$

$$\sum_{i \in I} z_i(\lambda - y_i) = 0. \qquad (3)$$

If P_1 is feasible and has a finite minimum, then starting with an optimal solution for P for some value of λ and solving P for successively larger values of λ, an optimal solution for P will eventually be reached where $z = 0$, and therefore this solution will also be optimal for P_1.

Any feasible solution for the dual of P_1, satisfying the complementarity condition for P and D with some λ, can be used to construct a joint solution to begin with the method. If no such solution is given or if it cannot be found by inspection, then it may immediately be obtained by a technique as described by van de Panne and Whinston (1964). In Section 4 we refer again to this case.

Let $(\bar{x}_j)$, $(\bar{y}_i)$, $(\bar{v}_j)$ and $(\bar{s}_j)$ be such a solution. Of course, these values will form also an optimal solution for P and D. We assume that the parametrization is performed by only adding artificial variables where they are necessary, i.e., if $\bar{z}_i \neq 0$, with

$$\bar{z}_i = b_i - \sum_{j \in J} a_{ij} \bar{x}_j \qquad \text{for } i \in I. \tag{4}$$

We may assume that $\bar{z}_i \geq 0$ for all i, which is not a loss of generality, since we can always multiply by (-1) each equation of P_1 for which this is not true.

2. The Restricted Problems

Let $(\bar{x}_j)$, $(\bar{y}_i)$, $(\bar{v}_j)$ and $(\bar{s}_j)$ be a joint solution to begin with the method, for $\lambda = \bar{\lambda}$. If all the values $(\bar{z}_i)$, as computed by (4), are vanishing, then this joint solution provides the optimal solution $(\bar{x}_j)$ of P_1; otherwise we have to compute other joint solutions for increasing values of the parameter. This will be performed in several iterations, constructing restricted problems, whose joint solutions provide corrections to find a new one of P and D.

For this purpose define the sets of indices

$$T_1 = \{j \mid \bar{x}_j = M_j, j \in J\},$$
$$N_1 = \{j \mid \bar{v}_j = 0, j \in T_1\},$$
$$T_2 = \{j \mid \bar{x}_j = m_j, j \in J\},$$
$$N_2 = \{j \mid \bar{s}_j = 0, j \in T_2\},$$
$$R = \{i \mid \bar{z}_i = 0, i \in I\},$$

with $\bar{z}_i$, for $i \in I$, defined by (4).

Now let us define the following quadratic program:

$$\text{RP} \begin{cases} \text{Minimize } \tfrac{1}{2} \sum\limits_{k,j \in J} r_{kj} \epsilon_k \epsilon_j + \sum\limits_{i \in R} \gamma_i \text{ subject to} \\[4pt] \sum\limits_{j \in J} a_{ij} \epsilon_j = 0 & \text{for } i \in R, \\[4pt] \sum\limits_{j \in J} a_{ij} \epsilon_j + \gamma_i = 0 & \text{for } i \in (I - R), \\[4pt] -\epsilon_j \geq 0 & \text{for } j \in N_1, \\[4pt] -\epsilon_j = 0 & \text{for } j \in (T_1 - N_1), \\[4pt] \epsilon_j \geq 0 & \text{for } j \in N_2, \\[4pt] \epsilon_j = 0 & \text{for } j \in (T_2 - N_2), \\[4pt] \gamma_i \text{ unrestricted} & \text{for } i \in (I - R). \end{cases}$$

It can easily be seen that $T_1 \cap T_2$ is empty. Thus problem RP can be written in the form:

$$\text{RP}_1 \begin{cases} \text{Minimize } \frac{1}{2} \sum_{j,k \in Q} r_{kj} \epsilon_k \epsilon_j + \sum_{j \in Q} d_j \epsilon_j \text{ subject to} \\ \sum_{j \in Q} a_{ij} \epsilon_j = 0 & \text{for } i \in R, \\ \epsilon_j \leq 0 & \text{for } j \in N_1, \\ \epsilon_j \geq 0 & \text{for } j \in N_2, \end{cases}$$

where

$$Q = J - (T_1 \cup T_2) \cup (N_1 \cup N_2),$$

and

$$d_j = - \sum_{i \notin R} a_{ij} \quad \text{for } j \in J. \tag{5}$$

The dual for RP_1 is

$$\text{RD}_1 \begin{cases} \text{Maximize } \left(-\frac{1}{2} \sum_{k,j \in Q} r_{kj} \epsilon_k \epsilon_j \right) \text{ subject to} \\ \sum_{i \in R} a_{ij} \sigma_i - \sum_{k \in Q} r_{kj} \epsilon_k \geq d_j & \text{for } j \in N_1, \\ \sum_{i \in R} a_{ij} \sigma_i - \sum_{k \in Q} r_{kj} \epsilon_k \leq d_j & \text{for } j \in N_2, \\ \sum_{i \in R} a_{ij} \sigma_i - \sum_{k \in Q} r_{kj} \epsilon_k = d_j & \text{for } j \in Q - (N_1 \cup N_2). \end{cases}$$

These subproblems depend on the joint solution considered for P and D, and are called the restricted primal RP_1 and the restricted dual RD_1. They have been constructed in order to provide feasibility and satisfaction of the complementary conditions for the new solution constructed in the following section.

3. Iterations of the Method

Let (ϵ_j^*), (σ_i^*) be a joint solution for the restricted problems RP_1 and RD_1, which can be computed by any simplex-like quadratic programming algorithm. It satisfies the complementary slackness conditions:

$$\sum_{j \in N_1} \epsilon_j \delta_j = 0,$$

$$\sum_{j \in N_2} \epsilon_j \mu_j = 0, \tag{6}$$

with (δ_j) and (μ_j) defined by (10) and (11), respectively.

18. QUADRATIC PROGRAMMING WITH BOUNDED VARIABLES

We then construct the new solution:
$$\begin{cases} (\bar{x}_j + \theta\epsilon_j), (\bar{y}_i + \theta\sigma_i), \\ (\bar{v}_j + \theta\delta_j), (\bar{s}_j + \theta\mu_j), \end{cases} \tag{7}$$

where
$$\begin{cases} \epsilon_j = \epsilon_j^* & \text{for } j \in Q, \\ \epsilon_j = 0 & \text{for } j \in (J - Q), \end{cases} \tag{8}$$

$$\begin{cases} \sigma_i = \sigma_i^* & \text{for } i \in R, \\ \sigma_i = 1 & \text{for } i \in (I - R), \end{cases} \tag{9}$$

$$\begin{cases} \delta_j = \sum_{i \in R} a_{ij}\sigma_i - \sum_{k \in Q} r_{kj}\epsilon_k - d_j & \text{for } j \in T_1, \\ \delta_j = 0 & \text{for } j \in (J - T_1), \end{cases} \tag{10}$$

$$\begin{cases} \mu_j = d_j + \sum_{k \in Q} r_{kj}\epsilon_k - \sum_{i \in R} a_{ij}\sigma_i & \text{for } j \in T_2, \\ \mu_j = 0 & \text{for } j \in (J - T_2), \end{cases} \tag{11}$$

and the value of θ is computed by
$$\theta = \min\{\theta_i; i = 1, \ldots, 6\}$$

with

$$\theta_1 = \begin{cases} \min(-\bar{x}_j/\epsilon_j) & \text{for } \epsilon_j < 0 \text{ (if any)}, \\ \infty & \text{otherwise,} \end{cases}$$

$$\theta_2 = \begin{cases} \min(-\bar{v}_j/\delta_j) & \text{for } \delta_j < 0 \text{ (if any)}, \\ \infty & \text{otherwise,} \end{cases}$$

$$\theta_3 = \begin{cases} \min(-\bar{s}_j/\mu_j) & \text{for } \mu_j < 0 \text{ (if any)}, \\ \infty & \text{otherwise.} \end{cases}$$

$$\theta_4 = \begin{cases} \min(-\bar{z}_i/\gamma_i) & \text{for } i \notin R \text{ and } \gamma_i < 0 \text{ (if any)}, \\ \infty & \text{otherwise,} \end{cases}$$

$$\theta_5 = \begin{cases} \min(M_j - \bar{x}_j)/\epsilon_j & \text{for } \epsilon_j > 0 \text{ (if any)}, \\ \infty & \text{otherwise,} \end{cases}$$

$$\theta_6 = \begin{cases} \min(m_j - \bar{x}_j)/\epsilon_j & \text{for } \epsilon_j < 0 \text{ (if any)}, \\ \infty & \text{otherwise.} \end{cases}$$

The quantities (γ_i) in RP are computed by

$$\gamma_i = -\sum_{j\in J} a_{ij}\epsilon_j \qquad \text{for } i \notin R. \tag{12}$$

Because of the way problem RP is constructed and the weighting factor, θ, is computed, it is easy to verify that this new solution, defined by (7) to (11), is feasible and satisfies the complementary slackness condition, (1) to (3), of P and D, for $\lambda = \bar{\lambda} + \theta$. Therefore, (7) represents a joint solution, which will be used to start the next iteration.

The algorithm terminates when at any iteration we have a value of θ such that $\bar{z}_i + \theta\gamma_i = 0$ for all i.

If at some iteration we have $\theta = \infty$, then we may conclude that P_1 has no optimal solution.

When the matrix (r_{kj}) is positive definite, the restricted problems have always finite optimal solutions. However, if (r_{kj}) is semi-definite, it may happen (see Gonçalves, 1971a) that, for some $\lambda = \bar{\lambda}$, the new solution yields a restricted dual with no feasible solution. Then, we have to find out whether problem D has any joint solution, for $\lambda = \bar{\lambda}$, yielding a feasible restricted dual (the number of such solutions is finite). If there is none, then P_1 is not feasible (see Gonçalves, 1971a). Most simplex-like quadratic programming algorithms offer alternatives to construct such joint solutions.

4. The Algorithm in Tableau Form

We can give to the above algorithm a tableau form identical to that presented by Gonçalves (1971b) for the primal-dual method for quadratic programming. The tables are transformed by simplex transformations, and we compute the optimal solutions for the restricted problems simultaneously with the other quantities defined by (10), (11) and (12).

The set-up table is of the form shown in Table 1, where A and R are the matrices of elements (a_{ij}) and (r_{ij}), respectively, as defined for P_1. The symbols p, h and d denote vectors with components defined by

$$p_j = c_j + \sum_{k\in J} r_{kj} m_k \qquad \text{for } j \in J, \tag{13}$$

$$h_i = b_i - \sum_{j\in J} a_{ij} m_j \qquad \text{for } i \in I, \tag{14}$$

and (5), respectively. By w and β we represent two vectors of components $w_j = s_j$ or v_j, and $\beta_j = \mu_j$ or δ_j, according to the trial solution and the corresponding restricted problems considered. In Table 1 we have $w_j = s_j$ and $\beta_j = \mu_j$ for all j. The last column stores the values, x_j, w_j and z_i, of the trial solution at each iteration, corresponding to the variables ϵ_j, β_j and γ_i appearing in the first column.

18. QUADRATIC PROGRAMMING WITH BOUNDED VARIABLES

The definition of p and h by (13) and (14), respectively, is equivalent to substituting x_j by $x_j + m_j$, for all j, in P_1. So, the bounds are transformed in $m'_j = 0$ and $M'_j = M_j - m_j$.

Without loss of generality we may assume the vector $p \geqslant 0$ (see van de Panne and Whinston, 1964, for a discussion of the starting procedure), and since $h \geqslant 0$ the table represents a solution for the dual to start with.

The corresponding restricted problems may be solved using the simplex method for quadratic programming by van de Panne and Whinston (1964), with a slight difference for the case when we have $\epsilon_j \leqslant 0$ and $\delta_j \leqslant 0$, for $j \in N_1$, instead of the usual non-negativity restriction.

The simplex transformations of the tableaux are extended to all its elements but those in the last column. As soon as an optimal solution is obtained for the restricted problems, the whole tableau will also give all the other values for δ_j, μ_j and γ_i as defined by (10), (11) and (12), respectively, which, with the optimal solution, will be used to compute the value of θ.

TABLE 1

Basic variables		x	y		Values of basic variables
w	β	$-R$	$-A^T$	d	p
z	γ	A	0	0	h

The last row is redefined, at each iteration, by the computation of the new solution, as defined by (7).

Once a variable γ_i leaves the basis, it may be crossed out with its corresponding column. The variables σ_i entering the basis can also be crossed out with their corresponding rows.

Because of the way the weighting factor, θ, is computed, the final tableau, at each iteration, is also an initial table to start with the next iteration (for details see Gonçalves, 1971a).

5. Efficiency of the Method

As is seen from the last section, the algorithm is fitted for the use of the product form of the inverse to carry out the simplex calculations.

Preliminary results from solving by hand some relatively small numerical examples are encouraging, but no experimental basis for a firm judgement is yet available.

6. Numerical Example

In this section we shall solve an example used by Whinston (1965) for demonstrating his method of quadratic programming with bounded variables.

$$\text{Minimize } 6x_1 + 2x_1^2 - 2x_1 x_2 + 2x_2^2$$

$$\text{Subject to } x_1 + x_2 = 2,$$

$$0 \leq x_1 \leq 2, \quad 0 \leq x_2 \leq 1,$$

whose dual is:

$$\text{Maximize } 2y - 2v_1 - v_2 - 2x_1^2 + 2x_1 x_2 - 2x_2^2$$

$$\text{Subject to } y - v_1 + s_1 - 4x_1 + 2x_2 = 6,$$

$$y - v_2 + s_2 + 2x_1 - 4x_2 = 0,$$

$$v_1, v_2, s_1, s_2 \geq 0.$$

Starting with the solution for the dual $x_1 = x_2 = 0$, $y = 0$, $v_1 = v_2 = 0$, $s_1 = 6$ and $s_2 = 0$, we have $T_1 = N_1 = \varnothing$, $T_2 = N_2 = \{1,2\}$ and $R = \{1\}$ since $z_1 = 2$. Then the restricted primal is

$$\text{Minimize } -\epsilon_1 - \epsilon_2 + \tfrac{1}{2}(\epsilon_1, \epsilon_2) \begin{bmatrix} 4 & -2 \\ -2 & 4 \end{bmatrix} \begin{bmatrix} \epsilon_1 \\ \epsilon_2 \end{bmatrix}$$

$$\text{Subject to } \epsilon_1 \geq 0, \quad \epsilon_2 \geq 0.$$

Using the tableau form we have the following succession of tableaux:

		ϵ_1	ϵ_2	σ		Values of basic variables
+	μ_1	-4^*	2	-1	-1	6
+	μ_2	2	-4	-1	-1	0
*	γ	1	1	0	0	2

		μ_1	ϵ_2	σ		
+	ϵ_1	$-\tfrac{1}{4}$	$-\tfrac{1}{2}$	$\tfrac{1}{4}$	$\tfrac{1}{4}$	0
+	μ_2	$\tfrac{1}{2}$	-3^*	$-\tfrac{1}{2}$	$-\tfrac{3}{2}$	0
*	γ	$\tfrac{1}{4}$	$\tfrac{3}{2}$	$-\tfrac{1}{4}$	$-\tfrac{1}{4}$	2

		μ_1	μ_2	σ		
+	ϵ_1	$-\tfrac{1}{3}$	$-\tfrac{1}{6}$	$\tfrac{1}{2}$	$\tfrac{1}{2}$	0
+	ϵ_2	$-\tfrac{1}{6}$	$-\tfrac{1}{3}$	$\tfrac{1}{2}$	$\tfrac{1}{2}$	0
*	γ	$\tfrac{1}{2}$	$\tfrac{1}{2}$	-1	-1	2

The third tableau provides the optimal solution for the restricted problems and all the other quantities defined by (10), (11) and (12): $\epsilon_1 = \tfrac{1}{2}$, $\epsilon_2 = \tfrac{1}{2}$, $\mu_1 = \mu_2 = 0$, $\delta_1 = \delta_2 = 0$, $\sigma = 1$ and $\gamma = -1$. So, $\theta = 2$ and an improved solution for the dual, as defined by (7), is: $x_1 = 1$, $x_2 = 1$, $v_1 = v_2 = 0$, $s_1 = 6$, $s_2 = 0$ and $y = 2$, with $z = 0$. Hence, $x_1 = x_2 = 1$ is also an optimal solution for the original problem.

Acknowledgements

This paper is based on Chapter III of the author's Ph.D. thesis, University of Birmingham, 1969. The author is indebted to Professor S. Vajda for valuable comments and stimulating discussions. The research was supported by NATO and the 'Instituto de Alta Cultura, Portugal'.

References

Eisemann, K. (1963). The primal-dual method for bounded variables. *Operat. Res.* **13**, 110–121.
Gonçalves, A. S. (1969). A Unified Theory of Primal-Dual Techniques in Linear Programming. Ph.D. thesis, University of Birmingham.
Gonçalves, A. S. (1971a). A generalized primal-dual technique for quadratic programming, in parametric form. *Revista Faculdade Ciências Universidade Coimbra* **47**, 1–23.
Gonçalves, A. S. (1971b). A primal-dual method for quadratic programming with economic applications. Paper presented at the CORS/AFCET conference of 'O.R. as applied to Banking and Finance'. Montréal.
Panne, C. van de, and Whinston, A. (1964). The simplex and the dual method for quadratic programming. *Operat. Res. Quart.* **15**, 355–388.
Whinston, A. (1965). The bounded variable problem—an application of the dual method for quadratic programming. *Naval Res. Log. Quart.* **12**, 173–180.

19. Quadratic Zero-One Programming by Implicit Enumeration

PIERRE HANSEN
*Institut d'Economie Scientifique et de Gestion,
Lille, France*

Summary

An implicit enumeration algorithm is proposed for the minimization of a quadratic function in zero-one variables under quadratic constraints; convexity or monotonicity is not required for the objective function or the constraints. The algorithm is a specialization of a previous algorithm for non-linear zero-one programming (Hansen, 1970) with new tests based on the idea of additive penalties. These penalties are computed for all free variables and added to yield tighter bounds than those of the other algorithms known to the author. The resolution of a few test problems on a computer is also discussed.

1. Introduction

1.1. *Zero-one Programming*

Because of their ability to express problems involving numerous interrelated decisions, programs in 0–1 variables have been studied very thoroughly over the last few years. This trend began when Balas proposed his 'additive algorithm' (1964a, 1964b, 1965) for solving linear 0–1 programs by implicit enumeration. Reformulations and improvements of this algorithm, due to Balas himself (Balas, 1967) and to Glover (1965, 1968, 1971), Petersen (1967), Geoffrion (1967, 1969) and other authors have rendered this approach probably the most efficient yet proposed for the resolution of linear 0–1 programs.

This success has motivated the attempts to design algorithms for solving non-linear 0–1 programs of certain classes: Ginsburgh and van Peetersen (1969) have considered non-linear programs with a non-decreasing objective function and non-decreasing non-linear constraints; Laughhunn (1970) has proposed an algorithm for the minimization of a positive definite quadratic function under linear constraints. The author has studied the minimization of unconstrained non-linear functions of 0–1 variables (Hansen, 1969) and has proposed an algorithm for the general case of the minimization of a non-linear objective function under non-linear constraints without restrictions such as convexity or monotonicity (Hansen, 1970). Hammer and Rubin (1969, 1970) have considered quadratic 0–1 programs with linear constraints and indefinite objective function; these authors have shown that any quadratic function in 0–1 variables may be replaced by an equivalent difference of a positive definite quadratic function in 0–1 variables and a linear function in 0–1 variables. Lawler (1963) had already used this convexification procedure in the particular case of the quadratic assignment problem; Abadie (1971) has shown that it could be used also for any non-linear program in 0–1 variables. Another implicit enumeration algorithm for quadratic 0–1 programming has been outlined in a recent book of Hillier (1969).

Several other approaches to the resolution of non-linear programs in 0–1 variables have been studied: Balas (1964b), Zangwill (1965) and Watters (1967) have noted that any polynomial program in 0–1 variables could be reduced to a linear program in 0–1 variables by introducing new variables and constraints. Boolean methods have been proposed by Hammer and Rudeanu (1968, 1969, 1970). Lawler and Bell (1966) and Dragan (1968), as well as Mao and Wallingford (1968) have presented algorithms using lexicographical enumeration of the feasible solutions.

Several algorithms have also been given for the resolution of non-linear convex programs with integer variables. These algorithms could be used to solve quadratic programs in 0–1 variables after convexification of the objective function and of the constraints. Kelley (1960), Künzi and Oettli (1963), Witzgall (1963), Dinkelbach (1965) and Courtillot (1965) have proposed cutting-plane algorithms. Implicit enumeration algorithms for pure or mixed non-linear programs in integer variables were proposed by Boot and Theil (1964), Balas (1969) and Abadie (1969, 1971).

1.2. *Statement of the Problem*

Quadratic programs in 0–1 variables with quadratic constraints are considered in this paper. Such problems may be written

$$\text{Minimize} f(x_1, x_2, \ldots, x_n) = \sum_j c_j x_j + \sum_{i<j} \sum_j c_{ij} x_i x_j \tag{1}$$

under the constraints

$$g_k(x_1, x_2, \ldots, x_n) = \sum_j a_{jk} x_j + \sum_{i<j} \sum_j a_{ijk} x_i x_j - b_k \geq 0, \tag{2}$$

$$x_j \in \{0, 1\}, \quad (i = 1, 2, \ldots, n-1; j = 1, 2, \ldots, n; k = 1, 2, \ldots, m), \tag{3}$$

where the c_j, c_{ij}, a_{jk}, a_{ijk} for all i, j and k are real numbers. No requirements such as convexity or monotonicity are made on the objective function or on the constraints.

Let us call a *candidate solution* or, more briefly, a *solution* any vector satisfying the constraints (3), that is every boolean n-vector. As usual in linear programming terminology, let us call a *feasible solution* any vector satisfying the constraints (2) and (3), and an *optimal solution* any vector satisfying (2) and (3) and minimizing (1).

1.3. *Implicit Enumeration: a Reminder*

Implicit enumeration algorithms use sequential separations of the set of solutions, thus defining subproblems associated to subsets of solutions. In 0–1 programming, the separations are carried out by choosing a free (not yet chosen or fixed) variable and fixing it to 0 or to 1. Separations are performed until subsets of solutions are obtained for which the tests of the algorithm may show:

(a) that a particular vector is the optimal solution of the subproblem under consideration, or
(b) that the subproblem has no feasible solution, or
(c) that the subproblem has no solution better than the best solution obtained so far.

In any of these cases, the subset of solutions is said to be *fathomed* (Geoffrion, 1967). After a subset of solutions has been fathomed the algorithm selects another subset of solutions (if any), which has not yet been tested. If a subset of solutions is not fathomed by the tests of the algorithm, it is separated.

According to the case (a) (b) or (c) in which they attempt to fathom a subset of solutions, the tests of an algorithm may be classified as *resolution tests*, *feasibility tests* or *optimality tests* (Hansen, 1971c); if the subset of solutions under consideration is not fathomed, a part of this subset as specified by a condition (in 0–1 programming, this condition is usually that a free variable takes the value 0 or 1) may be fathomable. To test this, different versions of the tests may be used. To most *direct tests* for fathoming the whole subset of solutions there correspond *conditional tests* for fathoming parts of the subset in question. In addition to direct and conditional tests, implicit enumeration algorithms sometimes use *ancillary tests*. These tests do not attempt to fathom a subset of solutions; they are designed to determine quickly whether a direct or conditional test will certainly or probably not yield any useful information. When this is the case, the direct or conditional test is omitted, and computation time is saved.

2. Additive Penalties

2.1. *Definition*

The tests of an implicit enumeration algorithm make use of lower bounds of the values taken by the objective function and of upper bounds of the values taken by the left-hand side of the constraints on subsets of solutions. At a current iteration, these bounds must hold for all feasible solutions of the subset under consideration; in the algorithms for non-linear or quadratic 0–1 programming known to the author, these bounds are valid for the larger set of all solutions under consideration. At a current iteration let K_0, K_1 and K_2 denote respectively the sets of the indices of the variables fixed to 0, to 1 or still free; let $I(K_0, K_1, K_2)$ denote the subset of solutions under consideration:

$$I(K_0, K_1, K_2) = \{X | x_j = 0, \forall j \in K_0; x_j = 1, \forall j \in K_1;$$
$$x_j \in \{0, 1\}, \forall j \in K_2\}. \quad (4)$$

A lower bound $\underline{f}$ will thus satisfy

$$\underline{f} \leqslant f(X), \quad \forall X \in I(K_0, K_1, K_2), \quad (5)$$

and an upper bound associated with the kth constraint will satisfy

$$\bar{g}_k \geqslant g_k(X), \quad \forall X \in I(K_0, K_1, K_2). \quad (6)$$

A *penalty* is an increment which may be added to the lower bound $\underline{f}$ (or subtracted from the upper bound $\bar{g}_k$) when a condition is imposed on the free variables; in linear and in non-linear 0–1 programming, this condition is usually that a free variable must take the value 0 or 1. Penalties have been used in mixed-integer programming (Beale and Small, 1965; Driebeek, 1966; Beale and Tomlin, 1969; Tomlin, 1970); although the term has not been used before, all conditional tests of linear 0–1 programming also use penalties. Let us consider the simultaneous fixation of several free variables, whose indices are in K_0' to 0, and of several free variables, whose indices are in K_1', to 1. If the penalties associated with these fixations may be added up, yielding a valid new penalty, they will be called *additive penalties*. More precisely we may state:

Definition 1

The penalties P_j^0 and P_j^1, $\forall j \in K_2$ are additive if, and only if,

$$\forall K_0', K_1' \subset K_2 : K_0' \cap K_1' = \emptyset \; \exists \underline{f}':$$

$$\underline{f} + \sum_{j \in K_0'} P_j^0 + \sum_{j \in K_1'} P_j^1 \leqslant \underline{f}' \quad (7)$$

and

$$\underline{f}' \leqslant f(X), \forall X \in I[K_0 \cup K_0', K_1 \cup K_1', K_2/(K_0' \cup K_2')]. \quad (8)$$

Definition 2

The penalties P_{jk}^0 and P_{jk}^1, $\forall j \in K_2$ are additive if, and only if

$$\forall K_0', K_1' \subset K_2 : K_0' \cap K_1' = \emptyset \ \exists \bar{g}_k':$$

$$\bar{g}_k - \sum_{j \in K_0'} P_{jk}^0 - \sum_{j \in K_1'} P_{jk}^1 \geqslant \bar{g}_k' \tag{9}$$

and

$$\bar{g}_k' \geqslant g_k(X), \ \forall X \in I[K_0 \cup K_0', K_1 \cup K_1', K_2/(K_0' \cup K_1')]. \tag{10}$$

When additive penalties are available, new tests may be used in an implicit enumeration algorithm.

2.2. *Additive Penalties for Quadratic 0–1 Programming*

The quadratic 0–1 programming problem may be restated at a current iteration, by introducing in (1), (2) and (3) the values of the fixed variables. Then the problem is

$$\text{Minimize } f(x_1, x_2, \ldots, x_n) = C + \sum_{j \in K_2} c_j' x_j + \sum_{i<j, i \in K_2} \sum_{j \in K_2} c_{ij} x_i x_j \tag{11}$$

under the constraints

$$g_k(x_1, x_2, \ldots, x_n) = A_k + \sum_{j \in K_2} a_{jk}' x_j + \sum_{i<j, i \in K_2} \sum_{j \in K_2} a_{ijk} x_i x_j \geqslant 0 \tag{12}$$

$$x_j \in \{0, 1\} \quad (i \in K_2, j \in K_2, k = 1, 2, \ldots, m) \tag{13}$$

where

$$C = \sum_{j \in K_1} c_j + \sum_{i<j, i \in K_1} \sum_{j \in K_1} c_{ij}, \tag{14}$$

$$c_j' = c_j + \sum_{i<j, i \in K_1} c_{ij} + \sum_{i>j, i \in K_1} c_{ji}, \tag{15}$$

$$A_k = \sum_{j \in K_1} a_{jk} + \sum_{i<j, i \in K_1} \sum_{j \in K_1} a_{ijk} - b_k, \tag{16}$$

and

$$a_{jk}' = a_{jk} + \sum_{i<j, i \in K_1} a_{ijk} + \sum_{i>j, i \in K_1} a_{jik}. \tag{17}$$

As in the other algorithms for quadratic 0–1 programming cited above, a lower bound $\underline{f}$ may be obtained by adding C and all the negative coefficients in (11):

$$\underline{f} = C + \sum_{j \in K_2} \min(0, c_j') + \sum_{i<j, i \in K_2} \sum_{j \in K_2} \min(0, c_{ij}). \tag{18}$$

Theorem 1

The penalties

$$P_j^0 = -\min(0, c_j') - \sum_{i<j, i\in K_2} \min\left(0, \frac{c_{ij}}{2}\right)$$
$$- \sum_{i>j, i\in K_2} \min\left(0, \frac{c_{ji}}{2}\right) \tag{19}$$

and

$$P_j^1 = \max(0, c_j') \tag{20}$$

are additive.

Proof

Formula (18) shows that $\underline{f}$ is equal to C plus the sum of the negative coefficients in the linear and quadratic terms of (11). When the free variables whose indices are in K_0' and K_1' are fixed, the lower bound $\underline{f}$ will be increased to $\underline{f}'$, because some terms with positive coefficients have the index of their variable, or both the indices of their variables, in K_1' and because some terms with negative coefficients have the index of their variable, or both the indices of their variables, in K_0'. The increments due to the linear terms are exactly taken into account by the first term on the right-hand side of (19) and by (20). The increments due to quadratic terms with positive coefficients are not taken into account by the penalties, as they depend on the values taken by two variables. The increments due to quadratic terms with negative coefficients are taken into account by the two last terms on the right-hand side of (19), where the factor $\frac{1}{2}$ avoids a double count. Therefore the sum of the penalties will be less than or equal to the increment of $\underline{f}$, so that (7) holds.

The non-strict inequality (7) becomes an equality when no quadratic term with a positive coefficient has both the indices of its variables in K_1' and no quadratic term with a negative coefficient has one of the indices of its variables in K_0' and the other not.

An upper bound of the values taken by the left-hand side of the constraint k on $I(K_0, K_1, K_2)$ may be obtained by adding A_k and all the positive coefficients in (12), whence

$$\bar{g}_k = A_k + \sum_{j\in K_2} \max(0, a_{jk}') + \sum_{i<j, i\in K_2} \sum_{j\in K_2} \max(0, a_{ijk}). \tag{21}$$

Theorem 2

The penalties

$$P^0_{jk} = \max(0, a'_{jk}) + \sum_{i<j, i\in K_2} \max\left(0, \frac{a_{ijk}}{2}\right) + \sum_{i>j, i\in K_2} \max\left(0, \frac{a_{jik}}{2}\right) \quad (22)$$

and

$$P^1_{jk} = -\min(0, a'_{jk}) \quad (23)$$

are additive. The proof is similar to that of Theorem 1 and is therefore omitted here.

3. Algorithm

3.1. Instructions

An initial solution X_h is first obtained by a heuristic algorithm. The indices of the variables fixed to 0 or to 1 are recorded in a memory vector M.

3.1.1. Initialization
Obtain a good initial solution X_h of value $f(X_h)$. Set $X_{opt} = X_h$ and $f_{opt} = f(X_h)$. Set $K_0 = K_1 = \emptyset$, $K_2 = \{1, 2, \ldots, n\}$, and $M = \emptyset$.

3.1.2. Tests
T1. (Direct optimality test.)
Compute $\underline{f}$ and $\forall j \in K_2, P^0_j, P^1_j$ and $P_j = \min(P^0_j, P^1_j)$.
Compute $\underline{f}^1 = \underline{f} + \sum_{j\in K_2} P_j$.

(a) If $\underline{f}^1 \geq f_{opt}$ go to 3.1.4 (regression step).
(b) If $\underline{f}^1 < f_{opt}$ go to test T2.

T2. (Conditional optimality test.)
Compute $\forall j \in K_2$, $D_j = |P^0_j - P^1_j|$.
$\forall j: (\underline{f}^1 + D_j \geq f_{opt}$ and $P^0_j > P^1_j)$ set $x_j = 1$, add j to M by the right and underline j.
$\forall j: (\underline{f}^1 + D_j \geq f_{opt}$ and $P^1_j > P^0_j)$ set $x_j = 0$, add j to M by the right and underline j.

(a) If one variable at least has been fixed in test T2, go back to test T1.
(b) If no variable has been fixed in test T2, go to test T3.

T3. (Resolution test.)
If $K_2 \neq \emptyset$, go to test T4.
Compute $\forall k = 1, 2, \ldots, m, g_k(x_1, x_2, \ldots, x_n) = A_k$.

(a) If $\exists\, g_k < 0$ go to 3.1.4.
(b) If $g_k > 0$, $\forall k = 1, 2, \ldots, m$, set $X_{opt} = X, f_{opt} = f(X)$ and go to 3.1.4.

T4 (Direct feasibility test.)
Consider the first constraint or, if coming from test T5, the current constraint or the next constraint as indicated.
Compute $\bar{g}_k$ and $\forall j \in K_2$, P_{jk}^0, P_{jk}^1 and $P_{jk} = \min(P_{jk}^0, P_{jk}^1)$.
Compute $\bar{g}_k^1 = \bar{g}_k - \sum_{j \in K_2} P_{jk}$.

(a) If $\bar{g}_k^1 < 0$ go to 3.1.4.
(b) If $\bar{g}_k^1 \geq 0$ go to test T5.

T5. (Conditional feasibility test.)
Compute $\forall j \in K_2$ $D_{jk} = |P_{jk}^0 - P_{jk}^1|$.
$\forall j : (\bar{g}_k^1 - D_{jk} < 0$ and $P_{jk}^0 > P_{jk}^1)$ set $x_j = 1$, add j to M by the right and underline j.
$\forall j : (\bar{g}_k^1 - D_{jk} < 0$ and $P_{jk}^1 > P_{jk}^0)$ set $x_j = 0$, add j to M by the right and underline j.

(a) If one variable at least has been fixed in test T5 and one variable at least remains free, iterate test T4 on the same constraint.
(b) If no variable has been fixed in test T5 and all constraints have not yet been tested, iterate test T4 on the next constraint.
(c) If no variable has been fixed in test T5, all constraints have been tested and one variable at least remains free, go to test T1.
(d) If no variable remains free, go to test T1.

3.1.3. *Choice step*
Compute $\forall j \in K_2$, P_j^0, P_j^1 and $D_j = |P_j^0 - P_j^1|$.
Select $D_{j*} = \max_{j \in K_2} D_j$.
If $P_{j*}^0 > P_{j*}^1$, set $x_j = 1$ and add j to M by the right.
If $P_{j*}^1 > P_{j*}^0$, set $x_j = 0$ and add j to M by the right.
Go to test T1.

3.1.4. *Regression step*
Seek from right to left in M an index j^* not underlined; if no such index exists, terminate.
Set free all x_j whose indices are to the right of j^* in M and erase those indices.
Set x_{j*} to 0 if x_{j*} was equal to 1, or set x_{j*} to 1 if x_{j*} was equal to 0.
Underline j^* in M.
Go to test T1.

3.2. *Justification of the Algorithm*
Test T1: all free variables must take a boolean value in any solution of the subset under consideration. Therefore, for every free variable a penalty of at least P_j must be incurred. As these penalties are additive, $\underline{f}^1$ is a valid lower

bound for f on every subset of $I(K_0, K_1, K_2)$ consisting of a unique vector and hence $\underline{f}^1$ is a valid lower bound for f on $I(K_0, K_1, K_2)$ itself.

Test T2: this test is made when $I(K_0, K_1, K_2)$ has not been fathomed in test T1: an attempt is made to fathom a part of $I(K_0, K_1, K_2)$. If a free variable x_j were fixed at the value corresponding to the largest of the penalties P_j^0 and P_j^1 and thereafter the test T1 were applied, the new bound would be at least as large as $\underline{f}^1 + D_j$. Therefore $\underline{f}^1 + D_j$ is a valid lower bound for f on the subset of $I(K_0, K_1, K_2)$ for every vector of which x_j takes the fixed value. If $\underline{f}^1 + D_j$ is greater or equal to f_{opt} that subset is fathomed (case c).

Test T3: this test is only applied when there are no free variables: $I(K_0, K_1, K_2)$ reduces to a unique vector and the test T3 determines whether this vector is feasible or not. If this vector is feasible, as test T1 has been passed, it gives to the objective function a value lower than f_{opt} and $I(K_0, K_1, K_2)$ is fathomed (case a). If this vector is not feasible, $I(K_0, K_1, K_2)$ is fathomed (case b).

The justification of the tests T4 and T5 is similar to that of the tests T1 and T2, except that the subsets of solutions are fathomed in case b instead of case c.

Convergence: the convergence of the algorithm follows immediately from the facts that (a) the number of solutions is finite, (b) at every iteration of the tests T1 to T5, one variable at least is fixed or the algorithm proceeds to the regression step or to the choice step, (c) the regression rule prohibits cycling, and (d) all tests take a finite computation time.

4. Computational Experience and Possible Extensions

4.1. *Resolution of Test Problems on a Computer*

The only published quadratic programming problem in 0–1 variables known to the author is a problem with eight variables and linear constraints due to Mao and Wallingford (1968). Thirteen new test problems have been designed; the data of these problems are given in Table 1. Three codes have been written in FORTRAN IV: MAOWAL, BNL 101 and BNL 102. MAOWAL is a code for the lexicographical enumeration algorithm of Mao and Wallingford (1968); BNL 101 is a code for quadratic 0–1 programming by implicit enumeration without additive penalties, based on Hansen (1970); BNL 102 is a code for the algorithm of the present paper. The boolean method of Hammer and Rudeanu (1968) has not been programmed because

(a) the constrained problems must first be reduced to unconstrained problems, and very long expressions may be obtained for problems with more than 10 variables;
(b) experiments by Gaspar (1967) have shown that memory requirement problems appeared already for problems with 15 variables;

274 PIERRE HANSEN

TABLE 1
Data of the test problems

c_j	2	1	8	7	6	8	5	9	4	5	8	1	2	1	8	5	3	5	1	1	6	9	2	6	9	7	8	2	7	8	3
i	2	4	2	5	1	3	2	6	4	1	8	6	9	3	9	3	7	4	3	9	17	20	6	2	8	4	4	15	17	26	
j	1	1	2	3	5	6	7	8	9	0	11	12	13	14	15	16	18	19	20	21	22	23	24	25	26	27	28	29	30		
c_{ij}	3	-1	8	-4	-7	7	4	-3	4	-3	9	-6	3	9	9	-4	-3	9	-4	-4	-6	-7	4	-6	5	-7	-7	2	5	3	
a_{j1}	3	-9	7	-5	-8	3	-7	-9	5	7	-9	-9	9	-3	9	-9	-9	2	-6	-7	8	9	4	-4	-3	2	-7	-7	4	-7	1
i	3	4	5	1	6	5	1	5	5	8	2	2	3	5	5	13	5	10	15	4	14	16	4	4	16	8	21	4	29	19	
j	1	2	3	5	6	7	8	9	10	11	12	13	14	15	17	18	19	20	21	22	23	24	25	26	27	28	29	30			
a_{ij1}	8	-8	5	-7	2	6	3	-6	-8	9	-2	-6	-3	1	-8	-3	-3	9	-4	-6	-8	-4	-4	-7	8	-6	6	-6	4	-5	
a_{j2}	5	1	6	4	7	8	9	-7	9	9	6	6	9	-1	9	7	6	6	4	2	5	4	6	-6	-8	6	-9	7	4	9	
i	5	3	2	6	5	2	3	7	3	8	5	9	8	8	4	5	15	3	11	17	6	6	4	9	9	13	24	7	29	30	
j	1	2	3	5	6	7	8	9	10	11	13	14	15	16	17	18	19	20	21	22	23	24	25	26	27	28	29	30			
a_{ij2}	9	-4	1	-6	9	5	9	-2	1	6	5	-6	-7	5	3	-6	-3	4	-3	4	8	-4	-8	-7	-5	7	-5	1	-5		
a_{j3}	-3	4	5	1	2	3	1	-9	8	7	3	5	5	5	1	7	1	10	6	6	5	-6	4	4	-8	2	7	4	7	7	
i	4	5	4	2	1	4	1	3	6	4	4	2	10	2	3	3	10	7	6	15	16	4	14	25	4	25	21	7	29	3	
j	2	1	2	3	5	6	7	8	9	10	11	13	14	15	17	18	19	20	22	23	24	25	26	27	28	29	30				
a_{ij3}	-7	7	-8	6	-1	3	4	3	-4	2	-7	-3	-4	9	-3	6	5	-4	-6	4	-9	1	-5	8	-7	9	7	-7	-7		
a_{j4}	4	4	-3	4	2	7	-6	5	5	-5	5	3	7	8	3	9	8	6	6	9	6	5	6	4	7	-8	-8	9	5	5	
i	2	5	4	2	1	3	4	3	2	2	6	4	10	1	4	9	15	7	13	11	5	16	22	4	4	13	13	12	26	
j	1	2	3	5	6	7	8	9	11	13	14	15	17	18	19	20	22	23	24	25	26	27	28	29	30						
a_{ij4}	9	-8	5	-2	6	-9	4	-7	1	4	-1	8	-4	2	-1	5	-5	1	-5	-5	-9	4	-8	3	-1	-2					
b_1			5	10	20	-5	30	0	45	25	40	0	35	45	50																
b_2			5	5	15	30	15	10	25	20	5	15	25	20	30																
b_3			5	5	25	-5	25	5	15	25	5	-5	30	10	20																
b_4			5	15	0	5	0	15	15	10	5	35	25	35	30																

The first 6, 8, ..., 30 columns give the data of the 1st, 2nd, ..., 13th problem.

19. QUADRATIC ZERO-ONE PROGRAMMING

TABLE 2

Number of the problem	Number of variables	Indices of the variables fixed at 1 in X_{opt}	Value of f_{opt}	Time MAOWAL	Time BNL101	Time BNL102
1	6	1, 6	10	0·139	0·165	0·112
2	8	1, 2, 3, 7	27	0·849	0·530	0·353
3	10	3, 4, 7, 9	28	2·035	0·975	0·340
4	12	1, 7, 9, 10, 11	21	9·578	5·522	1·156
5	14	1, 3, 6, 9, 12, 13	12	30·652	3·618	0·551
6	16	9, 13	5	103·235	6·811	0·521
7	18	1, 3, 6, 9, 12, 13, 17	17	483·231	7·854	1·901
8	20	1, 3, 6, 12, 13, 18, 19	7	>1,000·000	21·188	1·300
9	22	1, 3, 6, 12, 13, 18, 19, 22	2	—	22·161	0·757
10	24	2, 8, 13, 16, 18, 19	7	—	127·690	6·785
11	26	1, 3, 6, 10, 12, 13, 18, 19	9	—	413·126	8·298
12	28	1, 3, 6, 12, 13, 18, 19, 21	16	—	1,041·966	42·729
13	30	1, 3, 6, 12, 13, 18, 19, 30	10	—	>1,500·000	9·735

The times are in seconds CPU on a CDC 6400; input and output times are excluded. For problem 11 BNL101 gives a different optimal solution than that of Table 2: the indices of the variables fixed at 1 are 1, 3, 6, 10, 12, 13, 17, 18, 19, 22.

(c) computation time as reported by Gaspar seems high (from 3 to 15 minutes on a very unsophisticated MECIPT-1 computer for unconstrained problems with 8 to 12 variables).

Linearization was not tested for lack of time and an adequate code; as every quadratic term yields a new variable and two new constraints (or, on average, slightly more than one constraint when using equivalent expressions for the constraints as proposed by Glover and Woolsey, 1971) the tests problems 4 and following would have given linear programs in 0–1 variables too large to be solved entirely in core.

Computation time for the test problems on a CDC 6400 computer are given in Table 2. It is seen that the order of these times is always $t_{\text{Maowal}} > t_{\text{BNL101}} > t_{\text{BNL102}}$ except for the problem 1. These results show that implicit enumeration is more efficient than lexicographical enumeration for the resolution of our test problems and that the additive penalties improve the bounds enough to reduce the computation times significantly.

4.2. *Possible Extensions*

Additive penalties may be obtained for the general case of non-linear 0–1 programming in a similar fashion to above; the only difference is that if a term contains k variables its coefficient must be divided by k instead of by 2 in formulas (19) and (22). With these modified penalties and obvious modifications of formulas (18) and (19), the algorithm may be applied in a straightforward manner.

One referee has suggested that the algorithm could be extended to mixed-integer 0–1 quadratic programming (where the constraints are quadratic in the 0–1, but linear in the continuous variables) by the procedure of Balas (1969). Additive penalties may also be used in other contexts than non-linear 0–1 or mixed-integer programming; in particular they provide an improved algorithm for the simple plant or warehouse location problem (Hansen, 1971b).

References

Abadie, J. (1969). Une méthode arborescente pour les programmes non linéaires partiellement discrets. *Revue fr. Informat. Rech. opér.* **3**, 24–50.

Abadie, J. (1971). Une méthode de résolution des programmes non linéaires partiellement discrets sans hypothèses de convexité. *Rev. fr. Informat. Rech. opér.* **5**, 23–28.

Balas, E. (1964a). Un algorithme additif pour la résolution des programmes linéaires en variables bivalentes. *C.r. hebd. Séanc. Acad. Sci. Paris*, **258**, 3817–3820.

Balas, E. (1964b). Extension de l'algorithme additif à la programmation en nombres entiers et à la programmation non linéaire. *C.r. hebd. Séanc. Acad. Sci. Paris*, **258**, 5136–5139.

Balas, E. (1965). An additive algorithm for solving linear programs with zero-one variables. *Operat. Res.* **13**, 517–546.

Balas, E. (1967). Discrete programming by the filter method. *Operat. Res.* **15**, 915–957.

Balas, E. (1969). Duality in discrete programming: II. The quadratic case. *Managmt Sci.* **16**, 14–32.

Boot, J. J. G., and Theil, H. (1964). A procedure for integer maximization of a definite quadratic function. *In*: 'Proceedings of the Third International Conference on Operations Research', (G. Kreweras and G. Morlat, eds.), pp. 667–682. Dunod, Paris, and English Universities Press.

Beale, E. M. L., and Small, R. (1965). Mixed integer programming by branch-and-bound technique. *In*: 'Proceedings IFIP Congress, New York', (W. A. Kalenick, ed.), Vol. 2, pp. 450–451. Spartan Press, Washington.

Beale, E. M. L., and Tomlin, J. (1970). Special facilities in a general mathematical programming system for non-convex problems using ordered sets of variables. *In*: 'Proceedings 5th International Conference on Operational Research'. John Wiley, New York.

Courtillot, M. (1965). Sur la résolution des programmes à solutions entières. *Revue. fr. Traitmt. Informat.Chiffres*, **8**, 81–94.

Dinkelbach, W. (1965). On convex integer programming. *In*: 'Colloquium on Applications of Mathematics to Economics', (A. Prekopa, ed.), pp. 75–78. Publishing House of the Hungarian Academy of Sciences, Budapest.

Dragan, I. (1968). A lexicographic algorithm for the solution of polynomial programs in binary variables (in Roumanian). *Studii Cercet. Mat.* **20**, 1135–1146.

Driebeek, N. J. (1966). An algorithm for the solution of mixed integer programming problems. *Managmt. Sci.* **12**, 576–587.

Gaspar, T. (1967). Programming the algorithm for the minimization of pseudo-boolean functions on a MECIPT-1 computer (in Roumanian). *Studii Cercet. Mat.* **19**, 1135–1148.

Geoffrion, A. (1967). Integer programming by implicit enumeration and Balas' method. *SIAM Rev.* **9**, 178–190.

Geoffrion, A. (1969). An improved implicit enumeration approach for integer programming. *Operat. Res.* **17**, 437–454.

Ginsburgh, V., and van Peetersen, A. (1969). Un algorithme de programmation quadratique en variables binaires. *Revue fr. Informat. Rech. opér.* **3**, 57–74.

Glover, F. (1965). A multiphase-dual algorithm for the zero-one integer programming problem. *Operat. Res.* **13**, 879–919.

Glover, F. (1968). Surrogate constraints. *Operat. Res.* **16**, 741–749.

Glover, F. (1971). Flows in arborescences. *Managmt Sci.* **17**, 568–586.

Glover, F., and Woolsey, E. (1971). Further reduction of zero-one polynomial programming problems to zero-one linear programming problems. To appear in *Operat. Res.*

Hammer, P. L., and Rubin, A. A. (1969). Quadratic programming with 0–1 variables. Operational Research Statistics and Economics Mimeograph Series, No. 53, Technion, Haifa, Israel.

Hammer, P. L., and Rubin, A. A. (1970). Some remarks on quadratic programming with 0–1 variables. *Revue fr. Informat. Rech. opér.* 67–79.

Hammer, P. L., and Rudeanu, S. (1968). 'Boolean Methods in Operations Research and Related Areas'. Springer Verlag, Berlin, Heidelberg, New York.

Hansen, P. (1969). Un algorithme S.E.P. pour les programmes pseudo-booléens non linéaires. *Cahiers Centre Etudes Rech. opér.* **11**, 26–44.

Hansen, P. (1970). Nonlinear 0–1 programming by implicit enumeration. Paper presented at the 7th Mathematical Programming Symposium 1970, The Hague, To be published in *Cahiers Centre Etudes Rech. Oper.*
Hansen, P. (1971a). Pénalités additives pour les programmes en variables zéro-un *C.r. hebd. Séanc. Acad. Sci. Paris* **273**, 175–177.
Hansen, P. (1971b). Pénalités additives pour le problème de la localisation des entrepôts. *C.r. hebd. Séanc. Acad. Sci. Paris* **273**, 252–253.
Hansen, P. (1971c). Les procédures d'optimisation par séparation: présentation générale. *Revue Belge Stat. Rech. opér.* **11**, 35–60.
Hillier, F. S. (1969). 'The Evaluation of Risky Interrelated Investments'. North Holland Publishing Co., Amsterdam and London.
Kelley, E. J. (1960). The cutting-plane method for solving convex programs. *J. SIAM* **8**, 703–712.
Künzi, H. P., and Oettli, W. (1963). Integer quadratic programming. *In*: 'Recent Advances in Mathematical Programming' (R. Graves and P. Wolfe, eds.), pp. 303–308. McGraw-Hill, New York.
Laughhunn, D. J. (1970). Quadratic binary programming with applications to capital budgeting problems. *Operat. Res.* **18**, 454–461.
Lawler, E. (1963). The quadratic assignment problem. *Managt. Sci.* **9**, 586–599.
Lawler, E., and Bell, M. D. (1966). A method for solving discrete optimization problems. *Operat. Res.* **14**, 1098–1112.
Mao, J. C. T., and Wallingford, B. A. (1968). An extension of Lawler and Bell's method of discrete optimization with examples from capital budgeting. *Managmt Sci.* **15**, 51–60.
Petersen, C. (1967). Computational experience with variants of the Balas algorithm applied to the selection of R and D projects. *Managmt Sci.* **13**, 736–750.
Rudeanu, S. (1969). Irredundant optimization of a pseudo-boolean function. *J. Optim. Theory Applns* **4**, 253–259.
Rudeanu, S. (1970). An axiomatic approach to pseudo-boolean programming, 1. *Matemat. Vest.* **7**, 403–414.
Tomlin, J. (1970). Branch and bound methods for integer and non-convex programming. *In*: 'Integer and Nonlinear Programming' (J. Abadie, ed.), Chapter 21. North Holland Publishing Co., Amsterdam.
Watters, L. J. (1967). Reduction of integer polynomial programming problems to zero-one linear programming problems. *Operat. Res.* **15**, 1171–1174.
Witzgall, C. (1963). An all-integer algorithm with parabolic constraints. *J. SIAM.* **11**, 855–871.
Zangwill, W. (1965). Media selection by decision programming. *J. Ad. Res.* **5**, 30–36.

20. Minimizing General Functions Subject to Linear Constraints

R. FLETCHER

Theoretical Physics Division, U.K.A.E.A. Research Group,
Atomic Energy Research Establishment, Harwell, England

1. Introduction

The problem under consideration is to

$$\text{Minimize } F(\mathbf{x}) \tag{1a}$$

$$\text{Subject to } \bar{C}^T \mathbf{x} \geq \bar{\mathbf{d}} \tag{1b}$$

where $F(\mathbf{x})$ is a differentiable function of n continuous variables $\mathbf{x}$, whose gradient vector of first derivatives will be denoted by $\mathbf{g}$ or by ∇F, where $g_i = \partial F/\partial x_i$, and whose Hessian matrix of second derivatives by G or by $\nabla^2 F$, where $G_{ij} = \partial^2 F/(\partial x_i \partial x_j)$. The constraints (1b) represent p linear inequalities, where $\bar{C}$ is an $n \times p$ matrix and $\bar{\mathbf{d}} \in E^p$. If $\mathbf{c}_i$ is one of the columns of $\bar{C}$, then $\mathbf{c}_i$ is the vector of coefficients in the ith constraint $\mathbf{c}_i^T \mathbf{x} \geq d_i$, and is the normal vector to the n-dimensional hyperplane $\mathbf{c}_i^T \mathbf{x} = d_i$. Methods to be considered will all solve a more general formulation in which additional linear equalities are allowed in (1b), but the modifications required are trivial. It will be assumed that there is at least one 'feasible point' $\mathbf{x}$; that is, a point which satisfies all the constraints (1b). Two fundamental problems of this nature are 'linear' and 'quadratic' programming, in which $F(\mathbf{x})$ is respectively linear and quadratic. Special techniques exist for these problems, which can be solved in a finite number of computer operations. It is, however, beyond the scope of this paper to discuss such methods in detail, although many of the fundamental ideas will be relevant to the general problem.

The methods which will be discussed owe their motivation both to developments in unconstrained optimization which took place in the early 1960s, and also to techniques for handling linear inequalities which were developed perhaps a little earlier. Surprisingly enough, very little was learned from early studies of quadratic programming, and the opposite situation is now true, that new methods of quadratic programming have been introduced, based on ideas developed whilst solving the general problem. Although different approaches have been suggested in a number of papers, they have a lot in common, and numerous hybrid methods would be possible, tackling say the linear constraints by A's method, but using the optimization strategy of B. In this paper therefore, I will try to present the fundamentals of the subject as I see it, and point out where each author has used a certain idea. I shall also try to indicate comparisons between methods as I go along. In this case this is difficult, as there is little numerical information around, and the best methods have relatively little to choose between them. Consequently, I shall be commenting on matters like: how thoroughly information is represented, whether it has to be rejected periodically, the number of operations per iteration, accuracy, and to what extent convergence results are available.

The majority of methods are based on results from the minimization subject to linear equality constraints and this topic is covered in Sections 2 and 3. The generalization of these results to inequality problems is considered in Section 4. Finally, in Section 5 a number of methods will be described in which linear or quadratic programs are solved as sub-problems. It will be noticed that no mention is made of penalty functions. In my opinion penalty functions (as they are generally understood) constitute a very inefficient way of attacking problems in which all the constraints are linear.

Before considering the methods in more detail a few important concepts will first be described. Firstly, all the methods are iterative and work with approximations $\mathbf{x}^{(k)}$ (superscripts denoting iterations) which are feasible points. If $\mathbf{x}^{(k)}$ satisfies a constraint in (1b) exactly, then that constraint will be described as being 'active'. Many of the methods involve a 'linear search subproblem' in which, for any approximation $\mathbf{x}^{(k)}$, a direction vector $\mathbf{s}^{(k)}$ is calculated and $\mathbf{x}^{(k+1)}$ is chosen as the point

$$\mathbf{x}^{(k+1)} = \mathbf{x}^{(k)} + \alpha \mathbf{s}^{(k)} \tag{2}$$

where α is chosen to minimize $F(\mathbf{x}^{(k+1)})$ subject to feasibility being maintained.

Another important concept concerns the so-called Kuhn-Tucker conditions which are necessary conditions for a constrained minimum. Let $\boldsymbol{\xi}$ be such a minimum, and let C be an $n \times m$ matrix ($m \leqslant n$) whose columns correspond to the normal vectors of the active constraints at $\boldsymbol{\xi}$. The matrix C can be assumed to be of full rank without loss of generality. Consider the projection matrices

$$P = C(C^T C)^{-1} C^T = CC^+ \tag{3a}$$

where $C^+ = (C^T C)^{-1} C^T$ denotes the full rank generalized inverse of C, and

$$\hat{P} = I - P. \tag{3b}$$

The matrix P projects vectors into the space spanned by the columns of C, and $\hat{P}$ projects vectors so that they are orthogonal to these columns, and hence lie in the intersection of the constraint hyperplanes. Thus any vector $\boldsymbol{\xi} - \epsilon \hat{P} \mathbf{g}(\boldsymbol{\xi})$ for $\epsilon > 0$ will retain equality in the active constraints, will not violate any non-active constraints for sufficiently small ϵ, and will have a lower objective

20. MINIMIZING GENERAL FUNCTIONS

function than $F(\xi)$. Hence a necessary condition for minimality of F is

$$\hat{P}g(\xi) = 0. \tag{4a}$$

An alternative statement of this by virtue of (3) is that there exists a vector $\lambda \in E^m$ such that

$$g = C\lambda \tag{4b}$$

and the elements of λ are given by the components of the vector

$$\lambda = C^+ g. \tag{4c}$$

The elements of λ are called variously Lagrange multipliers, Kuhn-Tucker multipliers, dual variables, reduced costs, etc.

Assume ξ satisfies (4a): then ξ might still not be optimum because it might be possible to reduce F by moving in a feasible direction which is in the space

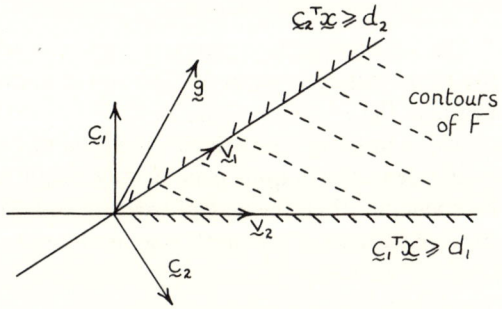

FIG. 1. The Kuhn-Tucker conditions.

spanned by the columns of C. Now, the set of feasible directions are convex linear combinations of the vectors $v_1, v_2, \ldots, v_m$, where $v_1^T, v_2^T, \ldots, v_m^T$ are the rows of C^+ (see Fig. 1). Hence if $g^T v_i \geq 0$ for all i, then there are no feasible descent directions in C-space, and hence ξ cannot be increased in this way. Thus by (4c) another necessary condition for optimality is that

$$\lambda \geq 0. \tag{4d}$$

Furthermore, if for the ith constraint $\lambda_i < 0$, then a feasible downhill direction of search is v_i, that is the row of C^+ corresponding to the constraint which has a negative multiplier. The effect of moving along v_i is that the ith constraint is no longer active.

An alternative statement of the Kuhn-Tucker condition (4d) is to assign a multiplier λ_i to each constraint, whence $\lambda_i \geq 0$ for any constraint whose residual $= 0$ and $\lambda_i = 0$ for any constraint whose residual ≥ 0: this is the so-called complementarity condition.

2. Equality Constraints—Elimination by Transformation

It is important first of all to look at methods for dealing with equality constraints on the variables, because many methods for inequality constraints rely heavily on these techniques. Assume then that the function (1a) is being minimized subject to m constraints

$$C^T \mathbf{x} = \mathbf{d} \tag{5}$$

where C is an $n \times m$ matrix such that the columns $\mathbf{c}_1, \mathbf{c}_2, \ldots, \mathbf{c}_m$ are the normal vectors of the corresponding equations, and where $m \leqslant n$.

Elimination methods can be considered as methods which set up new variables $\mathbf{y} \in E^n$ by an affine transformation from $\mathbf{x}$, such that the constraints (5) can be satisfied merely by setting $y_1, y_2, \ldots, y_m$ to zero. Clearly, the set of variables can be divided into two classes, $\mathbf{y}_1 \in E^m$, defined by

$$\mathbf{y}_1 = C^T \mathbf{x} - \mathbf{d} \tag{6}$$

and a set $\mathbf{y}_2 \in E^{n-m}$. The way in which the set $\mathbf{y}_2$ is defined is arbitrary to some extent, and gives rise to the different methods. The aim of methods for dealing with equality constraints will then be to minimize the function (1a) with respect to the variables $\mathbf{y}_2$, whilst holding the variables $\mathbf{y}_1$ equal to zero.

If the variables $\mathbf{y}_2$ are defined in a quite general way, requiring only that the complete set $\mathbf{y}$ is independent, then the way of treating equality constraints is that used in the 'reduced gradient method'—see Wolfe (1967) for instance. Assume then that

$$\mathbf{y}_2 = D^T \mathbf{x} - \mathbf{e} \tag{7}$$

where D is a general $n \times (n - m)$ matrix. In fact, when solving inequality problems, convenient choices for D and $\mathbf{e}$ are constraints from (1b) which do not correspond to active constraints. Then the equations (6) and (7) can be collected, giving

$$\begin{pmatrix} \mathbf{y}_1 \\ \mathbf{y}_2 \end{pmatrix} = [C:D]^T \mathbf{x} - \begin{pmatrix} \mathbf{d} \\ \mathbf{e} \end{pmatrix} \tag{8}$$

where the $n \times n$ matrix $[C:D]$ is non-singular. Clearly, given any value of the variables $\mathbf{y}_2$, and because $\mathbf{y}_1 = \mathbf{0}$, and given the matrix $[C:D]^{-1}$, then the variables $\mathbf{x}$, and hence $F(\mathbf{x})$, can be calculated. Thus the constrained minimization of a function $F(\mathbf{x})$ of n variables has been reduced to the unconstrained minimization of a function $F(\mathbf{y}_2)$ of $n - m$ variables.

Now it is most important in minimization methods to be able to isolate the gradient of the objective function, and in this case the gradient of F with respect to $\mathbf{y}_2$ must be isolated from the vector $\mathbf{g}$. This is fairly simple, however, in that

$$\frac{\partial y_i}{\partial x_j} = [C:D]^T_{ij} \tag{9}$$

by virtue of (8). Because

$$\frac{\partial F}{\partial x_j} = \sum_i \frac{\partial F}{\partial y_i} \frac{\partial y_i}{\partial x_j} \tag{10}$$

there follows

$$\mathbf{g_x} = [C\!:\!D]\mathbf{g_y} \tag{11}$$

or

$$\mathbf{g_y} = [C\!:\!D]^{-1}\mathbf{g_x} \tag{12}$$

where the subscripts **x** and **y** indicate the set of variables with respect to which differentiation has been taken. Of course, because the objective function is being considered as a function $F(\mathbf{y}_2)$, only the components $\mathbf{g}_{\mathbf{y}_2}$ represent the derivatives of F with respect to the $\mathbf{y}_2$ variables which occur in the now unconstrained problem.

Although when solving equality constrained problems the values of the Lagrange multipliers may not be of interest, they are when generalizing to solve inequality problems; and it is then important to have established simple formulae for computing these multipliers. Consider therefore that **x** is the point which solves the equality constrained problem, so that the required multipliers $\boldsymbol{\lambda}$ can be obtained from equation (4b), which can be written here as

$$\mathbf{g_x} = [C\!:\!D]\begin{pmatrix}\boldsymbol{\lambda}\\ \mathbf{0}\end{pmatrix}.$$

Clearly, on premultiplying by $[C\!:\!D]^{-1}$ and using (12),

$$\mathbf{g_y} = \begin{pmatrix}\mathbf{g}_{\mathbf{y}_1}\\ \mathbf{g}_{\mathbf{y}_2}\end{pmatrix} = \begin{pmatrix}\boldsymbol{\lambda}\\ \mathbf{0}\end{pmatrix}. \tag{13}$$

This tells us that not only has the problem been solved, by virtue of $\mathbf{g}_{\mathbf{y}_2} = \mathbf{0}$, but also that the Lagrange multipliers are given quite simply by $\mathbf{g}_{\mathbf{y}_1}$, the other part of the transformed gradient vector. Unfortunately, it is usually desirable to obtain estimates of the multipliers when **x** is not the solution point. In this case the vector $\mathbf{g}_{\mathbf{y}_1}$ can still be used as an estimate of the multiplier, but is in error by $O(\|\mathbf{h}\|)$, where $\|\mathbf{h}\|$ is the distance of **x** from the solution. Such an estimate will be called a 'first-order' estimate of the multiplier; it will be seen in a later section that second-order estimates can be obtained.

A special case of this reduced gradient method is simply the technique of using constraints to eliminate variables. Assume that the variables **x** are partitioned into $\mathbf{x}_1 \in E^m$ and $\mathbf{x}_2 \in E^{n-m}$, and C into a non-singular $m \times m$ matrix C_1 and an $(n-m) \times m$ matrix C_2, so that (5) can be written

$$\begin{pmatrix}C_1\\ C_2\end{pmatrix}^T \begin{pmatrix}\mathbf{x}_1\\ \mathbf{x}_2\end{pmatrix} = \mathbf{d}, \tag{14}$$

or
$$C_1^T \mathbf{x}_1 + C_2^T \mathbf{x}_2 = \mathbf{d}, \tag{15}$$

then $\mathbf{x}_1$ can be regarded as dependent upon $\mathbf{x}_2$, because

$$\mathbf{x}_1 = C_1^{-T}(\mathbf{d} - C_2^T \mathbf{x}_2) \tag{16}$$

and the elements of $\mathbf{x}_2$ can be regarded as the independent variables of an unconstrained minimization problem. In the context of the reduced gradient method, this is the special case in which $\mathbf{y}_2 = \mathbf{x}_2$, whence

$$D = \begin{pmatrix} O \\ I \end{pmatrix} \quad \text{and} \quad \mathbf{e} = \mathbf{0},$$

so that the transformation being used is

$$\begin{pmatrix} \mathbf{y}_1 \\ \mathbf{y}_2 \end{pmatrix} = \begin{pmatrix} C_1 & O \\ C_2 & I \end{pmatrix}^T \begin{pmatrix} \mathbf{x}_1 \\ \mathbf{x}_2 \end{pmatrix} - \begin{pmatrix} \mathbf{d} \\ \mathbf{0} \end{pmatrix}. \tag{17}$$

By virtue of the identity

$$\begin{pmatrix} C_1 & O \\ C_2 & I \end{pmatrix}^{-1} = \begin{pmatrix} C_1^{-1} & O \\ -C_2 C_1^{-1} & I \end{pmatrix} \tag{18}$$

and equation (12), the transformed gradient can be written as

$$\mathbf{g}_{\mathbf{y}_2} = \mathbf{g}_{\mathbf{x}_2} - C_2 C_1^{-1} \mathbf{g}_{\mathbf{x}_1}. \tag{19}$$

Similarly, the first-order estimate of the Lagrange multiplers is given by

$$\boldsymbol{\lambda} \approx \mathbf{g}_{\mathbf{y}_1} = C_1^{-1} \mathbf{g}_{\mathbf{x}_1}. \tag{20}$$

Another special case occurs when the set of variables represented by $\mathbf{y}_2$ are mutually orthonormal and orthogonal to those represented by $\mathbf{y}_1$, so that the variables $\mathbf{y}_2$ lie in the intersection of the equality constraint hyperplanes. In this case the normal vectors whose columns are C and D are related by

$$\left.\begin{array}{l} C^T D = O \\ D^+ = D^T \end{array}\right\} \tag{21}$$

Then the identity

$$[C:D]^{-1} = \begin{bmatrix} C^+ \\ D^+ \end{bmatrix} \tag{22}$$

can be used, where C^+ and D^+ again refer to full rank generalized inverses. Thus the first-order estimate of the Lagrange multipliers is

$$\boldsymbol{\lambda} \approx \mathbf{g}_{\mathbf{y}_1} = C^+ \mathbf{g}, \tag{23}$$

and the gradient vector of the independent variables is

$$\mathbf{g}_{y_2} = D^+ \mathbf{g}. \tag{24}$$

However, it is not necessary to go to the trouble of explicitly choosing an orthogonal set D, because if the direction vector

$$\begin{pmatrix} \mathbf{0} \\ \mathbf{g}_{y_2} \end{pmatrix}$$

is transformed back into x-space, it generates the direction vector

$$\bar{\mathbf{g}} = [C:D]^{-T} \begin{pmatrix} \mathbf{0} \\ \mathbf{g}_{y_2} \end{pmatrix} = DD^+ \mathbf{g}. \tag{25}$$

By virtue of the identity $[C:D][C:D]^{-1} = I$, and equations (22) and (3),

$$DD^+ = I - CC^+ = \hat{P}. \tag{26}$$

Thus $\bar{\mathbf{g}}$ is given simply by

$$\bar{\mathbf{g}} = \hat{P}\mathbf{g} \tag{27}$$

so that the direction of the gradient vector in the reduced variables is equivalent to the vector obtained by taking the projection of the gradient vector which is orthogonal to the constraint normals. This, of course, is the way in which equality constraints are handled in the 'gradient projection method' of Rosen (1960).

Having derived various methods of removing equality constraints, so as to leave unconstrained minimization problems in $n - m$ variables, and established expressions for derivatives, the problem of finding the unconstrained minimum is merely that of applying standard techniques. The original algorithms (Wolfe, 1967; Rosen, 1960) merely used the gradient vectors as directions of search in steepest-descent-like methods. Nowadays the importance of incorporating second-order information into the iteration is appreciated, and conjugate gradient versions of all the above techniques can be constructed, using the Fletcher and Reeves (1964) formula for example. Such methods give much improved rates of convergence with negligible extra effort: for instance, Abadie and Guigou (1970) report experiments of this nature in conjunction with the reduced gradient method. Termination for quadratic functions is assured in only $n - m$ iterations, as against n when the problem is unconstrained, and this point should be taken into account when resetting periodically to the (transformed) gradient.

Another possibility is to build up approximations to Hessian or inverse Hessian matrices in the $n - m$ independent variables to be used in a quasi-Newton method, and such an approach has been considered by McCormick (1970) in his 'variable reduction method', which uses the elimination of variables approach outlined in (14) to (20). Using the reduced gradient approach in a similar way would also be quite feasible. It is not so obvious however, how the gradient projection method ought to be generalized because here the transformed gradient is not formed explicitly. It is important therefore to look at the role played by second derivatives in equality constrained problems much more closely, and this is the subject of the next section.

3. Equality Constraints—Second-Order Analysis

The fundamental way to use knowledge of the Hessian G_x (i.e. the Hessian in which differentiation is carried out with respect to the **x** variables) in conjunction with the transformation and elimination schemes of the previous section, is merely to write down the transformed expression for the Hessian G_y in the transformed variables, given by

$$G_y = [C:D]^{-1} G_x [C:D]^{-T} \qquad (28)$$

following similar reasoning to that above. The $(n-m) \times (n-m)$ partition of this, corresponding to the $\mathbf{y}_2$ variables can then be used in a Newton method in which a step $\boldsymbol{\delta}$ given by

$$\boldsymbol{\delta} = -[G_{y_2,2}]^{-1} \mathbf{g}_{y_2} \qquad (29)$$

is taken in the $\mathbf{y}_2$ variables. In such a procedure it is also possible to get second-order estimates of the Lagrange multipliers because of the knowledge of second derivatives. However, I have not heard of any use of this approach in connection with either the reduced gradient or elimination of variables method, and the reason for this may be that it is not possible to simplify equation (29), with the result that the resulting housekeeping is rather lengthy. On the other hand, when the variables D are defined to be orthogonal to C, as in (21), then this is the basis of the quadratic programming method of Murray (1969, 1970). In this case the equations (28) and (29) simplify considerably, and it is also possible to write down the second-order estimate of the Lagrange multipliers quite readily. Although these ideas have so far only been suggested by Murray for quadratic programming, there would seem to be no objection to using them for the minimization of a general function subject to linear constraints.

However, as in the previous section, it is not necessary to define the set of variables corresponding to the variables D explicitly, but to proceed by a projection-like approach. As with the projection method, all the arithmetic is carried out in the **x** variables alone, and the transformation is implicit. To generate this method it is recognized that a Newton step for the constrained problem is that which would minimize a general quadratic function

$$q(\mathbf{x}) = \tfrac{1}{2}\mathbf{x}^T G \mathbf{x} - \mathbf{b}^T \mathbf{x} \qquad (30)$$

subject to the equality constraints (5). Note that $q(\mathbf{x})$ has been defined so that its Hessian matrix is G. The solution of this problem $\hat{\mathbf{x}}$, and the associated Lagrange multipliers $\hat{\boldsymbol{\lambda}}$ are then given by finding the stationary point of the Lagrangian function

$$\mathscr{L}(\mathbf{x}, \boldsymbol{\lambda}) = q(\mathbf{x}) - \boldsymbol{\lambda}^T(C^T \mathbf{x} - \mathbf{d}) \tag{31}$$

and hence by solving the linear equations in $(n + m)$ variables

$$\begin{bmatrix} G & -C \\ -C^T & 0 \end{bmatrix} \begin{pmatrix} \hat{\mathbf{x}} \\ \hat{\boldsymbol{\lambda}} \end{pmatrix} = \begin{pmatrix} \mathbf{b} \\ -\mathbf{d} \end{pmatrix}. \tag{32}$$

However, it is possible to rearrange this equation in a different way, using the relationship

$$\begin{pmatrix} G & -C \\ -C^T & 0 \end{pmatrix}^{-1} = \begin{pmatrix} G^{-1} - G^{-1}C(C^T G^{-1} C)^{-1} C^T G^{-1} & -G^{-1}C(C^T G^{-1} C)^{-1} \\ -(C^T G^{-1} C)^{-1} C^T G^{-1} & -(C^T G^{-1} C)^{-1} \end{pmatrix}$$

$$= \begin{pmatrix} H & -C^{*T} \\ -C^* & -(C^T G^{-1} C)^{-1} \end{pmatrix}. \tag{33}$$

Furthermore, if $\mathbf{x}$ is a point which satisfies the constraints (5), and $\mathbf{g}$ is the corresponding gradient, then the solution to (32) can be written, using (33), as

$$\hat{\mathbf{x}} = \mathbf{x} - H\mathbf{g} \tag{34}$$

and

$$\hat{\boldsymbol{\lambda}} = C^* \mathbf{g}. \tag{35}$$

These relationships are used by Fletcher (1971) as the basis of a quadratic programming algorithm, and the detailed analysis can be found there. The formulae can be thought of as projection in a metric G rather than a metric I as in Section 2; that is, if all the G^{-1} terms are replaced by I, then $H \to \hat{P}$, $C^* \to C^+$, and the Rosen gradient projection method is recovered. Equation (34) also has a clear relationship to Newton's method. The matrix H is of rank $n - m$ and contains the inverse curvature information corresponding to the independent variables. It can therefore be considered as a projected inverse Hessian. Fletcher (1970a) shows in fact that H is the generalized inverse of the projected Hessian; that is, $H = (\hat{P}G\hat{P})^+$, showing that H is being used to fulfill the functions of both inverse Hessian and projection. It will be noted that in general $H \neq \hat{P}G^{-1}\hat{P}$, as might have been expected on cursory inspection.

The equations so far derived in this section have concerned quadratic functions only, but are readily modified for general functions. An iterative method has obviously to be set up in which $G^{(k)}$ is evaluated at each iterate $\mathbf{x}^{(k)}$ and used to replace G in solving (32). Equally, (34) is replaced by the linear search process

$$\mathbf{x}^{(k+1)} = \mathbf{x}^{(k)} - \alpha H^{(k)} \mathbf{g}^{(k)}. \tag{36}$$

Finally, use of (35) now gives second-order estimates of Lagrange multipliers, rather than their exact values.

Two methods have also been suggested which use these equations in setting up quasi-Newton like methods for constrained minimization and which only require first derivatives of F to be computed. These are the methods of Goldfarb (1969) and of Murtagh and Sargent (1969, Method 2). Goldfarb uses the fact that if $H^{(0)}$ is a positive semidefinite matrix of rank $n - m$, but otherwise arbitrary, and if $\mathbf{x}^{(0)}$ satisfies the constraints (5), then application of the well-known DFP method (Fletcher and Powell, 1963) will in general minimize the quadratic function $q(\mathbf{x})$, subject to the constraints (5) in $n - m$ iterations. Furthermore, the ultimate matrix $H^{(n-m)}$ is the partition H in equation (33). A suitable $H^{(0)}$ is readily available: for instance, the matrix $H^{(0)} = \hat{P}$ which is equivalent to choosing $H^{(0)} = I$ in unconstrained optimization.

The other method due to Murtagh and Sargent, is again of quasi-Newton type, but instead of storing H directly, it is built up out of the components $(C^T G^{-1} C)^{-1}$, G^{-1}, and C as dictated by equation (33). The matrix H is then used to generate directions of search in the usual way. Murtagh and Sargent update an approximation to G^{-1} by a change of rank 1 at each iteration, and use the fact that a corresponding rank 1 formula can be derived by which to update $(C^T G^{-1} C)^{-1}$. Because the information is available to compute C^* as defined in (33), second-order estimates of the Lagrange multipliers are available in this method by using (35). Murtagh and Sargent show that their method minimizes a quadratic function subject to the constraints (5) in, at most, $n - m + 1$ iterations.

4. Inequality Constraints—the Active Set Strategy

In this section, generalizations will be described of the techniques of Sections 2 and 3 to deal with inequality constraints. These generalizations are all motivated in a very similar way, by what will be called the 'active set strategy'. The problem under consideration is now the general problem (1), and it will be assumed that an initial feasible point of this problem is available. Various techniques for doing this have been described, but I prefer those on the lines of Wolfe (1965), and also Fletcher (1970b) who gives a FORTRAN subroutine. The set of indices of the constraints (1b) for which equality holds, is collected and is the 'active set'. In the active set strategy the problem is treated as an equality problem, with equality constraints derived from the inequality constraints indexed by the active set. Minimization is carried out by any of the methods described in Sections 2 or 3. If the active set chosen corresponds to the set of constraints which are active at the solution of (1) then, hopefully, minimizing on this active set will cause convergence to the solution of (1).

However, usually the initial active set will not correspond to the final active set. Thus there have to be rules in the strategy in which the solution of one particular equality problem is discontinued, and constraint indices are either added to or deleted from the active set. This is done in such a way that the current approximation $\mathbf{x}^{(k)}$ is always feasible, and satisfies the constraints in the current active set as equalities. Of course, if the formulation (1) is extended to permit equality constraints in the first instance, then such constraints remain in the active set on every iteration. The rules for changing the active set will be discussed in the next few paragraphs. It is also important, especially with methods which do not require calculation of second derivatives, that the various operators (inverse transformation matrices, projection matrices, etc.) can be calculated efficiently when a change in the active set is made. Discussion of this topic will close the section.

The rule for adding a constraint to the active set is the obvious one. If the point $\mathbf{x}^{(k+1)}$ which would be obtained in a linear search process is not an unconstrained minimum but is restricted by the presence of a previously inactive constraint, then the index of the constraint which is causing this restriction is added to the active set. It will be noticed that the new $\mathbf{x}^{(k+1)}$ thus obtained satisfies the new constraint as an equality. This strategy generally works well, but has one disadvantage. When conjugate direction methods are in use, then any conjugacy relations which have been built up in one active set are no longer relevant in the new active set, and so the cycle of conjugate directions has to be restarted when a new constraint is added. This corresponds to throwing away useful information, and so is disadvantageous. The disadvantage is not shared by the methods which approximate inverse Hessians.

The choice of rule for removing constraints from the active set is much less apparent, and a bad choice can have quite disastrous effects on the rate of convergence. The simplest choice is merely to examine an estimate of the Lagrange multiplier vector, and by virtue of the Kuhn-Tucker condition (4d), to remove constraints (if any) for which $\lambda_i < 0$. Such a rule is suggested by Wolfe (1967) in connection with the reduced gradient method. In some methods it might be convenient to remove only one constraint, in which case the one with the most negative multiplier would be chosen. This is essentially the rule used in the simplex method for linear programming. However, when minimizing general functions subject to linear constraints, it has been noticed that simple rules like this can cause the phenomenon of 'zigzagging', in which the current active set oscillates between two (or more) different sets of integers. Wolfe (see Zoutendijk, 1970) has given an example with two constraints in which zigzagging between the constraints can cause an algorithm to fail to converge to a Kuhn-Tucker point; however, this example is somewhat unrepresentative, because the function is not twice differentiable along the intersection of the constraints. However, it is certainly true that methods which would normally converge at a superlinear rate if the correct active set were imposed, can converge at a linear rate due to the iteration zigzagging infinitely between two subsets of the correct active set (see again Zoutendijk, 1970).

Zigzagging occurs because the conditions under which a constraint is removed are not sufficiently stringent. Indeed, it could not possibly occur if one insisted on minimizing the function for the given active set before considering removal of any constraints. (Clearly so because there would then be no points with lower function value for which the active set were the same.) This feature is used in a number of quadratic programming algorithms to guarantee finite termination of the algorithm. However, it would be very wasteful when minimizing general functions to iterate to the minimum for a certain active set when it might become apparent at a much earlier stage that the active set was likely to be incorrect. Various *ad-hoc* rules have therefore come into being in an attempt to compromise between the two extremes above. For instance, Zoutendijk (1970) suggests that if a constraint enters the active set, then it should stay in the set for a number of iterations. Another strategy is to force a constraint to stay in the active set if it enters, leaves, and then re-enters the active set on successive iterations. Yet another strategy is only to examine multipliers with a view to removing constraints if an unconstrained minimum is found in the linear search. Although crude, strategies such as these work well in practice. However, I know of no proofs that zigzagging cannot occur in methods in which these strategies are used.

A more sophisticated approach is used by Rosen (1960), Goldfarb (1969), and Murtagh and Sargent (1969). They all follow a similar strategy, which is typified by the following. Consider first of all the reduction in the function likely to be achieved by keeping the same active set, and then consider the reduction in the function which could be achieved by removing a constraint from the active set. On neglecting terms of more than second order, and assuming that a minimum exists and is finite if the ith constraint is removed, then the corresponding reductions are

$$\tfrac{1}{2}\mathbf{g}^T H \mathbf{g} \tag{37}$$

and

$$\tfrac{1}{2}\mathbf{g}^T H \mathbf{g} + \tfrac{1}{2}\lambda_i^2 u_{ii} \tag{38}$$

where λ_i (<0) is the second-order estimate of the Lagrange multiplier, and u_{ii} is the ith diagonal element of $(C^T G^{-1} C)^{-1}$. Hence a suitable strategy is to find the integer q such that $\lambda_q^2 u_{qq}$ maximizes $\lambda_i^2 u_{ii}$ subject to $\lambda_i < 0$ and, assuming q exists, to drop the corresponding constraint only if

$$\mathbf{g}^T H \mathbf{g} < \lambda_q^2 u_{qq}. \tag{39}$$

I have no personal experience with this sort of strategy, but the authors who use it seem to have no zigzagging problems. However, neither of the above strategies have to my knowledge figured in proofs that zigzagging cannot occur.

Another strategy is suggested by McCormick (1969; 1970), in which the significant feature is that an iteration is not terminated when a direction of search meets a constraint. In the simple form (1969) of the algorithm, on meeting a constraint the linear search continues along what is essentially a projection of the search direction into the augmented active set. This process is continued either until a local minimum along a line is reached, or until the active set constraint matrix C has full rank. Only at this stage is the iteration terminated. McCormick finds it necessary only to examine first-order estimates of the multipliers to decide what constraints to drop from the active set. McCormick is able to prove that zigzagging cannot occur with this algorithm; that is to say, if the iteration is infinite, then there exists an iteration k' such that for all iterations $k \geqslant k'$, the active set is the same.

The final part of this section concerns techniques for manipulating the various operators such as inverse matrices, projection matrices, projected inverse Hessians, and so on which arise in the equality constraint problem. When changes occur in the active set, columns are either added to or deleted from the matrix C in (5), and it is important to examine how easily the operators corresponding to the new active set can be calculated from the old. For instance, in the reduced gradient method, if a constraint with normal c enters the active set, then c replaces one of the columns of D. To update the matrix $[C:D]^{-1}$ is then the well-known pivot operation of linear programming. In fact, it seems invariably possible to write down recurrence relations which allow the operators to be updated by making a change of rank 1 to the previous operator. The reader is referred to the source papers for details of such recurrence relations. These recurrence relations are an important feature of the algorithms, in that they enable the housekeeping to be done in $O(n^2)$ rather than $O(n^3)$ operations per iteration.

One feature of these formulae will be considered in more detail here. Consider the operator H used in Goldfarb's method, for instance, in which second-order information is represented. When a constraint with normal c is added to the active set, then a recurrence relation exists for H involving only H and c which preserves the second-order information in a satisfactory way. However, when a constraint is removed from the active set then the feasible region of the active set is extended, and H and c do not contain sufficient information to determine the new H in a satisfactory way. To determine H, therefore, information from other partitions of the inverse (33) is required (see Fletcher, 1971 for instance), and such information is not available in the implementation used by Goldfarb. Thus a more arbitrary rank 1 change must be made to H, and in fact it is usual to add in a rank 1 term which would give the correct H if G were a unit matrix. A similar problem exists with McCormick's method, which he avoids in a rather unsatisfactory way by resetting the reduced inverse Hessian to a unit matrix every time the active set changes. Murtagh and Sargent on the other hand, have an estimate of G^{-1}, and they have no difficulty in defining the correct recurrence relations whether constraints are added to or removed from the active set. Unfortunately, they restrict G to being positive definite, which is unrealistic, and this should be borne in mind when assessing the value of their approach.

The relative number of housekeeping operations required in the various methods is difficult to judge, because the comparisons differ according to whether one is adding or removing constraints from the basis, and whether rank 1 or rank 2 correction formulae are in use for updating Hessian matrices. Although all the methods lead to some multiple of n^2 operations per iteration, the methods which eliminate variables require the least housekeeping. Amongst the second-order methods the Murtagh and Sargent method seems to be the most expensive (assuming that a rank 1 updating formula is used by all methods). However, I would normally expect any such differences to be dominated by the cost of function and derivative evaluation and not to be of any great significance.

The effect of errors in the first-order methods has been studied indirectly, as it is now known how to represent and update inverse operators in linear programming by using a triangular decomposition of the basis matrix—see Gill and Murray (1970), for instance. This feature also makes the gradient-like methods suitable for dealing with problems in which the constraint matrix $\bar{C}$ in (1b) is sparse. The effect of errors in second-order methods has not been studied, and because inverse matrices are being represented directly, it is possible that the effect might be severe. Similarly, these second-order methods do not take into account any sparsity in either $\bar{C}$ or A. However, Murray (1970) has considered the application of triangular factors in quadratic programming, and it may be that his work has implications for algorithms which minimize a general function subject to linear constraints, especially in regard to preserving sparsity.

5. Methods Requiring the Repeated Solution of Linear or Quadratic Programs

In this section an approach will be described which attempts to dispense with the generalization of equality constraint methods through an active set strategy, and instead solves an inequality linear or quadratic program at each iteration. The reason for doing this is essentially to determine the active set automatically, rather than by using an exchange-type algorithm as in the active set strategy. The first two methods of this type are similar to the gradient-like methods of Section 2, as they apply to inequality constrained problems. It is intuitively attractive to choose the 'steepest feasible descent direction' as a direction of search; that is at any approximation $\mathbf{x}^{(k)}$, to find the vector $\mathbf{s}^{(k)}$ which solves

$$\max \{-\mathbf{g}^{(k)T}\mathbf{s}; C^T\mathbf{s} \geqslant \mathbf{0}, \mathbf{s}^T\mathbf{s} \leqslant 1\} \qquad (40; a, b)$$

where the columns of C correspond to the constraints of (1b) which are satisfied as equalities by $\mathbf{x}^{(k)}$. Equivalently (apart from a constant of proportionality) this can be written as

$$\min \{\mathbf{s}^T\mathbf{s}; C^T\mathbf{s} \geqslant \mathbf{0}, -\mathbf{g}^{(k)T}\mathbf{s} \geqslant 1\} \qquad (41; a, b)$$

which is a quadratic programming problem; although if those constraints in C of (41a) for which $C^T\mathbf{s}^{(k)} = \mathbf{O}$ were known beforehand, then the quadratic programming problem would reduce to a simple projection problem of the type described in Section 2. This result has been pointed out by a number of people, including Dennis (1959), Zoutendijk (1960) and Lemke (1961). Essentially, the difference between this approach and the gradient projection approach for inequalities is that the active set is determined directly for $\mathbf{x}^{(k)}$ by solving (41), rather than on the past history of the iterative process. As in Section 2, conjugacy of the Fletcher-Reeves type can be introduced into the scheme when the active set remains unchanged. Unfortunately, zigzagging can occur with this method, and anti-zigzagging precautions must be taken. This requires certain constraints to be forced to be equalities in an *ad-hoc* way, and therefore there seems to be no advantage in going to the expense of solving a quadratic programming problem to determine an 'optimum' active set.

A similar type of approach is described by Zoutendijk (1960; 1970), in which solution of (41) for $\mathbf{s}^{(k)}$ is replaced by solution of

$$\min \{\|\mathbf{s}\|; C^T\mathbf{s} \geqslant \mathbf{0}, -\mathbf{g}^{(k)T}\mathbf{s} \geqslant 1\} \qquad (42; a, b)$$

where $\|\cdot\|$ now refers to $\|\cdot\|_1$ or to $\|\cdot\|_\infty$. This is now a linear programming problem rather than the quadratic program (41). Conjugacy can be imposed by adding the conditions $-\mathbf{g}^{(j)T}\mathbf{s} = 1$ to (42b), where $j = k - 1, k - 2, \ldots$ are iterations on which the active set is unchanged. The method is such that a quadratic function subject to linear equality constraints is solved in $O(n^2)$ operations per iteration, and so the cost of having to solve a linear program is not prohibitive. Unfortunately the method has again to be supplemented by an anti-zigzagging device, and so the comments made in the previous paragraph, about imposing this *ad-hoc* strategy on what would otherwise be an optimum choice, still apply. In general the method is similar to those based on the ideas of Section 2, and I can see no special advantages in this type of approach. For instance, the need to throw away conjugacy conditions when the active set changes still applies. The method permits of savings when $\bar{C}$ and the vectors $\mathbf{g}^{(j)}$ are sparse. However, methods based on the reduced gradient or elimination of variables approach, together with conjugacy, require only $\bar{C}$ to be sparse to achieve similar savings.

Another class of methods arises by approximating the objective function at each $\mathbf{x}^{(k)}$ by a quadratic function $q^{(k)}(\mathbf{x})$, chosen so that

$$q^{(k)}(\mathbf{x}^{(k)}) = F^{(k)}, \qquad \nabla q^{(k)}(\mathbf{x}^{(k)}) = \mathbf{g}^{(k)}, \qquad \nabla^2 q^{(k)}(\mathbf{x}^{(k)}) = \Gamma^{(k)} \qquad \text{(43a, b)}$$

where $\Gamma^{(k)}$ is either the Hessian matrix $G^{(k)}$ if available, or an approximation to it. An obvious approach is then to find the point $\mathbf{y}$ which solves

$$\min \{q^{(k)}(\mathbf{x}); \bar{C}^T \mathbf{x} \geqslant \bar{\mathbf{d}}\}. \qquad (44)$$

This is a quadratic programming problem. The direction $\mathbf{s}^{(k)} = \mathbf{y} - \mathbf{x}^{(k)}$ can then be used as a search direction to determine the next approximation $\mathbf{x}^{(k+1)}$. I do not know if this approach has been tried, but it would seem to be worthy of consideration. It might be, however, that zigzagging would take place, because of the possibility of large changes being made to $\mathbf{x}^{(k)}$ on each iteration leading to oscillation. Again, if an anti-oscillation device had to be imposed upon such a method, it would contradict the philosophy behind the method of choosing the active set automatically.

The difficulties with methods which have been discussed so far is that (with one exception), the taking of large steps seems to lead to zigzagging, and furthermore the usual problems in carrying out a linear search have to be overcome. A quite different class of methods arises if the linear search and active set strategies are dispensed with. This can be done by making an approximation like (43) and by restricting the size of the correction to $\mathbf{x}^{(k)}$ so that $\mathbf{x}^{(k+1)}$ lies in a region in which $q^{(k)}(\mathbf{x})$ is a satisfactory approximation to $F(\mathbf{x})$. One way to do this is to add the constraint

$$\|\mathbf{x} - \mathbf{x}^{(k)}\|_\infty \leqslant h \qquad (45)$$

to the sub-problem (44), minimizing $q^{(k)}(\mathbf{x})$ subject to both (1b) and (45). This is still a quadratic sub-problem, and is the basis of the 'method of hypercubes' described by Fletcher (1970c). An important feature of the algorithm is that h is varied from one iteration to the next, so that the region defined by (45) is the largest possible region in which $q^{(k)}(\mathbf{x})$ and $F(\mathbf{x})$ agree to some prescribed extent. Convergence can be proved for this algorithm if $\nabla^2 q^{(k)} = G^{(k)}$ or if a certain rank 2 correction formula is used to maintain an approximation $\Gamma^{(k)}$ to $G^{(k)}$. In the latter case the method is advantageous, in that it is never necessary to reset $\Gamma^{(k)}$ to some arbitrary matrix. A most important feature of the algorithm however is that it is only necessary to bound $\|\Gamma^{(k)}\|$ to prove that zigzagging cannot occur with the algorithm. Hence no anti-zigzagging strategy need be imposed, and so no *ad-hoc* rules concerning the choice of active constraints need be made.

Similar considerations motivate extensions of Marquardt-Levenberg methods to solve linearly constrained problems. In these methods the restriction on the step is imposed in a different way, by adding a 'damping term' $\frac{1}{2}\lambda \|\mathbf{x} - \mathbf{x}^{(k)}\|_2^2$ to the function in the sub-problem (44). The size of the step

$\mathbf{x}^{(k+1)} - \mathbf{x}^{(k)}$ is now controlled by varying λ on each iteration. Some theorems about the method as it applies to least squares problems are given by Schrager (1970). However, control of the iteration using λ is more difficult than by using h as in Fletcher's scheme. Both implementations of methods in which steplength is restricted lead to quadratic programming sub-problems, and this necessitates using $O(n^3)$ housekeeping operations per iteration, whether or not second derivatives are available. Now if the methods based on the active set strategy take exact second derivatives into account, then they too use $O(n^3)$ operations per iteration, and so the restricted step methods are competitive in this aspect. However, if second derivatives are approximated using first derivative information, then the restricted step methods still require $O(n^3)$ housekeeping operations per iteration whilst those based on the active set strategy only $O(n^2)$. Thus whilst I prefer the restricted step methods in most respects the disadvantage over operations counts may limit their use in problems in which only first derivatives are available.

There are many other methods which solve linear programs at each iteration; for instance MAP of Griffith and Stewart (1961), which is similar to Fletcher's method except that a linear approximation rather than a quadratic approximation is made at each iteration. Also there are the 'decomposition' and 'cutting plane' methods—see Wolfe (1967) for example. However, none of these methods converge at a rate which is better than linear, and it is felt that this is a considerable disadvantage. Some of them also require very restrictive convexity conditions. All the other methods which have been discussed here in more detail converge in practice at a rate which is at least superlinear, and are quite general in their application.

References

Abadie, J., and Guigou, J. (1970). Numerical experiments with the GRG method. *In*: 'Integer and Non-linear Programming' (J. Abadie, ed.). North-Holland Publishing Co., Amsterdam.

Dennis, J. B. (1959). 'Mathematical Programming and Electrical Networks'. Wiley, New York.

Fletcher, R., and Powell, M. J. D. (1963). A rapidly convergent descent method for minimization. *Comput. J.* 6, 163–168.

Fletcher, R., and Reeves, C. M. (1964). Function minimization by conjugate gradients. *Comput. J.* 7, 149–154.

Fletcher, R. (1970a). Generalized inverses for non-linear equations and optimization. *In*: 'Numerical Methods for Non-linear Algebraic Equations' (P. Rabinowitz, ed.). Gordon and Breach, London.

Fletcher, R. (1970b). 'The Calculation of Feasible points for Linearly Constrained Optimization Problems'. UKAEA Research Group report, A.E.R.E.-R.6354.

Fletcher, R. (1970c). 'An Efficient, Globally Convergent, Algorithm for Unconstrained and Linearly Constrained Optimization Problems'. Paper presented at the 7th International Mathematical Programming Symposium, The Hague, 1970, and in report T.P. 431.

Fletcher, R. (1971). A general quadratic programming algorithm. *J. Inst. Maths. Applics* **7**, 76–91.

Gill, P. E., and Murray, W. (1970). 'A Numerically Stable Form of the Simplex Algorithm'. National Physical Lab. report Maths 87.

Goldfarb, D. (1969). Extension of Davidon's variable metric algorithm to maximization under linear inequality and equality constraints. *SIAM J. appl. Maths.* **17**, 739–764.

Griffith, R. E., and Stewart, R. A. (1961). A non-linear programming technique for the optimization of continuous processing systems. *Managmt Sci.* **7**, 379–392.

Lemke, C. E. (1961). The constrained gradient method of linear programming. *SIAM J.* **9**, 1–17.

McCormick, G. P. (1969). Anti zigzagging by bending. *Managmt Sci.* **15**, 315–320.

McCormick, G. P. (1970). The variable reduction method for nonlinear programming. *Managmt. Sci.* **17**, 146–160.

Murray, W. (1969). 'Indefinite Quadratic Programming'. National Physical Lab. report Ma 76.

Murray, W. (1970). 'An Algorithm to find a Local Minimum of an Indefinite Quadratic Program'. Paper presented at the 7th International Mathematical Programming Symposium, The Hague, 1970, and in NPL report DNAC 1 (1971).

Murtagh, B. A., and Sargent, R. W. H. (1969). A constrained minimization method with quadratic convergence. *In*: 'Optimization' (R. Fletcher, ed.). Academic Press, London.

Rosen, J. B. (1960). The gradient projection method for non-linear programming, Part I, linear constraints. *SIAM J.* **8**, 181–217.

Schrager, R. I. (1970). Non-linear regression with linear constraints: an extension of the magnified diagonal method. *J. ACM.* **17**, 446–452.

Wolfe, P. (1965). The composite simplex algorithm. *SIAM. Rev.* **7**, 42–54.

Wolfe, P. (1967). Methods of non-linear programming. *In*: 'Non-linear Programming' (J. Abadie, ed.). North-Holland Publishing Co., Amsterdam.

Zoutendijk, G. (1960). 'Methods of Feasible Directions'. Elsevier, Amsterdam.

Zoutendijk, G. (1970). Non-linear programming, computational methods. *In*: 'Integer and Non-linear Programming' (J. Abadie, ed.). North-Holland Publishing Co., Amsterdam.

21. A Gradient Projection Algorithm for Non-Linear Constraints

J. B. ROSEN AND J. KREUSER
*University of Minnesota, Minneapolis, Minnesota, U.S.A.
and University of Wisconsin, Madison, Wisconsin, U.S.A.*

Summary

Among the most appealing methods for non-linear constraint problems, from a theoretical standpoint, are various versions of the cutting plane and penalty function methods. These methods, however, tend to give computational difficulties as an optimum point is approached, due to ill-conditioning of the matrices involved. Such difficulties appear to be largely eliminated by the algorithm summarized below, even though it makes use of both cutting planes and a penalty function. This algorithm is designed primarily for convex problems where a convex function is minimized over a convex domain defined by linear and non-linear constraints. It is motivated by the existence of efficient computational algorithms for convex linearly constrained problems (Fletcher, 1971), and reduces the original problem to a sequence of such problems (major iterations). A major iteration is generated by linearizing each non-linear constraint about the current (infeasible) point, and adding to the objective function a linear external penalty for each violated non-linear constraint. The resulting function is essentially the Lagrangian corresponding to these violated constraints.

A similar penalty is used by Kelley and Speyer (1970) in their improvement of an earlier method (Rosen, 1961), and by Lill (1971). For the algorithm summarized below, a Kantorovich-type theorem is given, showing quadratic convergence in terms of major iterations. Computational results verify this quadratic convergence even when some of the assumptions of the theorem are not satisfied.

1. Algorithm

The problem considered may be stated as

$$\min_{x} \{\phi_0(x) | \phi_j(x) \leq 0, j = 1, \ldots, m\} \tag{1}$$

where $x \in E^n$, the $\phi_j: E^n \to E^1$, are convex and differentiable, $j = 0, 1, \ldots, m$, and an optimal solution x^* to (1) is assumed to exist. For any fixed x^k we define the linearization of $\phi_j(x)$ about x^k as

$$h_j^k(x) \equiv \phi_j(x^k) + \nabla\phi_j'(x^k)[x - x^k], \qquad j = 1, \ldots, m, \qquad (2)$$

where $\nabla\phi$ denotes the gradient (column) vector and $\nabla\phi'$ is its transpose. Also, define the set of indices $I(x) \equiv \{j | \phi_j(x) \geq 0\}$, and the corresponding Jacobean matrix

$$J = J(x) = [\nabla\phi_j(x)]_{j \in I(x)}. \qquad (3)$$

Finally, we denote by $J^+(x)$ the generalized inverse of $J(x)$, where $J^+ = (J'J)^{-1}J'$, provided $J'J$ is non-singular. A somewhat simplified statement of the algorithm is as follows:

Step 0. Start with arbitrary point x^0 (feasible if known).
Step 1. Given x^k compute $\lambda^k = -J^+(x^k) \nabla\phi_0(x^k)$.
Step 2. Compute x^{k+1} as optimum point of

$$\min_x \left\{ \phi_0(x) + \sum_{\lambda_j^k > 0} \lambda_j^k \phi_j(x) | h_j^k(x) \leq 0, j = 1, \ldots, m \right\}. \qquad (4)$$

Step 3. Are Kuhn-Tucker conditions for (1) satisfied?
    No: $x^{k+1} \to x^k$, go to step 1.
    Yes: $x^* = x^{k+1}$.

To prove convergence, the following additional definitions and assumptions are needed. We let $f(x)$ denote a column vector with components $\phi_j(x)$, $j \in I(x)$, and let $H_j(x)$ denote the Hessian matrix of $\phi_j(x)$. We assume that in an appropriate neighbourhood of x^* the quantities $\|H_j\|$, $\|J\|$, $\|(J'J)^{-1}\|$ and $\text{cond}(H_j)$ are uniformly bounded. For simplicity we also assume that $\phi_0(x)$ is linear ($H_0 \equiv 0$).

2. Convergence Theorem

Let x^0 be an exterior point with $\|x^0 - x^*\| \leq \beta_1$, and $\|f(x^0)\| \leq 1/2\beta_2$. Then $\{x^k\} \to x^*$, with

$$\|f(x^k)\| \leq \beta_2 \|f(x^{k-1})\|^2 \leq 2\|f(x^0)\|(0 \cdot 5)^{2k}$$

and

$$\|x^{k+1} - x^*\| \leq \|x^{k+1} - x^k\| \leq \beta_3 \|x^k - x^{k-1}\|^2 \leq \beta_4 \|f(x^0)\|(0 \cdot 5)^{2k}$$

for $k = 1, 2, \ldots$. The constants $\beta_1, \beta_2, \beta_3, \beta_4$ are given explicitly in terms of the previously assumed uniform bounds.

3. Experience

The algorithm has been tested computationally on a number of quadratic constraint test problems. A code (Kreuser, 1971) based on Goldfarb's synthesis of Rosen's linear constraint gradient projection and Davidon-Fletcher-Powell unconstrained minimization was used to solve (4) at each major iteration. The largest such problem solved so far consisted of 30 variables and 30 quadratic constraints, of which 20 were active at the optimum point. All the assumptions of the convergence theorem were not necessarily satisfied by the test problems. In particular the starting point was farther away from x^* than required by the theorem, but quadratic convergence was obtained nevertheless. This behaviour is similar to that of Newton's method for which one often gets convergence even when the sufficient conditions are not satisfied.

Typical computational behaviour is illustrated by the following tabulated results. The problem consisted of 15 variables and 15 ellipsoidal constraints, of which 10 were active at the optimum.

k	Constraints violated	$\|f(x^k)\|$	$\|x^k - x^*\|$	$\|\lambda^k - \lambda^*\|$	$\phi_0^k - \phi_0^*$	Gradient evaluations	Standard time
0	0	0	0·387 + 1	0·344 + 1	0·920 + 4	1	0·000
1	15	0·477 + 5	0·155 + 2	0·344 + 1	−0·368 + 5	17	0·019
2	15	0·114 + 5	0·620 + 1	0·275 + 1	−0·147 + 5	44	0·147
3	15	0·243 + 4	0·191 + 1	0·212 + 1	−0·453 + 4	59	0·222
4	15	0·329 + 3	0·315 + 0	0·114 + 1	−0·747 + 3	73	0·296
5	10	0·953 + 1	0·118 − 1	0·259 + 0	−0·281 + 2	98	0·372
6	3	0·141 − 1	0·444 − 4	0·105 − 1	−0·427 − 1	116	0·433
7	0	0·101 − 2	0·408 − 4	0·755 − 4	−0·195 − 2	119	0·441

An interior starting point ($x^0 = 0$) was chosen. The linearization about an interior point for convex constraints gives x^1 as an exterior point (in fact, all 15 constraints are violated). The sequence of points then converges from the exterior of the domain as shown. The optimal function value is 9203, so that the relative error in ϕ_0^7 is approximately $1 \cdot 5 \times 10^{-7}$. These calculations were done on a Univac 1108, for which we have time (sec) = 26 × (standard time). A gradient evaluation consists of computing the gradient of the Lagrangian function, i.e., the objective function in (4).

Details and proofs, together with more comprehensive test results, will appear in Kreuser's thesis (1972).

References

Fletcher, R. (1971). Minimizing general functions subject to linear constraints, in this volume (pp. 279).

Kelley, H. J., and Speyer, J. L. (1970). Accelerated gradient projection. *In*: 'Proceedings of Colloquium on Optimization'. Lectures on Mathematics 132. Springer-Verlag.

Kreuser, J. (1971). Gradient Projection Code Manual. University of Wisconsin Computing Center.

Kreuser, J. (1972). Convergence rates for nonlinear constraint Lagrangian methods. Ph.D. Thesis, Computer Sciences Dept., University of Wisconsin.

Lill, S. A. (1971). Generalization of an exact method for solving equality constrained problems to deal with inequality constraints, in this volume (pp. 303).

Rosen, J. B. (1961). The gradient projection method for nonlinear programming. Part II, nonlinear constraints. *J. Soc. ind. appl. Math.* **9**, 514–553.

22. Pseudo-Complementary Algorithms for Mathematical Programming

ULRICH ECKHARDT

Zentralinstitut für Angewandte Mathematik,
Kernforschungsanlage Jülich, Jülich, Germany

Summary

Lemke's complementary pivot theory (Lemke, 1968; Cottle and Dantzig, 1968) turned out to be a useful tool in theory and practice of mathematical programming. This paper presents a generalized version of that theory and some numerical examples.

1. Statement of the Problem and Definitions

Consider the following problem:

$$\text{Maximize } \langle c, x \rangle$$
$$\text{Subject to } \langle a_t, x \rangle \geq b_t \quad \text{for all } t \in T. \tag{1}$$

Here $\langle \cdot, \cdot \rangle$ denotes the scalar product in R^d, $x \in R^d$, $c \in R^d$, $a_t \in R^d$, $b_t \in R$ for all $t \in T$, where T is any index set (not necessarily finite).

The dual problem in the sense of Charnes et al. (1969) is given by

$$\text{Minimize } \sum_{j=1}^{k} y_{t_j} b_{t_j}$$
$$\text{subject to } \sum_{j=1}^{k} y_{t_j} a_{t_j} = c, \quad y_{t_j} \geq 0 \quad \text{for all } j = 1, \ldots, k, k \text{ finite.} \tag{2}$$

Definition 1

A basis of the problem pair (1), (2) is any d-tuple $B = (t_1, \ldots, t_d)$, $t_j \in T$, such that the vectors $a_{t_1}, \ldots, a_{t_d}$ are linearly independent.

Definition 2

Let B be a basis. The basic solution corresponding to B is the unique solution x of

$$\langle a_{t_j}, x \rangle = b_{t_j} \quad \text{for all } t_j \in B.$$

The dual basic solution corresponding to B is the unique solution $y_{t_1}, \ldots, y_{t_d}$ of

$$\sum_{t_j \in B} y_{t_j} a_{t_j} = c.$$

We define

$$y_t = \begin{cases} 0, & \text{if } t \in T - B, \\ y_{t_j}, & \text{if } t = t_j \in B. \end{cases}$$

In order to express the dependence on B we will sometimes denote the basic solutions corresponding to B by $x(B)$ and $y_t(B)$, respectively.

Definition 3

$P = \{x \in R^d | \langle a_t, x \rangle \geq b_t \text{ for all } t \in T\}$ is the set of *feasible* vectors of (1). Let $y_t \neq 0$ only for a finite set of values $t_1, \ldots, t_k$. The function y_t is called dual feasible if $y_t \geq 0$ for all $t \in T$ and

$$\sum_{j=1}^k y_{t_j} a_{t_j} = c.$$

Definition 4

$x \in R^d$ is termed a $\bar{b}_t$-feasible solution of (1) if

$$\langle a_t, x \rangle \geq b_t + \bar{b}_t \qquad \text{for all } t \in T.$$

2. The Algorithm

Given a basis B_r for any r with basic solutions $x_r = x_r(B_r)$ and $y_t^{(r)} = y_t^{(r)}(B_r)$, where $y_t^{(r)}$ is dual feasible. Let $\bar{x}_r$ be $\bar{b}_t^{(r)}$-feasible, $\{\gamma_r\}$ a sequence with $0 \leq \gamma_r \leq 1$ for all r.

(1) Define for any $x \in R^d$ the following:

$$d_x^{(r)}(t) = \langle a_t, x \rangle - b_t - \bar{b}_t^{(r)},$$

$$d_x(t) = \langle a_t, x \rangle - b_t,$$

and

$$V_\lambda(t) = d_{\bar{x}_r}^{(r)}(t) + \lambda \cdot [d_{x_r}(t) - d_{\bar{x}_r}^{(r)}(t)].$$

Select λ_r, $t_r \in T$ (if possible) such that

$$\left.\begin{array}{l} \text{(a) } 0 \leq \lambda_r \leq 1, \\ \text{(b) } V_{\lambda_r}(t) \geq 0 \qquad \text{for all } t \in T, \\ \text{(c) } V_{\lambda_r}(t_r) = 0. \end{array}\right\} \qquad (3)$$

If no such λ_r, t_r exist or if $t_r = t_{r+1} = \cdots$ the algorithm is said to be *impracticable*.
(2) Set
$$\bar{x}_{r+1} = \bar{x}_r + \gamma_r \lambda_r (x_r - \bar{x}_r) \tag{4}$$
$$\bar{b}_t^{(r+1)} = (1 - \gamma_r \lambda_r) \bar{b}_t^{(r)}.$$

Then $\bar{x}_{r+1}$ is $\bar{b}_t^{(r+1)}$-feasible and $\max_t |\bar{b}_t^{(r+1)}| \leq \max_t |\bar{b}_t^{(r)}|$. That means $\bar{x}_{r+1}$ is 'more feasible' than $\bar{x}_r$.

(3) Choose any dual feasible basis B_{r+1} by means of B_r and t_r according to a suitable selection rule. We give two examples for such a rule:

SR1: $B_{r+1} = B_r \cup \{t_r\} - \{t'\}$, $t' \in B_r$ chosen by the usual simplex pivot choice rule such that B_{r+1} is dual feasible.

SR2: B_{r+1} is the optimal basis for the finite linear programming problem:

Maximize $\langle c, x \rangle$

Subject to $\langle a_{t_j}, x \rangle \geq b_{t_j}$, $\quad j = 0, \ldots, r+1$.

The basis B_{r+1} gives us x_{r+1} and $y_t^{(r+1)}$.

(4) Apply an appropriate test for termination, for example, B_{r+1} feasible or $|\bar{b}_t^{(r+1)}|$ small enough. If it fails, go to (1).

Convergence, and even practicability, of the algorithm cannot be proved under these quite general assumptions. In the next section we consider some special cases.

For further reference the above algorithm is denoted by $A(\bar{x}_0, B_0, \gamma_r, i)$, when it is applied to the starting basis B_0, the approximate solution $\bar{x}_0$ with pivot selection rule SRi, $i = 1, 2$, and the relaxation coefficients γ_r.

3. Special Cases and Applications

3.1. *The Finite Case*

If T is finite, termination of the algorithm with SR1 can be proved by standard nondegeneracy arguments (Eckhardt, 1970). Termination means that the algorithm will stop after a finite number of steps with a solution of the problem pair (1), (2) or the indication that no solution exists.

We give a list of some special cases:

(1) If $\bar{x}_0 = \bar{x}_0(B')$ is a feasible basic solution for some basis B' with $B' - B_0 = \{i\}$, then $A(\bar{x}_0, B_0, 1, 1)$ generates the almost complementary path determined by B_0 and B' in the sense of Lemke (1968).

(2) The algorithm $A(\bar{x}_0, B_0, 1, 1)$ for any $\bar{x}_0$ has the following properties:

(a) After a finite number of steps $\bar{x}_r$ is feasible.
(b) After a finite number of steps the situation described in (1) occurs.

(3) When $\bar{x}_0$ is feasible, the algorithm $A(\bar{x}_0, B_0, \gamma_r, 1)$ has the following additional properties:

  (a) For each r we have:

$$\langle c, \bar{x}_r \rangle \leq \text{optimal value} \leq \langle c, x_r \rangle.$$

  (b) Any hyperplane $\langle a_{t_r}, x \rangle = b_{t_r}$ selected according to (3) is non-redundant to P.

  (c) The point $\bar{x}_r$ can be computed with the original data of the problem by means of numerically well behaved operations. Thus, rounding errors have only little influence on $\bar{x}_r$.

For further properties of the algorithm and complete proofs, see Eckhardt (1970).

If no dual feasible basis B_0 is known we define approximate dual feasibility as above and construct an algorithm which applies to this situation. So we can start with an approximate dual $\bar{c}$-feasible basis and iterate such that $\bar{c}$-feasibility is maintained until after a finite number of steps x_r is feasible. In a second phase the algorithm is applied to the dual problem, in order to obtain the solution. It is also possible to find a symmetric method such that at each step the basis B_r becomes 'more feasible' and 'more dual feasible'. An example of such a method is Lemke's 'Scheme I' (Lemke, 1968) or Ravindran's method (1970).

The algorithm can be formulated in a compact simplex version such that the mean iteration time in our experiments exceeded that of ordinary simplex method by less than 1%, or was even below it (see also Ravindran, 1970).

An extensive numerical investigation of Lemke's method has been carried out by Ravindran (1970). It turned out that Lemke's algorithm required only half to one-third of the number of iterations of the simplex method if the coefficient matrix has only non-negative elements. If the matrix has more general elements no such significant difference was found. Souami (1969) gives similar results for matrices with non-negative elements. For a set of 175 randomly generated test problems of various dimensions the algorithm $A(\bar{x}_0, B_0, 0, 1)$ with feasible $\bar{x}_0$ required on the average only one-third of the number of iterations of the simplex method.

Usually, a good approximate solution of the problem is not readily available. However, in many applications of linear programming in numerical analysis (see for example Rabinowitz, 1968) much *a priori* information on the problem is available which can be effectively used by a pseudo-complementary algorithm. In the next section we give an example for direct application of Lemke's method.

A quite different field of application is the quadratic programming problem (Eaves, 1971; Cottle and Dantzig, 1968). In this paper, however, we will not be concerned with that subject.

3.1.1. Finding Bounds for an Eigenvalue with Non-negative Eigenvector of a Matrix.

Let A be a (d,d)-matrix, x a d-vector, and ρ a number. We are looking for a solution ρ, x (if any) of

$$Ax - \rho x = 0, \qquad x \geq 0 \tag{5}$$

with $\langle e_0, x \rangle = 1$, $e_0 = (1, \ldots, 1) \in R^d$. Existence of such a solution is assured under certain conditions, namely when A has positive elements and is irreducible, by a well-known theorem of Frobenius.

Let N be the set of all ρ for which (5) has a solution. The matrix $B_\rho = (A - \rho I)^T(A - \rho I)$ is positive semi-definite. Using this fact, Motzkin's theorem (see Mangasarian, 1969) yields either the system

$$B_\rho x > 0, \; x \geq 0 \tag{6}$$

has a solution x, or

$$B_\rho y = 0, \; y \geq 0, \; \langle e_0, y \rangle = 1 \tag{7}$$

has a solution y, but never both.

The system (7) is equivalent to (5) by definition of B_ρ. Now, given $x \geq 0$ we introduce

$$I(x) = \{\rho \mid x \text{ is no solution of (6)}\}.$$

Then $N \subset I(x)$. The set $I(x)$ can easily be computed in the following way. Writing $A^T A x = (\alpha_i)_{i=1}^d$, $(A^T + A)x = (\beta_i)_{i=1}^d$, and $x = (x_i)_{i=1}^d$, (6) becomes

$$\alpha_i - \rho \beta_i + \rho^2 x_i > 0, \qquad i = 1, \ldots, d.$$

With $\gamma_i = \beta_i^2 - 4\alpha_i x_i$ define the interval

$$I_i = \begin{cases} \left[\dfrac{1}{2x_i}(\beta_i - \sqrt{\gamma_i}), \dfrac{1}{2x_i}(\beta_i + \sqrt{\gamma_i}) \right], & \text{if } x_i > 0, \gamma_i \geq 0, \\ (-\infty, +\infty), & \text{if } x_i > 0, \gamma_i < 0, \\ [\alpha_i/\beta_i, +\infty), & \text{if } x_i = 0. \end{cases}$$

Then $I(x) = \bigcup_{i=1}^d I_i$.

Now we proceed as follows. For $x_0 \geq 0$ compute $I(x_0)$. Select $\rho_0 \in I(x_0)$ and find a solution x_1 of (6) for $\rho = \rho_0$. Then $\rho_0 \notin I(x_1)$ and $N \subset I(x_0) \cap I(x_1)$.

In order to solve (6) for a fixed ρ we are led to the following linear programming problem:

$$\text{Maximize } (-\gamma)$$
$$\text{Subject to } B_\rho x + \gamma e_0 - v = 0,$$
$$\langle e_0, x \rangle = 1,$$
$$x \geq 0, \qquad v \geq 0.$$

The dual problem is

$$\text{Minimize } \delta$$
$$\text{Subject to } B_p y - \delta e_0 + u = 0,$$
$$\langle e_0, y \rangle = 1$$
$$y \geq 0, \quad u \geq 0.$$

For this pair an almost complementary solution in the sense of Lemke is:

$$x_1 = 1, \quad x_j = 0 \quad \text{for } j \neq 1,$$
γ minimal such that $v_k = 0, v_j \geq 0$ for all j,
$$y_k = 1, \quad y_j = 0 \quad \text{for } j \neq k,$$
δ minimal such that $u_i = 0, \quad u_j \geq 0$ for all j.

Starting with this almost complementary solution we get a solution with $\gamma < 0, \gamma = 0$ if, and only if, $\rho \in N$.

There are many methods for solving problem (5). It should be remarked, however, that the described method does not depend on the multiplicity of the eigenvalues $\rho \in N$ or the position of the other eigenvalues with respect to N.

We consider two examples:

Example 1

$$A = \begin{bmatrix} 1 & 0 & 0 & 1 \\ 2 & 1 & 0 & 0 \\ 0 & 2 & 1 & 0 \\ 0 & 0 & 2 & 1 \end{bmatrix}$$

Hall and Porsching (1968) gave this as an example for a matrix with bad behaviour. The unique solution is $\rho = 2\cdot 681\ 787$. If we put $x = \frac{1}{4} e_0$ this gives immediately the interval

$$I(x) = [1\cdot 381, 3\cdot 619].$$

With $\rho_1 = 3\cdot 619$, and x_1 the solution of (6) obtained as described above, we get

$$I(x_1) = [1\cdot 438, 3\cdot 279],$$

and $\rho_2 = 3.279$ yields

$$I(x_2) = [1\cdot 471, 3\cdot 121].$$

Now at each step we put ρ_r^l as the lower bound of the $(r-1)$st ρ-interval and ρ_r^u as the upper bound of it and perform the computation parallel for both values. The results for 12 steps are listed in Table 1.

TABLE 1
Bounds for the maximal eigenvalue of
Example 1

r	Lower bound	Upper bound
0	1·381	3·619
1	1·438	3·279
2	1·471	3·121
3	1·777	3·029
4	2·003	2·969
5	2·166	2·926
6	2·280	2·894
7	2·362	2·870
8	2·421	2·850
9	2·464	2·834
10	2·497	2·821
11	2·522	2·810

Example 2

$$A = [a_{ij}] = \left[\frac{1}{(i+j-1)!}\right]_{i,j=1}^{4}.$$

The solution is $\rho = 1 \cdot 258\ 319$. Here we started with $\rho_0 = 1$ and ρ_r as the upper bound of the $(r-1)$st ρ-interval. The results are listed in Table 2.

TABLE 2
Bounds for the maximal eigenvalue of
Example 2

r	Lower bound	Upper bound
0	1·115	1·439
1	1·183	1·343
2	1·219	1·300
3	1·238	1·279
4	1·248	1·269
5	1·253	1·264
6	1·255	1·261
7	1·257	1·260
8	1·257	1·259

3.2. *The Continuous Case—Infinite T*

The algorithm does not work in the general version given above. In order to assure practicability we make the following assumptions.

(1) The point $\bar{x}_0$ is *strictly* feasible, that means $d_{\bar{x}_0}(t) \geq a > 0$ for all $t \in T$. If $d_{\bar{x}_0}(t') = 0$ for any $t' \in T$ then it is possible that $t_0 = t_1 = \cdots = t'$. According to the fact that non-degeneracy arguments are not applicable for infinite T this situation must be avoided.

(2) It must be true that $0 \leq \gamma_r < 1$ for all r, to maintain strict feasibility for all x_r.

(3) The set T is compact and $d_x(t)$ is continuous in t for any feasible x. This assures existence of t_r in problem (3).

Some special conditions depending on the pivot selection rule are needed to assure convergence. The algorithm has the properties listed under (3) in Section 3.1.

3.2.1. *The Convex Programming Problem.* It is a well-known consequence of the separation theorem for closed convex sets that the convex programming problem

$$\text{Maximize } \langle c, x \rangle$$

$$\text{Subject to } x \in K,$$

where K is a closed convex set in R^d, can be formulated, under quite general conditions, as a linear programming problem with an infinite number of constraints (see Charnes *et al.*, 1969). Here, a dual feasible basis is defined by a set of d linearly independent hyperplanes meeting K at most at boundary points. Given a basis B_r, problem (3) means the construction of a supporting hyperplane at the intersection point $V(\lambda_r)$ of the line segment $V(\lambda)$ and the boundary of K. We consider two cases.

1. Complete pivoting—selection rule SR2. Under appropriate topological conditions convergence can be proved. Kleibohm (1966) and Veinott (1967) proved convergence for algorithm $A(\bar{x}_0, B_0, 0, 2)$. If $0 \leq \gamma_r = \gamma < 1$ for all r we have the MFD method of Zoutendijk (1966). The linear programming problem of SR2 may be solved by a pseudo-complementary algorithm.

2. Pivot choice rule SR1. This algorithm requires a simplex scheme of fixed size, in contrast to the methods described in Section 1, above. However, it does not converge in all cases. Numerically, the situation of non-convergence is indicated by the observation that $\langle c, x_r - \bar{x}_r \rangle$ does not converge to zero. It means that the dual iteration steps are nearly degenerated and that the points $\bar{x}_r$ tend to the boundary of P at the same time. This case seems to be rare in practice. It can be circumvented by taking some complete pivoting steps, for example, or by occasionally choosing $\bar{x}_r$ in a manner different from equation (4) so that it will leave the boundary. This can be achieved by means of an appropriate gradient method in the manner of Zoutendijk's anti-zigzagging precautions. Choosing $\gamma_r = \lambda_r$ seems to be a good strategy in order to circumvent this phenomenon and to speed up convergence.

Numerical experience indicated a good behaviour of the algorithm. We give three simple examples (see Collatz and Wetterling, 1971).

Example 1

$$\text{Maximize } -2x_1^2 - x_2^2 + 48x_1 + 40x_2 = z$$
$$\text{Subject to } x_1 + x_2 \leq 8,$$
$$x_1 \leq 6,$$
$$x_1 + 3x_2 \leq 18,$$
$$x_1, x_2 \geq 0.$$

Starting solutions:

$$\bar{x}_0 = \begin{pmatrix} 0 \\ 0 \end{pmatrix}, \quad \bar{z}_0 = -10,$$

$$x_0 = \begin{pmatrix} 6 \\ 2 \end{pmatrix}, \quad z_0 = 368.$$

Desired accuracy: $z_r - \bar{z}_r \leq 10^{-4}$.

TABLE 3
Number of iterations for different choices of γ_r (Example 1)

γ_r	Number of iterations
0·4	29
0·45	29
0·5	26
0·55	26
0·6	25
λ_r	20
λ_r^2	18
λ_r^3	18

Example 2

$$\text{Maximize } x_1 = z$$
$$\text{Subject to } \exp(x_1) - x_2 \leq 0,$$
$$\exp(x_2) - x_3 \leq 0,$$
$$x_3 - 10 \leq 0,$$
$$x_1, x_2, x_3 \geq 0.$$

Starting values:

$$\bar{x}_0 = \begin{pmatrix} 0 \\ 1{\cdot}05 \\ 2{\cdot}9 \end{pmatrix}, \quad x_0 = \begin{pmatrix} 2 \\ 3 \\ 0 \end{pmatrix}.$$

Desired accuracy: $z_r - \bar{z}_r \leqq 10^{-4}$.

TABLE 4
Number of iterations for different choices of γ_r (Example 2)

γ_r	Number of iterations
0·4	27
0·45	25
0·5	30
0·55	56
λ_r	15

Example 3

Maximize $0{\cdot}8x_1 - 0{\cdot}2x_3 = z$ under the same restrictions as in Example 2. The starting solutions and the desired accuracy are the same as in Example 2.

TABLE 5
Number of iterations for different choices of γ_r (Example 3)

γ_r	Number of iterations
0·4	16
0·5	13
0·6	12
0·65	11
0·7	10
0·75	9
0·8	9
0·85	12
0·9	318
λ_r	10
λ_r^2	10

3.2.2. *The Linear Chebyshev Approximation Problem.* Given a function $f(t)$ on a compact set and a finite set of functions $v_j(t)$, then the problem can be formulated as follows (Collatz and Wetterling, 1971).

Minimize ρ

Subject to $\sum x_j v_j(t) - \rho \leq f(t)$

$-\sum x_j v_j(t) - \rho \leq -f(t)$.

In this case the algorithm can be interpreted as a modified Remez-algorithm (Meinardus 1964). Convergence could not be proved, but the algorithm can be applied to improve a given approximate solution.

As an example we consider the approximation of $f(t) = |t^3|$ by a polynomial $\phi(t) = x_1 t^4 + x_2 t^2 + x_3$ in the interval $[-1, 1]$ (Meinardus, 1964). Starting with the mean square-error solution $\bar{x}_0$, a basis B_0 is given by the set of extrema of $d(t) = f(t) - \phi(t)$. The result of one iteration step of algorithm $A(\bar{x}_0, B_0, 1, 1)$ is listed in Table 6. The point x_1 is a solution to the problem with an accuracy of six decimals.

TABLE 6
Bounds for the maximal deviation of the linear Chebyshev approximation problem

r	$\bar{\rho}_r$	ρ_r	$\max|d(t)|$
0	0·015 625	0·008 829	0·009 041
1	0·008 993	0·008 883	0·008 884

If the set T is finite, we can apply a pseudo-complementary algorithm to this problem in a quite natural way. According to computational experience there seems no significant reduction of the number of iteration steps as compared with the usual simplex algorithm. In any case, however, we have the advantages of the algorithm described in Section 3.1.

4. Conclusions

The class of pseudo-complementary algorithms described in this paper contains some well-known methods for mathematical programming, especially the complementary pivot algorithm of Lemke (1968). In contrast to the pure combinatorial approach of Lemke the methods given here are motivated geometrically. This gives a better understanding of the properties of these methods. Relaxing the idea of complementarity in the described way leads to a rich variety of algorithms which can be well adapted to known properties of the given problem.

In the infinite case, the cutting plane methods for convex programming of Kleibohm (1966), Veinott (1967) and Zoutendijk (1966) are members of the class of pseudo-complementary algorithms.

References

Charnes, A., Cooper, W. W., and Kortanek, K. O. (1969). On the theory of semi-infinite programming and a generalization of the Kuhn-Tucker saddle point theorem for arbitrary convex functions. *Nav. Res. Log. Quart.* **16**, 41–51.

Collatz, L., and Wetterling, W. (1971). 'Optimierungsaufgaben'. Second edition. Springer-Verlag, Berlin, Heidelberg, and New York.

Cottle, R. W., and Dantzig, G. B. (1968). Complementary pivot theory of mathematical programming. *In*: 'Mathematics of the Decision Sciences, Part I.' Lectures in Applied Mathematics, Vol. 11. (G. B. Dantzig and A. F. Veinott, eds.), pp. 115–136. AMS, Providence.

Eaves, B. C. (1969). The linear complementary problem in mathematical programming. Technical Report No. 69–4, Stanford University.

Eckhardt, U. (1970). Fastkomplementäre Iterationspfade und Teilprobleme beim linearen Programmieren. *In*: 'Methods of Operations Research, VIII'. (R. Henn, H. P. Künzi, and H. Schubert, eds.), pp. 64–76. Verlag Anton Hain, Meisenheim.

Hall, C. A., and Porsching, T. A. (1968). Computing the maximal eigenvalue and eigenvector of a positive matrix. *SIAM J. Numer. Anal.* **5**, 269–274.

Kleibohm, K. (1966). Ein Verfahren zur approximativen Lösung von konvexen Programmen. Thesis, Zürich.

Lemke, C. E. (1968). Complementary pivot theory. *In*: 'Mathematics of the decision Sciences, Part I'. Lectures in Applied Mathematics, Vol. 11 (G. B. Dantzig and A. F. Veinott, eds.), pp. 95–114. AMS, Providence.

Mangasarian, O. L. (1969). 'Nonlinear Programming'. McGraw-Hill, New York.

Meinardus, G. (1964). 'Approximation von Funktionen und ihre Numerische Behandlung'. Springer-Verlag, Berlin, Göttingen, Heidelberg, and New York.

Rabinowitz, P. (1968). Applications of linear programming in numerical analysis. *SIAM Rev.* **10**, 121–159.

Ravindran, A. (1970). Computational aspects of Lemke's algorithm applied to linear programming. *Operat. Res.* **7**, 214–262.

Souami, B. (1969). Rechnerische Untersuchungen einiger Varianten des Simplex-Verfahrens. Diplomarbeit Technische Hochschule Aachen, Lehrstuhl für Unternehmensforschung.

Veinott, A. F. (1967). The supporting hyperplane method for unimodal programming. *Operat. Res.* **15**, 147–152.

Zoutendijk, G. (1966). Nonlinear programming: A numerical survey. *SIAM J. Control* **4**, 194–210.

23. A Survey of Methods for Solving Constrained Minimization Problems via Unconstrained Minimization

F. A. LOOTSMA
Philips Research Laboratories, Eindhoven, The Netherlands

1. Introduction

1.1. *Scope of the Present Survey*

The constrained minimization problem to be considered in this paper is defined as

$$\left. \begin{array}{l} \text{Minimize } f(x) \\ \text{Subject to the constraints } g_i(x) \geq 0, \quad i = 1, \ldots, m, \end{array} \right\} \quad (1)$$

where $f, g_1, \ldots, g_m$ denote real-valued functions of a vector x in the n-dimensional vector space E_n. For the time being, we shall restrict ourselves to a problem with inequality constraints, only; occasionally, we shall be concerned with problems which have equality as well as inequality constraints, or equality constraints only.

We shall be dealing with methods for solving (1) which reduce the computational process to unconstrained minimization of a *compound function* combining in a particular way the objective function f, the constraint functions $g_1, \ldots, g_m$, and possibly one or more *controlling parameters*.

Surveying the literature one can roughly distinguish three classes of methods. The *interior-point* methods operate in the set

$$R^0 = \{x \mid g_i(x) > 0, i = 1, \ldots, m\}. \quad (2)$$

The *exterior-point* methods, on the other hand, present an approach to a minimum solution $\bar{x}$ of (1) from outside the constraint set

$$R = \{x \mid g_i(x) \geq 0, i = 1, \ldots, m\}. \quad (3)$$

Finally, there are the *Lagrangian methods* based on the Lagrangian function associated with problem (1) and operating in E_n.

The interior-point and exterior-point methods can be subdivided into two classes; the *parametric* methods have one or more controlling parameters in the compound function (penalty function) to control the convergence towards a minimum solution; the *non-parametric* methods do not (explicitly) operate with controlling parameters. Lagrangian methods are always parametric, however.

Our mode of operation will be as follows. First, we shall consider the parametric interior-point and exterior-point methods. Next, we shall be dealing with their non-parametric counterparts and other non-parametric variants. Finally, the Lagrangian methods will be discussed.

In view of the abundance of interior-point and exterior-point methods we have been searching for a significant classification. Basically, these methods are designed to take into account the constraints of a minimization problem or, since almost none of the problems arising in practice have interior minima, to approach the boundary of the constraint set in a specifically controlled manner. We therefore classified these methods according to the behaviour of the compound function (penalty function) in the neighbourhood of the boundary.

Whenever possible, each section gives a brief sketch of the history of the methods to be discussed. For a more detailed description of the history the reader is referred to Fiacco (1967), and Fiacco and McCormick (1968).

1.2. *Problem Conditions*

The analysis of the interior-point and exterior-point methods is mainly carried out under conditions which imply local uniqueness of a minimum solution $\bar{x}$ of (1) and uniqueness of the vector $\bar{u}$ of associated Lagrangian multipliers.

We assume that the reader is familiar with the theory of necessary and sufficient conditions for minima of non-linear programming problems; see, for instance, John (1948), Kuhn and Tucker (1951), Arrow *et al.* (1961), Mangasarian and Fromowitz (1967), McCormick (1967), Fiacco and McCormick (1968). For reasons of convenience we introduce the following definition. An ordered pair $(\bar{x}, \bar{u})$ is a Kuhn-Tucker point of (1) if the requirements of (4) to (7) are simultaneously satisfied.

$$g_i(\bar{x}) \geqslant 0, \quad i = 1, \ldots, m, \tag{4}$$

$$\bar{u}_i \geqslant 0, \quad i = 1, \ldots, m, \tag{5}$$

$$\bar{u}_i g_i(\bar{x}) = 0, \quad i = 1, \ldots, m, \tag{6}$$

$$\nabla f(\bar{x}) - \sum_{i=1}^{m} \bar{u}_i \nabla g_i(\bar{x}) = 0. \tag{7}$$

Thus, a Kuhn-Tucker point $(\bar{x}, \bar{u})$ solves the relations

$$\left. \begin{array}{l} \nabla f(x) - \sum_{i=1}^{m} u_i \nabla g_i(x) = 0, \\ u_i g_i(x) = 0, \qquad i = 1, \ldots, m, \end{array} \right\} \qquad (8)$$

a system consisting of $m+n$ non-linear equations and involving $m+n$ variables. Let $\bar{J}$ denote the Jacobian matrix of (8), evaluated at $(\bar{x}, \bar{u})$. If $\bar{J}$ is non-singular, then there exists a neighbourhood of $(\bar{x}, \bar{u})$, where $(\bar{x}, \bar{u})$ is the unique solution of (8). This is the basic idea underlying the considerations to follow.

In order to distinguish the constraints which are active at a feasible solution x we introduce the set

$$A(x) = \{i \mid g_i(x) = 0, 1 \leq i \leq m\}.$$

Furthermore, we define

$$D(x, u) = \nabla^2 f(x) - \sum_{i=1}^{m} u_i \nabla^2 g_i(x).$$

We shall say that a Kuhn-Tucker point $(\bar{x}, \bar{u})$ of (1) satisfies the *Jacobian uniqueness conditions* if the following three conditions are simultaneously satisfied.

1. The multipliers $\bar{u}_i$, $i \in A(\bar{x})$, are positive.
2. The gradients $\nabla g_i(\bar{x})$, $i \in A(\bar{x})$, are linearly independent.
3. For any $y \in E_n$, $y \neq 0$, such that $\nabla g_i(\bar{x})^T y = 0$, $i \in A(\bar{x})$, it must be true that $y^T D(\bar{x}, \bar{u}) y > 0$.

For a detailed discussion of these conditions we refer the reader to Mangasarian and Fromowitz (1967), Fiacco and McCormick (1968), and Lootsma (1970). We have the following results. If the functions f, g_1, ..., g_m have continuous second derivatives in E_n, and if a Kuhn-Tucker point $(\bar{x}, \bar{u})$ of problem (1) exists satisfying the Jacobian uniqueness conditions, then the Jacobian matrix $\bar{J}$ is non-singular; furthermore, the point $\bar{x}$ is an isolated local minimum of (1) and the vector $\bar{u}$ of associated Lagrangian multipliers is uniquely determined.

We shall henceforth assume that problem (1) is a convex programming problem (i.e. f is convex and the functions g_1, ..., g_m are concave), and that R is compact and R^0 non-empty. Then a solution $\bar{x}$ of (1) exists. Furthermore, the sets R and R^0 have some important topological properties which will be explained in Section 2.3. Finally, $\bar{x}$ is the unique, global minimum of (1) if the Jacobian uniqueness conditions are satisfied at the Kuhn-Tucker point $(\bar{x}, \bar{u})$.

2. Parametric Interior-point and Exterior-point Methods

2.1. *Development and Classification of Interior-point Methods*

Parametric interior-point methods are based on penalty functions of the form

$$f(x) - r \sum_{i=1}^{m} \phi[g_i(x)]. \tag{9}$$

Here, r denotes a positive controlling parameter. The function ϕ is a function of one variable η, defined and continuously differentiable in the interval $\{\eta | \eta > 0\}$, and such that $\phi(0+) = -\infty$. Then the function of (9) is defined in R^0, but it has a positive singularity at every feasible point such that $g_i(x) = 0$ for some i, $1 \leq i \leq m$. Under mild conditions a point $x(r)$ exists minimizing (9) over R^0 for sufficiently small, positive r. This is due to the second term in (9) which presents itself as a barrier in order to prevent violation of the constraints. Following Murray (1967), we shall therefore briefly refer to interior-point penalty functions as *barrier functions*. Let $\{r_k\}$ denote a monotonic, decreasing null sequence as $k \to \infty$. Then any limit point of $\{x(r_k)\}$ is a minimum solution of (1).

There are three interior-point methods that have attracted considerable theoretical and computational attention (see also Davies, 1970). First, there is the *logarithmic programming method* using $\phi(\eta) = \ln \eta$, originally proposed by Frisch (1955). It was further developed by Parisot (1961) to solve linear programming problems, and later on the present author (1967; 1968a) gave a treatment of the method as a tool for solving non-linear problems. Second, we find the *sequential unconstrained minimization technique* (SUMT) with $\phi(\eta) = -\eta^{-1}$. It was originally suggested by Carroll (1961) and further developed by Fiacco and McCormick (1964a, b, 1966), Pomentale (1965), and Strong (1965). Lastly, there is an interior-point method with $\phi(\eta) = -\eta^{-2}$, described by Kowalik (1966), Box et al. (1969), and Fletcher and McCann (1969).

The classification that we have introduced is based on a property of the derivative ϕ' of ϕ: a barrier function is said to be of order λ if ϕ' is analytic and if it has a pole of order λ at $\eta = 0$. Illustrative examples are given by the cases where $\phi'(\eta) = \eta^{-\lambda}$ with a positive, integer λ. For $\lambda = 1$ we obtain the logarithmic barrier function of Frisch (1955) and Parisot (1961). For $\lambda = 2$ the function (9) is reduced to the inverse barrier function of SUMT proposed by Carroll (1961) and Fiacco and McCormick (1964). Finally, the inverse quadratic barrier function of Kowalik et al. (1966) is obtained for $\lambda = 3$.

2.2. Development and Classification of Exterior-point Methods

The parametric exterior-point methods can be treated in a similar manner. One is concerned with the penalty function

$$f(x) - s^{-1} \sum_{i=1}^{m} \psi[g_i(x)], \tag{10}$$

where s is a positive controlling parameter, and ψ a continuously differentiable function of one variable η such that

$$\begin{aligned}\psi(\eta) &= 0 \quad \text{for } \eta \geq 0, \\ \psi(\eta) &< 0 \quad \text{for } \eta < 0.\end{aligned} \tag{11}$$

The second term in (10) gives a (positive) contribution if, and only if, x is unfeasible. Constraint violation is progressively weighted as s decreases to 0. Under certain conditions a point $x(s)$ exists minimizing (10) over E_n for sufficiently small, positive values of s. Any limit point of the sequence $\{x(s_k)\}$, where $\{s_k\}$ is a monotonic, decreasing null sequence, is a minimum solution of (1). Following Fiacco and McCormick (1968) we shall briefly refer to penalty functions of the type (10) and (11) as *loss functions*.

The exterior-point methods have a somewhat longer history than interior-point methods. The first suggestion here was given by Courant (1943). Further developments came from Ablow and Brigham (1955), Camp (1955), Butler and Martin (1962), Pietrzykowski (1962), Fiacco and McCormick (1967a), and Beltrami (1967, 1969a). They were mainly concerned with the quadratic loss function obtained by substituting $\psi(\eta) = -[\min(0, \eta)]^2$ into (10). A more general treatment of the exterior-point methods was presented by Zangwill (1967), Fiacco (1967), Fiacco and McCormick (1968) and Roode (1968).

For classification purposes we have introduced a function ω such that

$$\omega(\eta) = \psi(\eta) \quad \text{for } \eta \leq 0.$$

Now a loss function is said to be of order μ if the derivative ω' of ω is analytic and if it has a zero of order μ at $\eta = 0$. Simple examples of loss functions are obtained by using $\omega'(\eta) = (-\eta)^\mu$ with positive, integer μ. For $\mu = 1$ we find the above named quadratic loss function.

2.3. A Mixed Penalty Function

In the present paper, parametric barrier functions will be represented by

$$B_r(x) = f(x) - r^\lambda \sum_{i=1}^{m} \phi[g_i(x)], \tag{12}$$

where λ denotes the order of the pole of ϕ' at $\eta = 0$. Raising r to the power λ

yields certain advantages when we are dealing with the Taylor series expansion of a minimizing trajectory associated with the barrier function. Similarly, a parametric loss function is given by

$$L_s(x) = f(x) - s^{-\mu} \sum_{i=1}^{m} \psi[g_i(x)], \tag{13}$$

where μ stands for the order of the loss function (the order of the zero of ω' at $\eta = 0$).

It is convenient to consider a mixed penalty function so that many properties of barrier-function and loss-function techniques can be simultaneously established. Apart from that, interior-point and exterior-point methods have particular advantages and suffer from particular disadvantages that will be explained in Section 2.7. Accordingly, combinations of these methods have been designed. The first ideas came from Fiacco and McCormick (1966) who proposed a penalty function for incorporating the inequalities as well as the equality constraints of a problem. *Mixed* penalty functions have been studied by Fiacco (1967), Fiacco and McCormick (1968) and the present author (1968b).

We shall think of the set $I = \{1, \ldots, m\}$ of constraint indices as being partitioned into two disjunct subsets I_1 and I_2; the partitioning is arbitrary and either I_1 or I_2 may be empty. The mixed penalty function to be treated here is then given by

$$M_{rs}(x) = f(x) - r^\lambda \sum_{i \in I_1} \phi[g_i(x)] - s^{-\mu} \sum_{i \in I_2} \psi[g_i(x)]. \tag{14}$$

This function reduces to the barrier function (12) if I_2 is empty, and to the loss function (13) if I_1 is empty. Furthermore, we introduce the sets

$$R_k = \{x \mid g_i(x) \geqslant 0, i \in I_k\}, \qquad k = 1, 2, \tag{15}$$

whence $R = R_1 \cap R_2$. We define R_1^0 by

$$R_1^0 = \{x \mid g_i(x) > 0, i \in I_1\}. \tag{16}$$

This is exactly the definition area of the mixed penalty function M_{rs} defined by (14). Under the convexity assumptions the sets R_1 and R_1^0 have the following topological properties if R_1^0 is non-empty (see Bui Trong Lieu and Huard, 1966; Tremolières, 1968; Evans and Gould, 1970a): the set R_1^0 is the interior of R_1; the boundary of R_1 is given by the set of points $x \in R_1$ such that $g_i(x) = 0$ for at least one $i \in I_1$; the set R_1 is the closure of R_1^0. These are important properties for establishing the convergence: the function M_{rs} generates a barrier at the boundary of R_1, and every point in R_1 *can be attained via a sequence of points in R_1^0*. Another important consequence of the convexity assumptions is that M_{rs} is a convex function in R_1^0. Hence, any local minimum of M_{rs} is a global minimum over R_1^0.

In order to simplify matters we shall restrict ourselves to functions ϕ and ψ such that

$$\left.\begin{aligned}\phi'(\eta) &= \eta^{-\lambda}, \\ \omega'(\eta) &= (-\eta)^{\mu}.\end{aligned}\right\} \tag{17}$$

For a more general treatment the reader is referred to the author's monograph, (Lootsma, 1970).

2.4. *Primal and Dual Convergence*

The convergence of the methods sketched in the previous sections is established in the next theorem. For the proof the reader is referred to Fiacco and McCormick (1968) and the present author, (Lootsma, 1970).

Theorem 1

If (a) problem (1) is a convex programming problem, (b) the constraint set R is compact, (c) the set $R_1^0 \cap R_2$ is non-empty, and (d) the sequences $\{r_k\}$ and $\{s_k\}$ are monotonic, decreasing null sequences as $k \to \infty$, then a point $x(r_k, s_k) \in R_1^0$ minimizing $M_{r_k s_k}$ over R_1^0 can be found for any k. Any limit point of the sequence $\{x(r_k, s_k)\}$ is a minimum solution of problem (1).

If the problem functions $f, g_1, \ldots, g_m$ have continuous first-order partial derivatives in E_n, then a feasible solution of the dual problem of (1) is given by the pair $[x(r, s), u(r, s)]$, where $x(r, s)$ denotes a point minimizing M_{rs} over R_1^0 for positive r and s, and $u(r, s)$ is taken to be the m-vector with components

$$\left.\begin{aligned}u_i(r, s) &= r^\lambda \phi'\{g_i[x(r, s)]\}, &&i \in I_1, \\ u_i(r, s) &= s^{-\mu} \psi'\{g_i[x(r, s)]\}, &&i \in I_2.\end{aligned}\right\} \tag{18}$$

The proof is simple. The mixed penalty function M_{rs} possesses continuous first-order derivatives in R_1^0. Then the gradient of M_{rs} vanishes at a minimizing point $x(r, s)$. Hence,

$$\left.\begin{aligned}\nabla f[x(r,s)] - \sum_{i=1}^{m} u_i(r,s)\, \nabla g_i[x(r,s)] &= 0, \\ u_i(r,s) &\geq 0, \quad i = 1,\ldots,m,\end{aligned}\right\} \tag{19}$$

in complete accordance with the definition of a dual-feasible solution.

This result appears to be particularly useful for theoretical investigations. Fiacco and McCormick (1964a) discovered that SUMT provides primal-feasible as well as dual-feasible solutions of the problem (1). In doing so, they made a connection between interior-point methods and the duality theory for non-linear programming developed in the preceding years by Dorn (1960a, b), Wolfe (1961), Huard (1962, 1963) and Mangasarian (1962). Duality and the Kuhn-Tucker relations are also the tools for investigation of a 'minimizing trajectory' and its Taylor series expansion about the Kuhn-Tucker point $(\bar{x}, \bar{u})$.

We shall conclude this section by stating a theorem concerning the dual convergence.

Theorem 2

If (a) problem (1) is a convex programming problem, (b) the constraint set R is compact, (c) the set $R_1^0 \cap R_2$ is non-empty, (d) the sequences $\{r_k\}$ and $\{s_k\}$ are monotonic, decreasing null sequences as $k \to \infty$, (e) the functions $f, g_1, \ldots, g_m$ have continuous second-order derivatives in E_n, and (f) a Kuhn-Tucker point $(\bar{x}, \bar{u})$ of problem (1) exists satisfying the Jacobian uniqueness conditions, then

$$\lim_{k \to \infty} [x(r_k, s_k), u(r_k, s_k)] = (\bar{x}, \bar{u}).$$

For a proof the reader is again referred to Fiacco and McCormick (1968).

2.5. *Series Expansion of the Minimizing Trajectory*

The appearance of controlling parameters in a penalty function poses the numerical question of how to choose appropriate values for them and how to use the information gathered during the computational process. One has to compromise between the desire for rapid convergence and the need to avoid minimization of extremely steep-valleyed penalty functions which may cause all kinds of numerical difficulties.

Acceleration of the convergence has been obtained by extrapolation according to the Richardson-Romberg principle, which is generally a powerful tool for approximating the limit of an infinitesimal process; we may, for instance, refer to Laurent (1963), Bulirsch (1964), Bulirsch and Stoer (1964, 1966). In the field of penalty-function techniques, a basis for extrapolation (the Taylor series expansion of a minimizing trajectory) was first derived by Fiacco and McCormick (1966) for SUMT, and later by the author (Lootsma, 1968a) for logarithmic programming.

So, let us now turn to the question of how the pair $[x(r, s), u(r, s)]$ behaves as a function of r and s in a neighbourhood of $(r, s) = (0, 0)$. For numerical purposes (extrapolation towards a minimum solution) it is desirable that $[x(r, s), u(r, s)]$ should be differentiable, preferably as many times as the problem

23. SOLVING CONSTRAINED MINIMIZATION PROBLEMS

functions admit. This is established in the next theorem (see Fiacco and McCormick, 1968; and Lootsma, 1970).

Theorem 3

If (a) problem (1) is a convex programming problem, (b) the constraint set R is compact, (c) the set $R_1^0 \cap R_2$ is non-empty, (d) the functions $f, g_1, \ldots, g_m$ have continuous $(k+1)$th order partial derivatives $(k \geqslant 1)$ in E_n, and (e) a Kuhn-Tucker point $(\bar{x}, \bar{u})$ of (1) exists satisfying the Jacobian uniqueness conditions, then the pair $[x(r,s), u(r,s)]$ is unique and has continuous kth order partial derivatives in a neighbourhood of $(r,s) = (0,0)$.

Let us now restrict ourselves to a mixed penalty function with only one controlling parameter. In the considerations to follow we shall accordingly be dealing with

$$M_r(x) = f(x) - r^\lambda \sum_{i \in I_1} \phi[g_i(x)] - r^{-\mu} \sum_{i \in I_2} \psi[g_i(x)]. \tag{20}$$

We take $x(r)$ to denote a point minimizing M_r over R_1^0, and introduce a vector $u(r)$ with components

$$u_i(r) = r^\lambda \phi'\{g_i[x(r)]\}, \quad i \in I_1,$$

$$u_i(r) = r^{-\mu} \psi'\{g_i[x(r)]\}, \quad i \in I_2.$$

We shall henceforth refer to the vector function $[(x(r), u(r)]$ as the *minimizing trajectory* associated with the mixed penalty function M_r of (20). A consequence of Theorem 3 is that the minimizing trajectory can be expanded in a Taylor series about $r = 0$. This provides, as an important numerical application, a basis for extrapolation towards $(\bar{x}, \bar{u})$.

It is interesting to consider a first-order approximation to the expression $f[x(r)] - f(\bar{x})$. We take A_1 and A_2 to denote the subsets of I_1 and I_2 respectively, corresponding to the active constraints at $\bar{x}$. Then

$$f[x(r)] - f(\bar{x}) = r \left[\sum_{i \in A_1} (\bar{u}_i)^{1-1/\lambda} - \sum_{i \in A_2} (\bar{u}_i)^{1+1/\mu} \right] + O(r^2). \tag{21}$$

Now, let I_2 be empty and let $\lambda = 1$. The mixed penalty function (20) then reduces to the logarithmic barrier function obtained by substituting $\phi(\eta) = \ln \eta$ into (12); formula (21) takes the simple form

$$f[x(r)] - f(\bar{x}) = \alpha r + O(r^2),$$

where α denotes the number of active constraints at $\bar{x}$. This is a slightly more workable result than the inequality

$$0 \leq f[x(r)] - f(\bar{x}) \leq mr,$$

derived by Parisot (1961) and the author (Lootsma, 1967) for the logarithmic barrier function when it is used to solve linear programs and convex programs respectively. Apparently, the parameter r can then be given such a value that $f(\bar{x})$ is approximated with a prescribed accuracy.

The discovery that penalty-function minima can be used for extrapolation (Fiacco and McCormick, 1966) has certainly increased the power of the penalty-function methods. Of course, an accurate location of these minima is necessary, although it is an open question how accurate the approximation should be. A striking coincidence was the development of variable-metric methods for unconstrained minimization (see Fletcher and Powell, 1963) with their fast ultimate convergence due to the property of quadratic termination (see Fletcher, 1970b). So the theoretical basis and the computational tools for using extrapolation techniques became available at roughly the same time.

The existence of a Taylor series expansion has been established under the condition that the derivative of ϕ has a pole and the derivative of ω a zero of *finite* order at the origin. Let us now consider two types of penalty functions where this condition is not satisfied.

(a) A class of barrier functions violating the above condition has recently been proposed by Osborne and Ryan (1970). The simplest member of the class is obtained by substituting.

$$\phi(\eta) = \ln(K - \ln \eta)$$

and $\lambda = 1$ into (20) with empty I_2; the symbol K denotes a positive constant. The associated minimizing trajectory cannot be expanded in a Taylor series about $r = 0$ (one only has to take a simple example and to compute the trajectory in order to see this), and we do not therefore immediately see the merits of the proposal.

(b) An intriguing loss function for solving problem (1) was proposed by Zangwill (1967). It is obtained by substituting $\psi(\eta) = \min(0, \eta)$ and $\mu = 1$ into (20) with empty I_1 so that it is given by

$$f(x) - r^{-1} \sum_{i=1}^{m} \min[0, g_i(x)]. \tag{22}$$

The derivative ψ' of ψ is discontinuous at $\eta = 0$ and ω' does not have a zero at the origin. The loss function (22) has the following remarkable property, however. There is a positive ρ_0 such that $x(r)$ is a minimum solution of the original problem (1) for any $0 < r < \rho_0$. This implies that a differentiable minimizing trajectory may not exist: if the minimum solution $\bar{x}$ is a boundary point of R, the loss function (22) is not differentiable at its minimizing point $x(r)$.

The computational process for solving (1) with a method based on the so-called *exact potential function* (22) would be as follows. Generate a sequence $x(r_1), x(r_2), \ldots$ of r-minima for a positive decreasing null sequence $r_1, r_2, \ldots$, until a point $x(r_k)$ is found which satisfies the constraints of (1). Such a number k exists, and $x(r_k)$ must be a required minimum solution. A serious drawback of (22) is that it is not differentiable at the boundary of the constraint set. Hence, it is doubtful whether the gradient methods for unconstrained minimization can be used to minimize (22). Recently, Pietrzykowski (1969) sketched a new algorithm to find the r-minima. It has been further developed by Conn (1971), and modified into a hybrid procedure which first identifies the active constraints at the minimum solution and then uses a quadratic loss function to solve the reduced, equality-constrained problem. The convergence proof of the method has not yet been published, however, and the numerical results are not very striking.

2.6. *Condition Number of Principal Hessian Matrix*

Numerically, problem (1) can be solved by unconstrained minimization of a penalty function for a sequence of positive, decreasing values of the controlling parameter. It is obvious that computational success depends critically on the power of unconstrained-minimization techniques. This, however, introduces the question of whether we can facilitate the computational process by an appropriate choice of the orders λ and μ of the barrier term and the loss term respectively. Several authors (Murray, 1967, 1969a, b; Fletcher and McCann, 1969; Lootsma, 1969) have accordingly been concerned with the Hessian matrix of a penalty function, and particularly its eigenvalues, in the limiting case where r decreases to 0. The motivation for the study was the idea that failures of unconstrained-minimization techniques may be due to ill-conditioning of the Hessian matrix at some iteration points, whereas Fletcher and McCann (1969) tried to exploit the behaviour of some Hessian matrices to accelerate the computational process.

We assume, amongst other things, that problem (1) is one of convex programming, and that a Kuhn-Tucker point $(\bar{x}, \bar{u})$ exists which satisfies the Jacobian uniqueness conditions. Our concern is the Hessian matrix $H(r)$ of the mixed penalty function M_r of (20), evaluated at the point $x(r)$ which minimizes M_r over R_1^0; in what follows we shall refer to $H(r)$ as the principal Hessian matrix of M_r. Since any method for minimizing M_r approaches $x(r)$, it is reasonable to assume that unconstrained minimization of M_r may be obstructed by ill-conditioning of $H(r)$.

Taking α to denote the number of active constraints at $\bar{x}$, one finds the following. There are α eigenvalues of $H(r)$ which vary with r^{-1}, whereas the remaining eigenvalues converge to finite, positive values as r decreases to 0.

Thus, if $1 \leq \alpha < n$, then the condition number $\chi(r)$ of $H(r)$ varies with r^{-1} for r small enough.

2.7. *A Mixed Penalty Function for Computational Purposes*

The methods using a penalty function of the type (20) show a remarkable similarity if ϕ' has a pole of finite order λ and ω' a zero of finite order μ at the origin. The minimizing trajectory $[x(r), u(r)]$ can be expanded in a Taylor series about $(\bar{x}, \bar{u})$, and the expansion is always one in terms of r. It is possible to show that

$$x(r) - \bar{x} = O(r).$$

On the other hand, the condition number $\chi(r)$ of the principal Hessian matrix varies with r^{-1} for small values of r. These results are valid for any choice of λ and μ and for any partitioning of the set of constraint indices into subsets I_1 and I_2. The behaviour of $\chi(r)$ is an indication that first-order penalty functions (with $\lambda = \mu = 1$) are not easier or harder to minimize than the higher-order ones, provided that the condition number is an appropriate measure of the degree of difficulty. The last hypothesis has not been thoroughly investigated and it is beyond the scope of the present paper to do so; a somewhat more extensive discussion may be found in Murray (1969b).

First-order and higher-order methods differ mainly in the *weight* r^λ attached to the barrier term and the weight $r^{-\mu}$ attached to the loss term in (20). Thus, extrapolation devices based on series expansions in terms of r can be used with higher-order penalty functions. The *weight factors* must then decrease more rapidly than with first-order methods (a phenomenon which is not always well understood). It is unlikely, however, that the more rapid decrease would cause disproportionate difficulties in minimizing the penalty function.

An obvious advantage of first-order penalty functions with respect to the higher-order ones is that their first and second derivatives are easier to evaluate. This is the argument put forward by Fiacco and McCormick (1968) to explain their preference for the logarithmic barrier function and the quadratic loss function.

Summarizing the results, we have seen that there is no obvious reason for not using a first-order penalty function: a logarithmic barrier function, a quadratic loss function, or a mixture of these functions. One might object that the second derivatives of the quadratic loss function are discontinuous at the boundary of the constraint set: a possible cause of difficulties during minimization (Murray, personal communication). Up to now, we have not seen convincing examples in the literature where minimization was obstructed by these discontinuities.

Barrier functions have the nice property that constraint violation during the computations is prevented by the barrier generated at the boundary of the constraint set. The starting point, however, has to be a point which satisfies the constraints with strict inequality sign. Hence, the author (1968b) has proposed to use a mixed penalty function with only those constraints in the loss term which are violated at the starting point or satisfied with strict equality sign.

The results obtained so far can be used in the case where both inequality and equality constraints are encountered. Let us consider the problem:

$$\left.\begin{array}{l} \text{Minimize } f(x) \\ \text{Subject to } g_i(x) \geq 0, \quad i = 1, \ldots, m, \\ \phantom{\text{Subject to }} h_j(x) = 0, \quad j = 1, \ldots, p. \end{array}\right\} \qquad (23)$$

If we incorporate the equalities in the loss term (incorporation in the barrier term is impossible), then a penalty function for solving (23) is given by

$$f(x) + rb(x) + r^{-1} l(x), \qquad (24)$$

where $b(x)$ denotes the barrier term

$$-\sum_{i \in I_1} \ln g_i(x),$$

and $l(x)$ the loss term

$$\sum_{i \in I_2} \{\min[0, g_i(x)]\}^2 + \sum_{j=1}^{p} h_j^2(x).$$

Let $x(r)$ be a minimizing point of (24) and let $u(r)$ and $w(r)$ denote vectors with components

$$u_i(r) = r/g_i[x(r)], \qquad i \in I_1,$$
$$u_i(r) = -2r^{-1} \min\{0, g_i[x(r)]\}, \qquad i \in I_2,$$
$$w_j(r) = -2r^{-1} h_j[x(r)], \qquad j = 1, \ldots, p,$$

respectively. The behaviour of the minimizing trajectory $[x(r), u(r), w(r)]$ has been thoroughly studied by Fiacco and McCormick (1968) so that we shall not go into details here.

2.8. *Concluding Remarks on Parametric Methods*

In order to describe the parametric interior-point and exterior-point methods we have introduced a classification of penalty functions according to their behaviour at the boundary of the constraint set. Furthermore, we have considered mixed penalty functions to avoid a separate treatment of interior-point and exterior-point methods. Our concern has mainly been the theoretical basis for acceleration by extrapolation, and its counterpart: ill-conditioning of the principal Hessian matrix of penalty functions as the controlling parameter converges to 0. Using these results we described the reasons for choosing the so-called first-order penalty functions for computational purposes. In addition to this, we have dealt with several methods which do not fit into the above classification.

3. Non-parametric Interior-point and Exterior-point Methods

3.1. *Moving Truncations and Non-parametric Interior-point Methods*

An interesting development was initiated by Rosenbrock (1960) and continued by Huard (1964), when he proposed the *method of centres*. It has been explored, theoretically and computationally, by Faure and Huard (1965, 1966), Bui Trong Lieu and Huard (1966), Huard (1967, 1968) and Tremolières (1968). The method of centres generates a sequence of points converging to a minimum solution of the problem. Each of these points (centres) is obtained by unconstrained *maximization* of a *distance function*: a particular combination of the objective function and the constraint functions. However, some distance functions may also be regarded as penalty functions *without controlling parameters*. Starting from this point of view, Fiacco and McCormick (1967b) presented a non-parametric version of SUMT, and Fiacco (1967) demonstrated that similar versions can be obtained for a large class of interior-point as well as exterior-point methods. Slightly earlier, a non-parametric exterior-point method was suggested by Kowalik (1966). Computational experience, however, prompted the author (1968c) to undertake a theoretical study of the rate of convergence of these methods as compared with the above-mentioned parametric techniques.

The methods to be treated here can be characterized more precisely in the following way. They do not *explicitly* operate with controlling parameters but with *moving truncations of the constraint set*. Convergence to a minimum solution of (1) is controlled by a sequence $\{t_k\}$ of *truncation levels* converging to the unknown minimum value $\bar{v}$ of the problem. In using a parametric technique, however, a null sequence $\{r_k\}$ of values assigned to the controlling parameter is employed.

Non-parametric methods are based on the following ideas. One operates in intersections of the constraint set R and the truncations

$$F(t) = \{x | f(x) \leqslant t; x \in E_n\}$$

for values of the truncation level t which are not less than the minimum value $\bar{v}$ of problem (1). Then the truncated constraint set

$$T(t) = R \cap F(t) = \{x | f(x) \leqslant t; x \in R\}$$

is non-empty. We define a non-parametric barrier function by

$$B_t^*(x) = p\phi[t - f(x)] + \sum_{i=1}^{m} \phi[g_i(x)], \tag{25}$$

with ϕ as given by (17). There is a positive weight factor p attached to the term which contains the objective function, for reasons that will be explained in Section 3.2.

We assume that R^0 is non-empty. If $t > \bar{v}$, then the set

$$T^0(t) = \{x | f(x) < t; x \in R^0\}$$

is also non-empty. Moreover, we assume that problem (1) is a convex programming problem. Then B_t^* is a concave function in $T^0(t)$, and there exists a point $c(t)$ which maximizes B_t^* over $T^0(t)$. By analogy with Huard's method of centres, the point $c(t)$ will be referred to as a *centre* of $T(t)$. Let $\{t_k\}$ denote a sequence of monotonic, decreasing truncation levels converging to $\bar{v}$ as $k \to \infty$. It will intuitively be clear that any limit point of the sequence $\{c(t_k)\}$ is then a minimum solution of (1).

In algorithms which operate along these lines [the method of centres (Huard, 1964) using (25) with $\phi(\eta) = \ln \eta$, some variants of it with relaxation facilities (Tremolières, 1968), and SUMT without parameters (Fiacco and McCormick, 1967b) using (25) with $\phi(\eta) = -\eta^{-1}$] a sequence $\{t_k\}$ as mentioned above is obtained as follows. The first step starts with a truncation level $t_1 = f(x_0)$, where x_0 is some feasible solution so that $t_1 \geqslant \bar{v}$. At the beginning of the kth step the truncation level t_{k-1} and the centre $c(t_{k-1})$ are available, whereas $t_{k-1} > f[c(t_{k-1})]$. The truncation level t_k is then taken to be

$$t_k = t_{k-1} - \rho\{t_{k-1} - f[c(t_{k-1})]\}. \tag{26}$$

Here, ρ stands for a relaxation factor such that $0 < \rho \leqslant 1$ in order to ensure that $t_{k-1} > t_k \geqslant \bar{v}$. The proof that the sequence $\{t_k\}$ converges to $\bar{v}$ is a straightforward generalization of the convergence proof for SUMT without parameters.

To illustrate the foregoing sketch we substitute $\phi(\eta) = \ln \eta$ into (25). A point maximizing the non-parametric logarithmic barrier function over $T^0(t)$ can also be found by the maximizing over $T(t)$ of the function

$$d_t(x) = [t - f(x)]^p \prod_{i=1}^{m} g_i(x). \tag{27}$$

This function is an example of the general distance function appearing in the method of centres: one of the properties of a distance function is that it vanishes at the boundary of the corresponding truncated constraint set; another is that it has a positive value at any interior point of the truncated constraint set under consideration. A point maximizing a distance function over this set was referred to as a *centre* of the set.

The precise relationship between parametric and non-parametric barrier-function techniques is expressed by the next theorem (see also Fiacco and McCormick, 1967b, 1968; Lootsma, 1968a, 1970).

Theorem 4

If (a) problem (1) is a convex programming problem, (b) R is compact and R^0 is non-empty, and (c) the functions f, g_1, ..., g_m have continuous first-order partial derivatives in E_n, then a centre $c(t_k)$ minimizes the parametric barrier function B_r of (12) over R^0 for r equal to

$$r_k = \{p\phi'(t_k - f[c(t_k)])\}^{-1/\lambda}. \tag{28}$$

If the sequence $\{t_k\}$ is generated by (26), then the sequence $\{r_k\}$ is a monotonic, non-increasing null sequence.

The proof of the theorem rests on the observation that the gradients of $B^*_{t_k}$ and B_{r_k} both vanish at the point $c(t_k)$.

Theorem 4 implies that a non-parametric barrier-function technique is equivalent to a parametric barrier-function technique *adjusting the controlling parameter automatically*. At first sight this is a welcome feature, and not only from a theoretical standpoint, for under certain conditions the parametric technique using the function B_r admits of a differentiable minimizing trajectory $[x(r), u(r)]$. The centres $c(t_k)$ can be written as

$$c(t_k) = x(r_k), \quad k = 1, 2, \ldots,$$

with r_k given by (28). Thus, we obtain a sequence $\{[x(r_k), u(r_k)]\}$ on the minimizing trajectory, and this sequence is clearly amenable to extrapolation towards $(\bar{x}, \bar{u})$. The crucial point, however, is the rate of convergence, and we shall accordingly be dealing with that subject in the next section.

3.2. Rate of Convergence of Non-parametric Interior-point Methods

In the next theorem the rate of convergence is established for the objective-function values $f_k = f[c(t_k)]$ at the successive centres $c(t_k)$, the truncation levels t_k, and the equivalent r-values r_k generated by (28). The proof has already been given by the author (1968c, 1970).

Theorem 5

If (a) problem (1) is a convex-programming problem, (b) the constraint set R is compact and R^0 is non-empty, (c) the problem functions have continuous second-order partial derivatives in E_n, (d) a Kuhn-Tucker point $(\bar{x}, \bar{u})$ of (1) exists satisfying the Jacobian uniqueness conditions, (e) the point $\bar{x}$ is a boundary point of R, and (f) the sequences $\{t_k\}$ and $\{r_k\}$ are generated by (26) and (28) respectively, then

$$\lim_{k\to\infty} \frac{f_k - \bar{v}}{f_{k-1} - \bar{v}} = \lim_{k\to\infty} \frac{t_k - \bar{v}}{t_{k-1} - \bar{v}} = \lim_{k\to\infty} \frac{r_k}{r_{k-1}} = 1 - \frac{\rho}{\bar{\beta}+1}, \quad (29)$$

with

$$\bar{\beta} = p^{-1/\lambda} \sum_{i \in A(\bar{x})} (\bar{u}_i)^{1-1/\lambda}. \quad (30)$$

Thus, convergence is linear, and the asymptotic error constant (the right-hand side of (29)) depends on the relaxation factor ρ, the order λ of the barrier function, the weight factor p, and the Lagrangian multipliers $\bar{u}_i$, $i \in A(\bar{x})$, which are unknown at the beginning of the computations for solving problem (1). There is one remarkable exception: in the first-order case where $\lambda = 1$ (Huard's method of centres) the asymptotic error constant depends only on ρ, p, and the number α of active constraints at the minimum solution $\bar{x}$, since $\bar{\beta}$ reduces to $p^{-1}\alpha$. Hence, one can roughly predict the rate of convergence (since α can mostly be estimated by the number m of constraints), and the weight factor p can be given an appropriate value (see Faure and Huard, 1966; Tremolières, 1968). For the higher-order methods, however, the asymptotic error constant is unpredictable.

It is convenient to use a relaxation ρ such that $0 < \rho < 1$. Then $t_k > f[c(t_{k-1})]$, and the search for $c(t_k)$ can immediately start from the previous centre $c(t_{k-1})$.

3.3. *Other Variants of the Method of Centres*

In order to solve problem (1) only two particular distance functions have been envisaged (see Huard, 1964, 1967, 1968). One of these is (27), and the other was given by

$$\min\{t - f(x), g_1(x), \ldots, g_m(x)\}, \quad (31)$$

which is clearly not a differentiable function. A centre $c(t)$, to be found by the maximization of (31) over $T(t)$, could equivalently be obtained by solving the problem:

$$\left.\begin{array}{l}\text{Maximize } \delta \\ \text{Subject to } t - f(x) \geq \delta, \\ \qquad g_i(x) \geq \delta, \quad i = 1, \ldots, m.\end{array}\right\} \quad (32)$$

Problem (32) is by no means easier to solve than (1). This is a sharp contrast with the penalty-function and distance-function techniques which reduce constrained minimization to the solution of a sequence of *unconstrained* problems each of which is *easier* to solve than the original problem. However, the method has been modified into an iterative procedure of the following kind. Suppose that an iteration point x_k has been obtained, and set $t_k = f(x_k)$. Then the functions $f, g_1, \ldots, g_m$ in (32) are replaced by their linear approximations

$$\tilde{f}(x) = f(x_k) + (x - x_k)^T \nabla f(x_k),$$
$$\tilde{g}_i(x) = g_i(x_k) + (x - x_k)^T \nabla g_i(x_k), \qquad i = 1, \ldots, m.$$

Now, let (δ_k, y_k) be a solution of the *linear programming* problem:

$$\left.\begin{aligned}
&\text{Maximize } \delta \\
&\text{Subject to } t_k - \tilde{f}(y) \geq \delta, \\
&\qquad\qquad \tilde{g}_i(y) \geq \delta, \qquad i = 1, \ldots, m.
\end{aligned}\right\} \qquad (33)$$

Then the next iteration point x_{k+1} is taken to be the point which maximizes (31) over the line segment connecting x_k and y_k. This method has been proposed by Huard (1968) in a more sophisticated form, and it is still under investigation at various places in the world.

3.4. *Non-parametric Exterior-point Methods*

The development of non-parametric exterior-point methods proceeds by analogy with the mode of operation in Sections 3.1 and 3.2. There are some minor differences: we have to deal with a monotonic, increasing sequence of truncation levels converging to $\bar{v}$ from below, in contrast to the convergence from above presented by non-parametric interior-point methods.

Let us now, first, describe the basic ideas and consider the non-parametric loss function

$$L_t^*(x) = p\psi[t - f(x)] + \sum_{i=1}^m \psi[g_i(x)]. \qquad (34)$$

We have again attached a positive weight factor p to the term which contains the objective function. It is obvious from the properties of ψ (which vanishes for non-negative values of its argument, and is negative elsewhere) that

$$L_t^*(x) \begin{cases} = 0 & \text{for all } x \in T(t), \\ < 0 & \text{for all } x \notin T(t). \end{cases}$$

The set $T(t)$ is non-empty if, and only if, the truncation level $t \geq \bar{v}$. If $t < \bar{v}$, it must be true that $L_t^*(x) < 0$ for all $x \in E_n$. A point $c(t)$ maximizing L_t^* over E_n exists under certain conditions for any t. If $t < \bar{v}$, it can be shown that

$$t < f[c(t)] \leq \bar{v},$$

so that $c(t) \notin F(t)$. Furthermore, it can readily be demonstrated that $c(t) \notin R$ in the (usual) case where f does not have an unconstrained minimum in R. We may, however, think of $c(t)$ as the common centre of the two disjunct sets R and $F(t)$. The basic idea will now be evident: if $\{t_k\}$ is a monotonic, increasing sequence of truncation levels converging to $\bar{v}$, then any limit point of the sequence $\{c(t_k)\}$ is a minimum solution of (1).

Further details will be briefly sketched. We consider the following construction of a sequence $\{t_k\}$ as mentioned above. In the first step the truncation level t_1 is taken to be a lower estimate of the minimum value $\bar{v}$ of problem (1). At the beginning of the kth step the truncation level t_k is generated in accordance with

$$t_k = t_{k-1} + \rho\{f[c(t_{k-1})] - t_{k-1}\}, \tag{35}$$

where ρ is a relaxation factor such that $0 < \rho \leq 1$. Under the conditions of Theorem 4 the successive centres $c(t_k)$ exist, and each of them is a point minimizing the parametric loss function L_s of (13) for s equal to the equivalent s-value

$$s_k = \{p\psi'(t_k - f[c(t_k)])\}^{1/\mu}. \tag{36}$$

Apparently, the relationship between the parametric and non-parametric exterior-point methods is similar to the relationship between the corresponding classes of interior-point methods.

The objective-function values $f[c(t_k)]$, the truncation levels t_k, and the equivalent s-values s_k generated by (36) converge *linearly* towards their respective limits ($\bar{v}$, $\bar{v}$ and 0) under the conditions (a)–(e) of Theorem 5 (see Lootsma, 1970). The asymptotic error constant is given by

$$1 - \frac{\rho}{\bar{\gamma} + 1} \tag{37}$$

with

$$\bar{\gamma} = p^{1/\mu} \sum_{i \in A(\bar{x})} (\bar{u}_i)^{1+1/\mu}, \tag{38}$$

which depends clearly on the Lagrangian multipliers for any order μ of the non-parametric loss function.

Some numerical experiences with the non-parametric, quadratic loss function—originally proposed by Kowalik (1966) and obtained by substituting $\psi(\eta) = -\{\min(0, \eta)\}^2$ into (34)—are reported by the author (Lootsma, 1968c). These experiments show the importance of the weight factor p to control the convergence. At the beginning of the computations, however, an appropriate choice of p cannot be made since the Lagrangian multipliers are then unknown. This is a disadvantage for any of the methods treated in this section.

3.5. *Convergence of Morrison and Tangent Parameters*

Kowalik et al. (1969) carried out some experiments with two methods, both using a non-parametric quadratic loss function, but different from the method of Section 3.4 only as far as the generation of truncation levels is concerned [formula (35)]. They considered an equality-constrained problem; it is possible, however, to rewrite the methods for the inequality-constrained problem (1).

Both methods start with a truncation level $t_1 \leq \bar{v}$. Let t_k be the current truncation level and c_k a point minimizing

$$F_k(x) = \{\min [0, t_k - f(x)]\}^2 + \sum_{i=1}^{m} \{\min [0, g_i(x)]\}^2$$

over E_n. In the method which is due to Morrison (1968), successive truncation levels are generated by

$$t_{k+1} = t_k + [F_k(c_k)]^{1/2}, \tag{39}$$

and these levels are accordingly known as *Morrison parameters*. The second method is due to a suggestion of Wolfe; here the truncation levels, known as the *tangent parameters*, are generated in accordance with

$$t_{k+1} = t_k + F_k(c_k)/(f(c_k) - t_k). \tag{40}$$

In both methods the truncation levels converge to the minimum value $\bar{v}$ of problem (1).

The experiments of Kowalik et al. (1969) were carried out to find which of the two methods has the faster convergence. Numerically, the method with the tangent parameters appeared to be superior. There is also a theoretical argument, however, to support these results. Reasoning along the lines of the proofs for Theorems 4 and 5 it is possible to show the following. The Morrison parameters and the objective-function values at the centres c_k converge *linearly* towards $\bar{v}$; the asymptotic error constant is given by

$$1 - \left(1 + \sum_{i \in A(\bar{x})} \bar{u}_i^2\right)^{-1/2}$$

The tangent parameters, however, and the objective-function values at the corresponding centres have *superlinear* convergence towards $\bar{v}$ since

$$\lim_{k \to \infty} \frac{f(c_k) - \bar{v}}{f(c_{k-1}) - \bar{v}} = \lim_{k \to \infty} \frac{t_k - \bar{v}}{t_{k-1} - \bar{v}} = 0.$$

Thus, the use of the tangent parameters will lead to a faster ultimate convergence.

3.6. *Concluding Remarks on Non-parametric Methods*

We have discussed the non-parametric versions of the interior-point and exterior-point methods classified according to the rules in Sections 2.1 and 2.2. As we have seen, these versions may be considered as parametric techniques with automatic adjustment of the controlling parameter. In applying the non-parametric techniques one finds a sequence of points (centres) which can be used to obtain points on the minimizing trajectory of the corresponding parametric methods. The adjusted value of the controlling parameter can be computed as soon as a centre has been located. Convergence of the objective-function values at the centres, and the adjusted parameter values is linear; the asymptotic error constant depends on the Lagrangian multipliers associated with the minimum solution of problem (1) for any order λ of barrier functions and any order μ of loss functions. There are two exceptions. The non-parametric first-order barrier-function method (Huard's method of centres) has indeed linear convergence, but the error constant does not depend on the Lagrangian multipliers; there is a variant of the non-parametric first-order loss-function method (the method of Morrison and Wolfe) which has superlinear convergence.

4. Lagrangian Methods

4.1. *Introductory Sketch*

Let us consider the classical constrained-minimization problem:

$$\left. \begin{array}{l} \text{Minimize } f(x) \\ \text{Subject to } h_j(x) = 0, \quad j = 1,\ldots,p, \end{array} \right\} \quad (41)$$

and let us suppose that a minimum solution $\bar{x}$ to (41) exists. It is well known that, assuming the functions $f, h_1, \ldots, h_p$ admit of continuous first derivatives in E_n and the Jacobian matrix of the functions $h_1, \ldots, h_p$ with respect to the variables $x_1, \ldots, x_n$ has full rank, then a vector $\bar{w} \in E_p$ can be found such that $(\bar{x},\bar{w})$ is a stationary point of the Lagrangian function

$$f(x) - \sum_{j=1}^{p} w_j h_j(x). \quad (42)$$

By analogy with the definitions in Section 1.2 we shall refer to $(\bar{x},\bar{w})$ as a Kuhn-Tucker point of (41). The stationary points of (42) are characterized by the $n + p$ non-linear equations

$$\left. \begin{array}{l} \nabla f(x) - \sum_{j=1}^{p} w_j \nabla h_j(x) = 0, \\ h_j(x) = 0, \quad j = 1,\ldots,p, \end{array} \right\} \quad (43)$$

with $n+p$ variables $x_1, \ldots, x_n, w_1, \ldots, w_p$. The technique of forming the Lagrangian function (42) and solving the system (43) to obtain a solution of (41) is called the *Lagrangian-multiplier technique*.

By this method the dimensionality of the problem is considerably increased: the original problem (41) is a problem in E_n but the system (43) presents a problem in E_{n+p}. This chapter is therefore concerned with the more recent methods which reduce the solution of a constrained-minimization problem to sequential minimization of a Lagrangian function as a function of x. Hence, the solution of problem (41) in E_n would be reduced to a sequence of problems in a vector space of the *same* dimension.

We shall start with a few simple observations and remarks. In the following sections the various methods will be sketched in more detail. Let us return to problem (1) and let us assume that $(\bar{x}, \bar{u})$ is a Kuhn-Tucker point of (1). Furthermore, let

$$\Lambda(x, u) = f(x) - \sum_{i=1}^{m} u_i g_i(x). \tag{44}$$

Then, $\bar{x}$ is a stationary point of the function $\Lambda(x, \bar{u})$. The computational implications are obvious: one could try to find a minimum solution of (1) by unconstrained minimization of $\Lambda(x, \bar{u})$.

This is a rather simplistic approach, however, and there are many pitfalls in it if the convexity conditions are not satisfied. Some problems which do have a minimum solution do not possess a Kuhn-Tucker point. Not every stationary point of $\Lambda(x, \bar{u})$ is a minimum solution of (1) and, moreover, it need not even be a feasible solution. Lastly, the vector $\bar{u}$ of Lagrangian multipliers associated with $\bar{x}$ is mostly unknown, so that the idea seems to be very impractical.

Nevertheless, the idea has an attractive feature. The function $\Lambda(x, \bar{u})$ is probably easier to minimize than the penalty functions treated before, since it does not tend to be ill-conditioned at the boundary of the constraint set.

A very simple result due to Everett (1963) has encouraged the study of Lagrangian methods. The result is expressed by a theorem stating that any point $x(u)$ minimizing the function $\Lambda(x, u)$ over E_n is a minimum solution of the perturbed problem:

$$\left. \begin{array}{l} \text{Minimize } f(x) \\ \text{Subject to } g_i(x) \geq g_i[x(u)], \quad i = 1, \ldots, m. \end{array} \right\} \tag{45}$$

Thus, the point $x(u)$ could be accepted as an approximation to a minimum solution of (1) if the perturbations $g_i[x(u)]$, $i = 1, \ldots, m$, are negligible, at least in the opinion of the user.

Recently, Gould and Howe (1971) considered in more detail the question of how to use the vector $\bar{u}$ if it is known *a priori*. In the non-convex case they propose to minimize the *exponential* penalty function

$$f(x) + k^{-1} \sum_{i=1}^{m} \bar{u}_i \{\exp[-kg_i(x)] - 1\}, \qquad (46)$$

which has, under mild conditions and for k large enough, a positive-definite Hessian matrix at $\bar{x}$. Hence, $\bar{x}$ is an unconstrained minimum of (46), even if the convexity conditions are not satisfied. It is not only a stationary point!

Nevertheless, we are still faced with the problem of how to find a vector $\bar{u}$ of Lagrangian multipliers associated with a minimum solution of (1). Several procedures have been proposed, one of them even before the theorem of Everett (1963) was published.

Before these methods can be explained, the original problem (1) must be extended. It will be assumed that we are restricted to a compact, convex set S. Thus, we consider the extended problem:

$$\left.\begin{array}{l} \text{Minimize } f(x) \\ \text{Subject to } g_i(x) \geq 0, \quad i = 1, \ldots, m, \\ x \in S. \end{array}\right\} \qquad (47)$$

Computationally, the set S could be regarded as being defined by a number of simple, linear constraints.

The set S has been introduced in order to avoid the danger of the Lagrangian function of (44) being minimized over E_n for some $u \geq 0$ which does not imply the existence of a minimum solution. The algorithms we are going to describe operate in S, and the Lagrangian function has a finite minimum over S for any $u \geq 0$.

There is, as we may note, a sharp contrast with the minimization of penalty functions. The interior-point methods, with an impenetrable barrier at the boundary of the constraint set, are very safe; under mild conditions a finite minimum exists for any positive value of r. The exterior-point methods are more dangerous: it may happen that constraint violations are not sufficiently penalized to compensate for the decrease of the objective function outside the constraint set; if the controlling parameter r is not small enough, a minimum may not exist. It will intuitively be clear, however, that Lagrangian functions are extremely dangerous to use if no additional constraints are imposed on the computational process for minimizing the function in question.

4.2. *A Column-generating Procedure*

Benders' decomposition method for mixed-integer programming (Benders, 1960) is well-known, but he also published a decomposition method to solve convex programming problems. This has gone unnoticed, and the method was accordingly rediscovered by Wolfe (1963) and Brooks and Geoffrion (1966).

The method of Benders (1960) for convex programming operates as follows. Suppose that f is convex and that the constraint functions are concave. Let $\{y_1, \ldots, y_p\}$ be a set of points in S, and let $\bar{\lambda}_1^{(p)}, \ldots, \bar{\lambda}_p^{(p)}$ be a solution of the linear programming problem

$$\begin{aligned}
\text{minimize } & \sum_{k=1}^{p} \lambda_k f(y_k) \text{ subject to} \\
& \sum_{k=1}^{p} \lambda_k = 1, \\
& \sum_{k=1}^{p} \lambda_k g_i(y_k) \geq 0, \quad i = 1, \ldots, m, \\
& \lambda_k \geq 0, \quad k = 1, \ldots, p.
\end{aligned} \quad (48)$$

Let

$$\bar{x}_p = \sum_{k=1}^{p} \bar{\lambda}_k^{(p)} y_k.$$

Using the convexity properties it is easy to show that $\bar{x}_p$ is a feasible solution of (47) such that

$$f(\bar{x}_p) \leq \bar{v}_p \stackrel{\text{def}}{=} \sum_{k=1}^{p} \bar{\lambda}_k^{(p)} f(y_k),$$

so that

$$\bar{v} \leq f(\bar{x}_p) \leq \bar{v}_p,$$

where $\bar{v}$ denotes the minimum value of (47). The dual problem of (48) is given by

$$\begin{aligned}
\text{maximize } & u_0 \text{ subject to the constraints} \\
& u_0 + \sum_{i=1}^{m} u_i g_i(y_k) \leq f(y_k), \quad k = 1, \ldots, p, \\
& u_i \geq 0, \quad i = 1, \ldots, m.
\end{aligned} \quad (49)$$

Let $\bar{u}_0^{(p)}, \bar{u}_1^{(p)}, \ldots, \bar{u}_m^{(p)}$ be a maximum solution of (49). The extreme values of (48) and (49) are equal whence

$$\bar{u}_0^{(p)} = \bar{v}_p.$$

Next, we have to answer the question of whether a point $y_{p+1} \in S$ exists which, if added to the set $\{y_1, \ldots, y_p\}$, would generate a 'better' approximation $\bar{x}_{p+1}$ to a minimum solution of (47). Considering the dual problem (49) we could ask whether a point $y_{p+1} \in S$ exists such that

$$\bar{u}_0^{(p)} + \sum_{i=1}^{m} \bar{u}_i^{(p)} g_i(y_{p+1}) > f(y_{p+1}).$$

Hence, let y_{p+1} be a point minimizing the Lagrangian function

$$f(y) - \sum_{i=1}^{m} \bar{u}_i^{(p)} g_i(y)$$

over S. Such a point exists by the compactness of S. Let

$$z_p = f(y_{p+1}) - \sum_{i=1}^{m} \bar{u}_i^{(p)} g_i(y_{p+1}).$$

It can readily be shown that z_p is not greater than the minimum value $\bar{v}$ of (47). Summarizing the results, we have the sequence of inequalities

$$z_p \leqslant \bar{v} \leqslant f(\bar{x}_p) \leqslant \bar{u}_0^{(p)} = \bar{v}_p.$$

If $z_p = f(\bar{x}_p)$, then $\bar{x}_p$ is a minimum solution of (47). Otherwise, the point y_{p+1} is added to the set $\{y_1, \ldots, y_p\}$, and the next approximation $\bar{x}_{p+1}$ is obtained by solving the linear programming problem (48) augmented by one column. Convergence of the sequence $\{\bar{x}_p\}$ has been established by Benders (1960) under the usual convexity conditions and the additional regularity condition that a feasible solution exists satisfying the (non-linear) inequalities with strict inequality sign.

Computationally, the above method is mainly concerned with minimization of the Lagrangian function over S. If S is described by linear inequalities, the method can be characterized as one of sequential linearly-constrained minimization.

4.3. *A General Approach*

Lagrangian methods were given a general framework by Falk (1967) who studied problem (47) under the usual convexity conditions and with strict convexity of the objective function. He introduced the function γ defined by

$$\gamma(u) = f[x(u)] - \sum_{i=1}^{m} u_i g_i[x(u)]. \tag{50}$$

where $x(u)$ denotes the point minimizing the Lagrangian function $\Lambda(x, u)$ over S for a fixed $u \geqslant 0$. Taking $D[\gamma]$ to denote the definition area of γ, Falk (1967) was primarily concerned with the properties of γ (concavity and differentiability) and the topological properties of $D[\gamma]$. He showed that a solution to (47) can be obtained by solving the auxiliary problem

$$\max \{\gamma(u) | u \in D[\gamma]\}. \tag{51}$$

If S is compact the set $D[\gamma]$ is the non-negative orthant in E_m. Then, the constraints of (51) are extremely simple. Evaluation of the objective function γ, however, involves minimization of the Lagrangian function over S.

The above ideas have been generalized by Roode (1968) to include a phenomenon first observed by Zangwill (1967). The exterior-point methods, using the loss function L_s of (13) to solve problem (1), can also be regarded as methods for solving the problem

$$\sup L_s(x) \text{ subject to} \\ \nabla L_s(x) = 0, \\ s > 0. \qquad (52)$$

It is easy to see why this is true under the convexity conditions; the constraints of (52) are precisely the necessary and sufficient conditions for a minimum $x(s)$ of the loss function L_s; furthermore, we have

$$L_s[x(s)] \leqslant \bar{v}$$

and

$$\lim_{s \downarrow 0} L_s(x(s)) = \bar{v}.$$

There is a remarkable similarity between problem (52) and the dual problem of (1) formulated by Wolfe (1961) as

$$\max \Lambda(x, u) \text{ subject to} \\ \nabla_x \Lambda(x, u) = 0, \\ u \geqslant 0, \qquad (53)$$

where Λ denotes the Lagrangian function (44). To unify the exterior-point methods and the Lagrangian methods, Roode (1968) introduced a generalized Lagrangian function and studied a straightforward generalization of the auxiliary problem (51). For a brief description we may also refer to Zoutendijk (1970).

The theoretical studies outlined above did not yield many results of computational value. Falk (1969) developed a method using a particular penalty function with one controlling parameter r, and derived an auxiliary problem of a very simple nature: maximization of an auxiliary function ρ of one variable r over its definition area $D[\rho]$, an interval in E_1. The bounds of the interval are unknown, however, and the evaluation of ρ involves the solution of a minimization problem. So, it is doubtful whether the method is workable, although its theoretical development has an undeniable elegance.

4.4. *Augmented Lagrangian Functions*

The approaches used in the previous sections are rather sophisticated as compared with the method of Fletcher (1970) for solving the equality-constrained problem (41). Consider the function

$$f(x) - \sum_{j=1}^{p} w_j(x) h_j(x) \qquad (54)$$

where the $w_j(x)$ are obtained as a least-squares solution of the problem

$$\min_{w_1,\ldots,w_p} \left\| \nabla f(x) - \sum_{j=1}^{p} w_j \nabla h_j(x) \right\|^2. \tag{55}$$

Thus, the $w_j(x)$ are the best approximations at x of the Lagrangian multipliers. Any feasible solution $\bar{x}$ of (41) with the property that multipliers $\bar{w}_1, \ldots, \bar{w}_p$ exist such that

$$\nabla f(\bar{x}) - \sum_{j=1}^{p} \bar{w}_j \nabla h_j(\bar{x}) = 0 \tag{56}$$

is a stationary point of (54). Computationally, however, the function (54) becomes interesting if a term (augment) is added which affects the Hessian matrix at $\bar{x}$. Consider the function

$$f(x) - \sum_{j=1}^{p} w_j(x) h_j(x) + K \sum_{j=1}^{p} h_j^2(x), \tag{57}$$

where the $w_j(x), j = 1, \ldots, p$, are again defined by (55), and K denotes a positive constant. It can easily be verified that a feasible solution $\bar{x}$ of (41) with the property (56) is a stationary point of (57). Under certain additional conditions [implying that $\bar{x}$ is an isolated local minimum of (41) and that $\bar{w}$ is uniquely determined] the Hessian matrix of (57) evaluated at $\bar{x}$ is positive definite for K sufficiently large, so that $\bar{x}$ is a point minimizing (57). In other words, (57) is a function with an unconstrained minimum at a solution of the constrained-minimization problem (41).

The augmented term suggests that (57) is some sort of quadratic loss function, but this is not true. If K is sufficiently large, the point $\bar{x}$ is obtained in one single minimization of (57). When a quadratic loss function of the type

$$f(x) + K \sum_{j=1}^{p} h_j^2(x)$$

is used, a solution of (41) is obtained in the *limiting case* where $K \to \infty$. Then a sequence of minimizations (with increasing values of K) is carried out until a suitable convergence criterion is satisfied.

A somewhat similar method has been proposed by Hestenes (1969), and Haarhoff and Buys (1970). They were also concerned with the equality-constrained problem (41) and considered the augmented Lagrangian function

$$f(x) - \sum_{j=1}^{p} w_j^{(k)} h_j(x) + \sum_{j=1}^{p} K_j h_j^2(x). \tag{58}$$

Let $w_j^{(0)} = 0, j = 1, \ldots, p$, and let x_k denote a point minimizing (58). Then the $w_j^{(k)}$ are adjusted according to

$$w_j^{(k+1)} = w_j^{(k)} - 2K_j h_j(x_k),$$

so that

$$\nabla f(x_k) - \sum_{j=1}^{p} w_j^{(k+1)} \nabla h_j(x_k) = 0. \tag{59}$$

Hence, the sequences $\{x_k\}$ and $\{w^{(k)}\}$ converge to $\bar{x}$ and $\bar{w}$ respectively if the weights $K_1, \ldots, K_p$ are sufficiently large.

It can readily be seen (Osborne, personal communication) that the above method is exactly the method of Powell (1968) using the function

$$f(x) + \sum_{j=1}^{p} K_j [h_j(x) + \theta_j]^2, \tag{60}$$

where the K_j are positive weight factors, and the θ_j are the shift parameters with a key function in the computational process. The procedure is as follows. Keep the K_j fixed, let $\theta^{(k)}$ denote the vector of current values assigned to the shift parameters, and let x_k denote a point minimizing (60) over E_n for $\theta = \theta^{(k)}$. The shift parameters are updated according to

$$\theta_j^{(k+1)} = \theta_j^{(k)} + h_j(x_k), \quad j = 1, \ldots, p.$$

If the weight factors $K_1, \ldots, K_p$ are sufficiently large, the sequence $\{x_k\}$ converges to a minimum solution of problem (41). Furthermore, the rate of convergence can be controlled by the choice of the weight factors: convergence of the shift parameters towards their optimum values is at least linear, and the asymptotic error constant can be made arbitrarily small by choosing the weight factors large enough. See also Pearson (1968) who studied the behaviour of the method if (41) is a quadratic programming problem.

Mostly—and this is an important feature of the above methods—large values need not be assigned to the weight factors in order to obtain a satisfactory rate of convergence.

4.5. *Concluding Remarks on Lagrangian Methods*

The Lagrangian methods that we have discussed deal with the following two types of problems:

(a) Inequality-constrained problems, under the additional condition that the vector x of variables is restricted to a compact, convex set S. Consequently, this involves operating in a set where the Lagrangian function $\Lambda(x,u)$ has a finite minimum for any $u \geqslant 0$. Methods for

solving the problem, and a general framework for these methods, have been proposed by Benders (1960), Falk (1967, 1969) and Roode (1968).
(b) Equality-constrained problems. Here, the Lagrangian function associated with the problem is augmented by a term which makes the Hessian matrix positive definite at a certain stationary point. Accordingly, a minimum solution of the constrained problem can be found by (sequential) unconstrained minimization of the augmented Lagrangian function (Powell, 1968; Hestenes, 1969; Fletcher, 1970; Haarhoff and Buys, 1970). To our knowledge, these methods have not yet been extended to solve inequality-constrained problems, which present the difficulty of *identifying* the active constraints at a minimum solution.

5. Evaluation and Conclusions

This survey was concerned with three classes of methods for solving a constrained-minimization problem:

(a) The *parametric interior-point and exterior-point methods* using a penalty function with parameters r and s (eventually one parameter r) to control the convergence towards a minimum solution. We introduced a classification of these methods based upon the behaviour of barrier and loss functions at the boundary of the constraint set. For barrier functions, the classification is based upon the behaviour of ϕ', the derivative of the function ϕ which describes the transformation of the *constraints* prior to their incorporation in the barrier term. For loss functions, the classification is based upon the behaviour of ω', the derivative of the function ω which describes the transformation of the constraint *violations*. If ϕ' has a pole and ω' a zero of *finite* order at the origin, the methods show a remarkable similarity in several aspects (basis for extrapolation, ill-conditioning of the principal Hessian matrix). Furthermore, extensive computational experience with these methods exists, although we did not mention it explicitly. The methods which do not possess the properties just sketched have not been thoroughly explored either. Their peculiar (and perhaps promising) features have been shown by means of a few examples.

(b) The *non-parametric interior-point and exterior-point methods* using a penalty function without controlling parameters. Here, convergence is controlled by moving truncations of the constraint set and by a sequence of truncation levels converging to the minimum value of the original problem. The methods under consideration are, in fact, the non-parametric counterparts of the parametric methods with the properties that ϕ' has a pole and ω' a zero of *finite* order at the origin. Non-parametric methods provide an automatic adjustment of the controlling parameter. For the rest, they behave like parametric methods, and extrapolation techniques could be employed successfully provided that the adjusted controlling parameter values have a satisfactory rate of convergence (linear convergence with an error constant $\approx \frac{1}{2}$). Our theoretical analysis shows that such a rate of convergence is difficult to obtain. We therefore hesitate to recommend non-parametric methods. Instead, we prefer the parametric methods where the rate of convergence can be freely chosen.

(c) *Lagrangian methods* using the Lagrangian function associated with the original problem, a generalized Lagrangian function, or an augmented Lagrangian function. A direct comparison with the interior-point and exterior-point methods is rather difficult. First, the Lagrangian methods cannot be applied to the inequality-constrained problem introduced at the start of the survey, but they need additional safeguards. Some Lagrangian methods have been developed for equality-constrained problems only. Finally, these methods suffer from a serious lack of computational experience, although the theoretical work has been impressive.

So far, the most general method seems to be the parametric method sketched in Section 2.7. It employs a mixed penalty function and can be used for problems with inequality as well as equality constraints.

We conclude this survey by mentioning some methods and ideas that have not been mentioned in the previous sections.

First, we may note that we have only been dealing with methods incorporating all the constraints in a penalty function regardless of whether they are linear or not. It might be attractive to treat the linear constraints separately. If only the non-linear constraints are included in a penalty function the computational method for solving (1) reduces to one of sequential *linearly-constrained* minimization. We indicated this approach when we were dealing with the method of Benders (1960). Unfortunately, numerical experience with an approach of this kind seems to be rather limited (see also Davies, 1970).

The idea of sequential minimization has been dropped by several authors (see Zoutendijk, 1966; Butz, 1967; Murray, 1969b). They presented a more direct approach and proposed to solve problem (1) by univariate searches each of which starts with a new, adjusted value of the controlling parameter. Computational experience with these methods does not exist, however, or is rather difficult to obtain.

A new class of penalty-function methods is constituted by the methods with *exponential penalty functions* (Allran and Johnson, 1970; Evans and Gould, 1970). There are no new features in these methods, however. This is also fully recognized by Evans and Gould (1970) so that a detailed description has been omitted.

This concludes our survey. It will be clear that there are still many directions for future research, particularly if one is concerned with the computational efficiency of these methods.

References

Abadie, J. (1967). 'Nonlinear Programming'. North-Holland Publishing Co., Amsterdam.
Abadie, J. (1970). 'Integer and Nonlinear Programming'. North-Holland Publishing Co., Amsterdam.
Ablow, C. M. and Brigham, G. (1955). An analog solution of programming problems. *Operat. Res.* **3**, 388–394.
Allran, R. R. and Johnson, S. E. J. (1970). An algorithm for solving nonlinear programming problems subject to nonlinear inequality constraints. *Comput. J.* **13**, 171–177.
Arrow, K. J., Hurwicz, L., and Uzawa, H. (1961). Constraint qualifications in maximization problems. *Naval Res. Log. Quart.* **8**, 175–191.
Bellmore, M., Greenberg, H. J., and Jarvis, J. J. (1970). Generalized penalty-function concepts in mathematical optimization. *Operat. Res.* **18**, 229–252.
Beltrami, E. J. (1967). A computational approach to necessary conditions in mathematical programming. *ICC Bulletin* **6**, 265–273.
Beltrami, E. J. (1969a). A constructive proof of the Kuhn-Tucker multiplier rule. *J. Math. Anal. Appls* **26**, 297–306.
Beltrami, E. J. (1969b). A comparison of some recent iterative methods for the numerical solution of nonlinear programs. *In*: 'Computing Methods in Optimization Problems' (M. Beckman, ed.), pp. 20–29. Springer, Berlin.
Benders, J. F. (1960). 'Partitioning in Mathematical Programming.' Avanti, Delft.
Box, M. J., Davies, D., and Swann, W. H. (1969). 'Nonlinear Optimization Techniques'. ICI Monograph No. 5, Oliver and Boyd, London.
Bracken, J., and McCormick, G. P. (1968). 'Selected Applications of Nonlinear Programming'. Wiley, New York.
Brooks, R., and Geoffrion, A. (1966). Finding Everett's Lagrange multipliers by linear programming. *Operat. Res.* **14**, 1148–1153.
Bui Trong, Lieu, and Huard, P. (1966). La méthode des centres dans un espace topologique. *Numer. Math.* **8**, 56–67.
Bulirsch, R. (1964). Bemerkungen zur Romberg-Integration. *Numer. Math.* **6**, 6–16.
Bulirsch, R., and Stoer, J. (1964). Fehlerabschätzungen und Extrapolation mit Rationalen Funktionen bei Verfahren vom Richardson-Typus. *Numer. Math.* **6**, 413–427.
Bulirsch, R., and Stoer, J. (1966). Asymptotic upper and lower bounds for results of extrapolation methods. *Numer. Math.* **8**, 93–104.
Butler, T., and Martin, A. V. (1962). On a method of Courant for minimizing functionals. *J. math. Phys.* **41**, 291–299.
Butz, A. R. (1967). Iterative saddle point techniques. *SIAM J. appl. Math.* **15**, 719–725.

Camp, G. D. (1955). Inequality-constrained stationary value problems. *Operat. Res.* **3**, 548–550.

Carpentier, J. (1962). Contribution à l'étude du dispatching économique. *Bull. Soc. Fr. des El.*, 8 ième série, **III**, 431–447.

Carroll, C. W. (1961). The created response surface technique for optimizing nonlinear restrained systems. *Operat. Res.* **9**, 169–184.

Colville, A. R. (1968a). 'Mathematical Programming Codes'. IBM New York Scientific Center, Techn. Rep. 320-2925.

Colville, A. R. (1968b). 'A Comparative Study of Nonlinear Programming Codes'. IBM New York Scientific Center, Techn. Rep. 320-2949.

Conn, A. R. (1971). 'A Gradient Type Method of Locating Constrained Minima'. Dept. of Appl. Anal. Comp. Sc. Research Report CSRR 2032, University of Waterloo, Ontario, Canada.

Courant, R. (1943). Variational methods for the solution of problems of equilibrium and vibrations. *Bull. Am. Math. Soc.* **49**, 1–23.

Davies, D. (1970). Some practical methods of optimization. *In*: 'Integer and Nonlinear Programming' (J. Abadie, ed.), pp. 87–117. North-Holland Publishing Co., Amsterdam.

Dorn, W. S. (1960a). Duality in quadratic programming. *Quart. Appl. Math.* **18**, 155–162.

Dorn, W. S. (1960b). A duality theorem for convex programming. *IBM J. Res. Dev.* **4**, 407–413.

Dorn, W. S. (1961). On Lagrange multipliers and inequalities. *Operat. Res.* **9**, 95–104.

Evans, J. P., and Gould, F. J. (1970a). Stability in nonlinear programming. *Operat. Res.* **18**, 107–118.

Evans, J. P., and Gould, F. J. (1970b). 'Stability and Exponential Penalty Function Techniques in Nonlinear Programming'. Dept. of Statistics, University of North Carolina, Chapel Hill, North Carolina.

Everett, H. (1963). Generalized Lagrange multiplier method for solving problems of optimum allocation of resources. *Operat. Res.* **11**, 399–417.

Falk, J. E. (1967). Lagrange multipliers and nonlinear programming. *J. Math. Anal. Appls* **19**, 141–159.

Falk, J. E. (1969), A relaxed interior approach to nonlinear programming. *Z. Wahrscheinlichk. verw. Geb.* **11**, 327–337.

Faure, P., and Huard, P. (1965). Résolution des programmes mathématiques à fonction nonlineaire par le méthode du gradient réduit. *Revue fr. Rech. opér.* **9**, 167–205.

Faure, P., and Huard, P. (1966). 'Résultats Nouveaux Relatifs à la Méthode des Centres'. Quatrième conférence de recherche opérationelle, Cambridge.

Fiacco, A. V. (1967). 'Sequential Unconstrained Minimization Methods for Nonlinear Programming'. Thesis, Northwestern University, Evanston, Illinois.

Fiacco, A. V. (1969). A general regularized sequential unconstrained minimization technique. *SIAM J. appl. Math.* **17**, 1239–1245.

Fiacco, A. V., and McCormick, G. P. (1964a). The sequential unconstrained minimization technique for nonlinear programming, a primal-dual method. *Managmt Sci.* **10**, 360–366.

Fiacco, A. V., and McCormick, G. P. (1964b). Computational algorithm for the sequential unconstrained minimization technique for nonlinear programming. *Managmt Sci.* **10**, 601–617.

Fiacco, A. V., and McCormick, G. P. (1966). Extensions of SUMT for nonlinear programming: equality constraints and extrapolation. *Managmt Sci.* **12**, 816–828.

Fiacco, A. V., and McCormick, G. P. (1967a). The slacked unconstrained minimization technique for convex programming. *SIAM J. appl. Math.* **15**, 505–515.

Fiacco, A. V., and McCormick, G. P. (1967b). The sequential unconstrained minimization technique without parameters. *Operat. Res.* **15**, 820–827.

Fiacco, A. V., and McCormick, G. P. (1968). 'Nonlinear Programming, Sequential Unconstrained Minimization Techniques'. Wiley, New York.

Fiacco, A. V., and Jones, A. P. (1969). Generalized penalty methods in topological spaces. *SIAM J. appl. Math.* **17**, 996–1000.

Fletcher, R. (1969). 'Optimization'. Academic Press, London.

Fletcher, R. (1970a). A class of methods for nonlinear programming with termination and convergence properties. *In*: 'Integer and Nonlinear Programming' (J. Abadie, ed), pp. 157–175. North-Holland Publishing Co., Amsterdam.

Fletcher, R. (1970b). A new approach to variable metric algorithms. *Comput. J.* **13**, 317–322.

Fletcher, R., and Powell, M. J. D. (1963). A rapidly convergent descent method for minimization. *Comput. J.* **6**, 163–168.

Fletcher, R., and McCann, A. P. (1969). Acceleration techniques for nonlinear programming. *In*: 'Optimization' (R. Fletcher, ed.), pp. 203–214. Academic Press, London.

Frisch, R. (1955). 'The Logarithmic Potential Method for Solving Linear Programming Problems'. Memorandum of the University Institute of Economics, Oslo.

Gould, F. J. (1969). Extensions of Lagrange multipliers in nonlinear programming. *SIAM J. appl. Math.* **17**, 1280–1297.

Gould, F. J. (1970). A class of inside-out algorithms for general programs. *Managmt. Sci.* **16**, 350–356.

Gould, F. J., and Howe, S. (1971). 'A New Result on Interpreting Lagrange Multipliers as Dual Variables'. Dept. of Statistics, University of North Carolina, Chapel Hill, North Carolina.

Graves, R. L., and Wolfe, P. (1963). 'Recent Advances in Mathematical Programming'. McGraw-Hill, New York.

Haarhoff, P. C., and Buys, J. D. (1970). A new method for the optimization of a nonlinear function subject to nonlinear constraints. *Comput. J.* **13**, 178–184.

Hadley, G. (1964). 'Nonlinear and Dynamic Programming'. Addison-Wesley, Reading, Mass.

Hestenes, M. R. (1969). Multiplier and gradient methods. *J. Optim. Theory Applns* **4**, 303-320.

Huard, P. (1962). Dual programs. *IBM J. Res. Dev.* **6**, 137–139.

Huard, P. (1963). Dual programs. *In*: 'Recent Advances in Mathematical Programming' (R. L. Graves and P. Wolfe, eds.), pp. 55–62. McGraw-Hill, New York.

Huard, P. (1964). 'Résolution de Programmes Mathématiques à Contraintes Nonlinéaires par la Méthode des Centres'. Note E.D.F., HR 5690/3/317.

Huard, P. (1967). Resolution of mathematical programming with nonlinear constraints by the method of centres. *In*: 'Nonlinear Programming' (J. Abadie, ed.), pp. 207–219. North-Holland Publishing Co., Amsterdam.

Huard, P. (1968). Programmation mathématique convexe. *R.I.R.O.* **7**, 43–59.

John, F. (1948). Extremum problems with inequalities as subsidiary conditions. *In*: 'Studies and Essays,' pp. 187–204. Courant Anniversary Volume. Interscience, New York.

Kowalik, J. (1966). Nonlinear programming procedures and design optimization. *Acta Polytech. Scand.* **13**, Trondheim.

Kowalik, J., and Osborne, M. R. (1968). 'Methods for Unconstrained Optimization Problems'. Elsevier, New York.

Kowalik, J., Osborne, M. R., and Ryan, D. M. (1969). A new method for constrained optimization problems. *Operat. Res.* **17**, 973–983.

Kuhn, H. W., and Tucker, A. W. (1951). Nonlinear programming. *In*: 'Proceedings of the Second Symposium on Mathematical Statistics and Probability' (J. Neyman, ed.), pp. 481–493. University of California Press, Berkeley.

Laurent, P. J. (1963). Un théorème de convergence pour le procédé d'extrapolation Richardson de. *C.r. hebd. Seanc. Acad. Sci. Paris* **256**, 1435–1437.

Lootsma, F. A. (1967). Logarithmic programming: a method of solving nonlinear-programming problems. *Philips Res. Repts.* **22**, 329–344.

Lootsma, F. A. (1968a). Extrapolation in logarithmic programming. *Philips Res. Repts.* **23**, 108–116.

Lootsma, F. A. (1968b). Constrained optimization via penalty functions. *Philips Res. Repts.* **23**, 408–423.

Lootsma, F. A. (1968c). Constrained optimization via parameter-free penalty functions. *Philips Res. Repts.* **23**, 424–437.

Lootsma, F. A. (1969). Hessian matrices of penalty functions for solving constrained-optimization problems. *Philips Res. Repts.* **24**, 322–331.

Lootsma, F. A. (1970). Boundary properties of penalty functions for constrained minimization. *Philips Res. Repts.* Suppl. No. 3.

Mangasarian, O. L. (1962). Duality in nonlinear programming. *Quart. Appl. Math.* **20**, 300–302.

Magasarian, O. L. (1965). Pseudo-convex functions. *SIAM J. Contr.* **3**, 281–290.

Mangasarian, O. L., and Fromowitz, S. (1967). The Fritz John necessary optimality conditions in the presence of equality and inequality constraints. *J. Math. Anal. Appls* **17**, 37–47.

McCormick, G. P. (1967). Second order conditions for constrained minima. *SIAM J. appl. Math.* **15**, 641–652.

McCormick, G. P., Mylander, W. C., and Fiacco, A. V. (1965). 'Computer Program Implementing the Sequential Unconstrained Minimization Technique for Nonlinear Programming'. Technical paper, RAC-TP-151, Research Analysis Corporation, McLean, Va., U.S.A.

Morrison, D. D. (1968). Optimization by least squares. *SIAM J. Numer. Anal.* **5**, 83–88.

Murray, W. (1967). 'Ill-conditioning in Barrier and Penalty Functions Arising in Constrained Nonlinear Programming'. Sixth symp. on math. programming, Princeton, N.Y.

Murray, W. (1969a). 'Behaviour of Hessian Matrices of Barrier and Penalty Functions Arising in Optimization'. Report Ma 77, National Physical Laboratory, Teddington, England.

Murray, W. (1969b). 'Constrained Optimization'. Report Ma 79, National Physical Laboratory, Teddington, England.

Osborne, M. R., and Ryan, D. M. (1970). On penalty function methods for nonlinear programming problems. *J. Math. Anal. Appls* **31**, 559–578.

Parisot, G. R. (1961). Résolution numérique approchée du problème de programmation linéaire par application de la programmation logarithmique. *Revue fr. Rech. opér.* **20**, 227–259.

Pearson, J. D. (1968). 'Powell's Method for Equality Constraints, Quadratic Programming'. Working paper RAC-S-1994, Research Analysis Corporation, McLean, Va., U.S.A.

Pietrzykowski, T. (1962). Application of the steepest descent method to concave programming. *In*: 'Proceedings of the IFIPS Congress, Munich, 1962', pp. 185–189. North-Holland Publishing Co., Amsterdam.
Pietrzykowski, T. (1969). An exact potential method for constrained maxima. *SIAM J. Numer. Anal.* **6**, 229–304.
Pomentale, T. (1965). A new method for solving conditioned maxima problems. *J. Math. Anal. Appls* **10**, 216–220.
Powell, M. J. D. (1968). A method for nonlinear constraints in minimization problems. *In*: 'Optimization' (R. Fletcher, ed.), pp. 283–298. Academic Press, London.
Roode, J. D. (1968). 'Generalized Lagrangian Functions in Mathematical Programming'. Thesis, Leiden, The Netherlands.
Rosenbrock, H. H. (1960). An automatic method for finding the greatest or least values of a function. *Comput. J.* **3**, 175–184.
Strong, R. E. (1965). A note on the sequential unconstrained minimization technique for nonlinear programming. *Managmt Sci.* **12**, 142–144.
Tremolières, R. (1968). 'La Méthode des Centres à Troncature Variable'. Thèse, Paris.
Wilde, D. J., and Beightler, C. S. (1967). 'Foundations of Optimization'. Prentice-Hall, Englewood Cliffs, N.J.
Wolfe, P. (1961). A duality theorem for nonlinear programming. *Quart. Appl. Math.* **19**, 239–244.
Wolfe, P. (1963). Methods of nonlinear programming. *In*: 'Recent Advances in Mathematical Programming' (R. L. Graves and P. Wolfe, eds.), pp. 67–86. McGraw-Hill, New York.
Zangwill, W. I. (1967). Nonlinear programming via penalty functions. *Managmt Sci.* **13**, 344–358.
Zoutendijk, G. (1960). 'Methods of Feasible Directions'. Elsevier, Amsterdam.
Zoutendijk, G. (1966). Nonlinear programming: a numerical survey. *SIAM J. Contr.* **4**, 194–210.
Zoutendijk, G. (1970). Nonlinear programming, computational methods. *In*: 'Integer and Nonlinear Programming' (J. Abadie, ed.), pp. 37–86. North-Holland Publishing Co., Amsterdam.

24. Non-linear Tolerance Programming

F. J. GOULD

Department of Statistics and Curriculum in Operations Research and Systems Analysis, University of North Carolina at Chapel Hill, U.S.A.

Summary

A continuously differentiable exact penalty function along with some numerical results is presented for solving non-linear programs with tolerances. Possible merits of tolerance programming, for problems with inexact data, are discussed.

1. Introduction

The purpose of this paper is, first, to suggest an approach different from conventional optimization theory for real-world problem solving and, second, to present a specific example of an implementation of the suggested approach. The implementation is in terms of the non-linear programming problem

$$P_0: \max f(x) \text{ subject to}$$

$$g_j(x) \leq 0, \quad j = 1, \ldots, m,$$

where it shall be assumed that all functions are real valued on R^n and k times continuously differentiable, i.e. all are elements of $C^k(R^n)$. Relative to this problem the following question is of theoretical interest and, seemingly, of computational importance. Is it possible to explicitly exhibit a function $\phi(x)$, $\phi \in C^k(R^n)$, such that ϕ attains an *unconstrained* maximum on R^n and any global (local) unconstrained maximum of ϕ is a global (local) solution to P_0? Such a function, $\phi(x)$, is sometimes termed an exact penalty function for P_0. To date it is not known whether or not this question can be answered in the affirmative. The major difficulty seems to be the stipulation, for cases of $k \geq 1$ that $\phi \in C^k(R^n)$. Roode (1968, p. 119), Pietrzykowski (1969), and Zangwill (1967) discuss a penalty function which is exact for concave programs P_0 but which is not everywhere differentiable. Evans *et al.* (1971) discuss exact penalty functions for more general problems P_0 but these functions also fail to be everywhere differentiable. This poses obvious difficulties in conducting the unconstrained optimization. If an exact penalty function in $C^k(R^n)$ can be discovered, then it is plausible that such knowledge could make it easier to solve P_0. (This question, however, would need to be further studied, for the function ϕ could be sufficiently complicated that it would in fact be more difficult to find an unconstrained maximum of ϕ than to find a constrained maximum of f by currently known asymptotic techniques.) The approach to be exposited in this paper yields such a function for problems P_0 where the right-hand sides are allowed to lie within certain tolerances. The behaviour of the derived functions, $\phi(x)$, has been studied on several test problems. The results are presented in Sections 3 and 4 and possible advantages are discussed.

2. An Approximation Seeking Approach

In describing the rationale of this method, it is first noted that in many practical problems sharp optimization may be unwarranted. That is, the problem P_0 is often a model builder's representation of some real-world process. For this reason alone most models are an inexact representation of reality. Furthermore, even given an exact model the data is often 'fuzzy'. The ideas discussed here do not appear to be fundamentally related to the 'fuzzy set' theory as, for example, exposited by Bellman and Zahdeh (1970), though it may be of interest to further explore possible relations. In applied work one often encounters problems where, though the data is fuzzy, it is known from past experience that the optimal policy or the optimal return is not highly sensitive to the tolerances in the data. As a consequence, the model builder may well be willing to replace the right-hand side of P_0 with any scalars b_j such that $-\epsilon_j \leqslant b_j \leqslant \epsilon_j$, where ϵ_j is some acceptable tolerance. (In solving problems with computer arithmetic one implicitly is always making such an assumption, where ϵ_j depends both on the hardware and the value of b_j. In many cases, ϵ_j is on the order of 10^{-6}. I am suggesting here that in many situations a 'larger' value of ϵ may also be acceptable to the practitioner.) Thus, in practice, we often see an exact optimization theory being applied to inexact models, where, in a sense, there is no justification for writing explicit numbers, and where all too frequently the theory is computationally hopeless for large-scale systems. What seems to be a possibility here is that a more powerful (computationally more useful) theory can be obtained by explicitly, and in a well-defined way, building a certain amount of inexactness into the optimization.

In essence, then, we consider situations where the following ground rules are in effect: it is acceptable to replace the 'conventional problem' P_0 with a continuum of surrogate problems, in the sense that one is willing to settle for an optimal solution to *any one* of the infinitely many problems, without caring which one, and perhaps even with indifference as to knowing which one has been solved. The contemplated replacement could be made in numerous ways; tolerances on right-hand sides are only one such way. For example, in linear programming (Gould, 1971) there might also be tolerances on objective function coefficients, or perhaps only on objective function coefficients. The point of making such a replacement would be the hope that certain advantages might be obtained in attempting to find an optimal solution to some unspecified member of the surrogate collection as opposed to the one specific problem P_0.

It has been pointed out by Powell, that Rosenbrock (1960) presented a perturbation of the objective function in such a way as to make it decrease to zero in a narrow band about the constraint region. Outside the band and inside the constraint region the function has its true (positive) value and outside the band outside of the constraint region the objective is identically zero. Unconstrained methods can then be applied to this perturbed function. Rosenbrock's determination of band width is related to the notion of tolerances discussed in this paper.

3. A Specific Illustration

I use the problem P_0 as a vehicle for illustrating the above ideas. Define the *multiplier functions* (Gould, 1970) $\lambda_j : R \to R \cup \{+\infty\}$ as follows:

$$\lambda_j(\xi) = \begin{cases} 0, & \xi \leq -\epsilon_j, \\ (\xi + \epsilon_j)^2, & -\epsilon_j < \xi \leq 0, \\ \dfrac{-2\epsilon_j^3}{\xi - \epsilon_j} - \epsilon_j^2, & 0 < \xi < \epsilon_j, \\ \infty, & \xi \geq \epsilon_j, \end{cases}$$

where ϵ_j is the user specified parameter discussed in Section 2. Now define the penalty function $\phi: R^n \to R \cup \{-\infty\}$ as

$$\phi(x) = f(x) - \sum_{j=1}^{m} \lambda_j(g_j(x)).$$

Finally, define

$$S_\epsilon = \{x : g_j(x) \leq \epsilon_j, j = 1, \ldots, m\},$$
$$S_\epsilon^\circ = \{x : g_j(x) < \epsilon_j, j = 1, \ldots, m\}.$$

Theorem

Assume S_ϵ is compact. Then:

(1) ϕ attains a global maximum (over R^n) and any such global maximizer is in the set S_ϵ°.
(2) If the functions in P_0 are all in $C^1(R^n)$, then ϕ is continuously differentiable on S_ϵ°.

(3) Let $\hat{x}$ be any element of S_ϵ° which is a global (local) maximizer of $\phi(x)$. Let $J_1 = \{i: -\epsilon_i < g_i(\hat{x}) < \epsilon_i\}$, $J_2 = \{i: g_i(\hat{x}) \leq -\epsilon_i\}$. Then $\hat{x}$ is a global (local) solution to

$$\max f(x) \text{ subject to}$$
$$g_i(x) \leq g_i(\hat{x}), \quad i \in J_1,$$
$$g_i(x) \leq -\epsilon_i, \quad i \in J_2.$$

Proof

The first conclusion is intuitively clear and is a standard result on barrier functions (penalty functions which are infinite on the constraint set boundary). The proof is a straightforward analogue of one appearing both in Fiacco and McCormick (1964) and in Gould (1970). The second conclusion is easily verified from the definition of $\lambda_j(\cdot)$ and $\phi(\cdot)$. To prove the third conclusion, suppose $\hat{x}$ is a global maximizer of $\phi(x)$. Then, noting that $J_1 \cup J_2 = \{i: 1 \leq i \leq m\}$,

$$\phi(\hat{x}) = f(\hat{x}) - \sum_{J_1} \lambda_i(g_i(\hat{x})) - \sum_{J_2} \lambda_i(g_i(\hat{x})) \geq f(x) - \sum_{J_1} \lambda_i(g_i(x)) - \sum_{J_2} \lambda_i(g_i(x))$$

for all x in R^n. Hence, since $\lambda_i(g_i(\hat{x})) = \lambda_i(-\epsilon_i) = 0$ for $i \in J_2$,

$$f(\hat{x}) - \sum_{J_1} \lambda_i(g_i(\hat{x})) - \sum_{J_2} \lambda_i(-\epsilon_i) \geq f(x) - \sum_{J_1} \lambda_i(g_i(x)) - \sum_{J_2} \lambda_i(g_i(x))$$

for all x in R^n. This implies

$$f(\hat{x}) \geq f(x) + \sum_{J_1} [\lambda_i(g_i(\hat{x})) - \lambda_i(g_i(x))] + \sum_{J_2} [\lambda_i(-\epsilon_i) - \lambda_i(g_i(x))]$$

for all x in R^n. Finally, this implies $f(x) \leq f(\hat{x})$ whenever $g_i(x) \leq g_i(\hat{x})$, $i \in J_1$, $g_i(x) \leq -\epsilon_i$, $i \in J_2$. The proof for the local result is analogous.

It can also be shown that the above penalty function $\phi(x)$ can be replaced with one which is k times continuously differentiable, provided the problem functions in P_0 are k times continuously differentiable, and results 1 and 3 still hold. Furthermore, simple modifications can be made to allow only left hand tolerances about zero (i.e. to force the approximate solution $\hat{x}$ to be feasible in P_0). Computational results for the above ϕ function are given in Section 4.

3.1. *An Alternative Form*

Somewhat simpler expressions for the multiplier functions are

$$\lambda_j(\xi) = \begin{cases} 0, & \xi \leq 0, \\ \dfrac{-1}{\xi^2 - \epsilon_j^2} - \dfrac{1}{\epsilon^2}, & 0 < \xi < \epsilon_j, \\ \infty, & \xi \geq \epsilon_j. \end{cases}$$

The above theorem also applies when ϕ is defined in terms of these functions.

3.2. *Handling Equality Constraints*

An equality constraint $g_i(x) = 0$ can be represented by two inequalities $g_i(x) \leq 0$ and $-g_i(x) \leq 0$. These two constraints can then be handled as in the above discussion. However, it is more efficient to define separate multiplier functions for equalities. Suppose, instead of P_0, we consider

$$P_0': \max f(x) \text{ subject to}$$
$$g_i(x) \leq 0, \quad i \in I_1,$$
$$g_i(x) = 0, \quad i \in I_2.$$

Let

$$S_\epsilon = \{x: g_i(x) \leq \epsilon_i, i \in I_1\} \cap \{x: -\epsilon_i \leq g_i(x) \leq \epsilon_i, i \in I_2\}$$

and

$$S_\epsilon^\circ = \{x: g_i(x) < \epsilon_i, i \in I_1\} \cap \{x: -\epsilon_i < g_i(x) < \epsilon_i, i \in I_2\}.$$

Define the multiplier functions $\lambda_i(\xi)$, for $i \in I_1$, as discussed above. For $i \in I_2$, define

$$\lambda_i(\xi) = \begin{cases} \dfrac{-1}{\xi^2 - \epsilon_i^2}, & -\epsilon_i < \xi < \epsilon_i, \\ \infty, & |\xi| \geq \epsilon_i. \end{cases}$$

Define the tolerance function $\phi(x)$ as

$$\phi(x) = f(x) - \sum_{i \in I_1} \lambda_i(g_i(x)) - \sum_{i \in I_2} \lambda_i(g_i(x)).$$

Then it can be shown, essentially as in the above theorem, that:

(1) ϕ attains a global maximum over R^n and any such global maximizer is in the set S_ϵ°.
(2) If the functions in P_0 are all in $C^1(R^n)$, then ϕ is continuously differentiable on S_ϵ°.
(3) Let $\hat{x}$ be any element of S_ϵ° which is a global (local) maximizer of $\phi(x)$. Define J_1 and J_2 as in the theorem. Then $\hat{x}$ is a global (local) solution to

$$\max f(x) \text{ subject to}$$
$$g_i(x) \leq g_i(\hat{x}), \quad i \in J_1 \subset I_1,$$
$$g_i(x) \leq -\epsilon_i, \quad i \in J_2 \subset I_1,$$

and

$$g_i(x) = g_i(\hat{x}), \quad i \in I_2,$$

where

$$-\epsilon_i < g_i(\hat{x}) < \epsilon_i, \quad i \in I_2.$$

3.3. Handling Non-Negativity Conditions

Considering the condition $x_j \geqslant 0$, we define a constraint $h_j(x)$ as $h_j(x) = -x_j \leqslant 0$ and define a corresponding multiplier function

$$\lambda(h_j(x)) = \begin{cases} 0, & h_j(x) \leqslant -2\epsilon, \\ (h_j(x) + 2\epsilon)^2, & -2\epsilon < h_j(x) \leqslant -\epsilon, \\ \dfrac{-2\epsilon^3}{h_j(x)} - \epsilon^2, & -\epsilon < h_j(x) < 0, \\ \infty, & h_j(x) \geqslant 0. \end{cases}$$

Alternatively, we could define

$$\lambda(h_j(x)) = \begin{cases} 0, & h_j(x) \leqslant -\epsilon, \\ \dfrac{-1}{(\xi + \epsilon)^2 - \epsilon^2} - \dfrac{1}{\epsilon^2}, & -\epsilon < \xi < 0, \\ \infty, & \xi \geqslant 0. \end{cases}$$

Either of the above forms prevents the variable x_j from becoming negative.

3.4. Relations to the Optimal Value for P_0

Maximizing the ϕ function gives a solution to some problem

$$P_b: \max f(x) \text{ subject to}$$
$$g_j(x) \leqslant b_j, \quad j = 1, \ldots, m,$$

where the values b_j are scalars such that $-\epsilon_j \leqslant b_j \leqslant \epsilon_j$, $j = 1, \ldots, m$. One naturally wonders how this obtained solution relates to the solution to P_0, especially when the values ϵ_j are chosen quite small. That is, will the optimal value of the objective function vary continuously with right-hand side changes? This question is discussed by Evans and Gould (1970). It can be shown that if

(i) S_0° non-empty, closure $(S_0^\circ) = S_0$ (where $S_0 = S_\epsilon$, $\epsilon = 0$),
(ii) $\exists \epsilon > 0$ such that S_ϵ is compact,

then the $f_{\sup}$ function (i.e. the optimal objective value as a function of the right-hand side) is continuous at zero.

Finally, as described in Gould (1969), the study of supports to the $f_{\sup}$ function enables one to obtain an upper bound on the optimal objective value in P_0. In particular, if x^* is an optimal global solution to P_0, and if $\hat{x}$ is a global maximizer of the tolerance function, then

$$f(x^*) \leqslant f(\hat{x}) + \sum_{j \in J_1} [\lambda_j(0) - \lambda_j(g_j(\hat{x}))] + \sum_{j \in J_2} [\lambda_j(0) - \lambda_j(-\epsilon_j)]$$
$$= f(\hat{x}) + \sum_{j \in J_1} [\epsilon_j^2 - \lambda_j(g_j(\hat{x}))] + \sum_{j \in J_2} \epsilon_j^2.$$

If P_0 is a concave program then the hyperplane

$$f(\hat{x}) + \sum_{j \in J_1} \lambda'_j(g_j(\hat{x}))[b_j - g_j(\hat{x})] + \sum_{j \in J_2} \lambda'_j(-\epsilon_j)[b_j - (-\epsilon_j)]$$

is a support to $f_{\sup}(b)$ and consequently we obtain the tighter bound

$$f(x^*) \leqslant f(\hat{x}) - \sum_{j \in J_1} \lambda'_j(g_j(\hat{x})) g_j(\hat{x}).$$

3.5. Relations to Other Penalty Functions

It is clear that the tolerance function can be interpreted as an exterior function in terms of the problem with right-hand sides $-\epsilon_j$ and an interior function in terms of the problem with right-hand sides ϵ_j. A major advantage of these attributes is the certainty that for the ϕ function *prespecified* tolerances can be achieved in a single iteration. Consider, by contrast, the quadratic loss function

$$Q(x, \alpha) = f(x) - \alpha \sum_{j=1}^{m} \tfrac{1}{4}(g_j(x) + |g_j(x)|)^2.$$

If $\hat{x}$ maximizes $Q(x, \alpha)$ then it is not difficult to show that $\hat{x}$ is a solution to the following relaxation of P_0:

$$\max f(x) \text{ subject to}$$

$$g_j(x) \leqslant 0 \quad \text{if } g_j(\hat{x}) \leqslant 0,$$
$$g_j(x) \leqslant g_j(\hat{x}) \quad \text{if } g_j(\hat{x}) > 0.$$

However, there is no guarantee that $g_j(\hat{x})$ will be within a prespecified tolerance of zero in a single iteration. Thus $g_j(\hat{x})$ may be too large and further iterations may be required. In general, considerable manipulation of the parameter α is required to make $g_j(\hat{x}) < \epsilon_j$. The interior attribute of the ϕ function guarantees that $g_j(\hat{x}) < \epsilon_j$ in a single iteration. On the other hand, consider an interior function such as

$$I(x, r) = f(x) + r \sum_{j=1}^{m} \ln(-g_j(x)).$$

If $\hat{x}$ maximizes $I(x, r)$ then $\hat{x}$ is a solution to the following subproblem:

$$\max f(x) \text{ subject to}$$

$$g_j(x) \leqslant g_j(\hat{x}), \quad j = 1, \ldots, m.$$

In this case $g_j(\hat{x}) < 0$, all j. However, again there is no guarantee that $g_j(\hat{x})$ will be within a prespecified tolerance of zero in a single iteration. In this case, $g_j(\hat{x})$ may be too small. The exterior attribute of the ϕ function guarantees that any $g_j(\hat{x})$ which is smaller than $-\epsilon_j$ can be replaced with $-\epsilon_j$. There are, of course, other termination criteria for interior methods, such as upper and lower bounds on the optimal objective value for P_0, but these criteria apply only to concave programs.

Finally, the tolerance function is more robust in that, as with other interior functions, a global maximum exists under milder conditions than with exterior methods (i.e. only continuity of the penalty function on S_ϵ° is required).

Computational performance of the tolerance function is compared with other methods in Section 4. It will be seen that for the examples studied the tolerance function compares quite favourably with both the interior and the exterior method.

4. Numerical Tests

4.1. Rosen–Suzuki Test Problem

This problem, a convex programming problem, is

$$\min f(x) = x_1^2 + x_2^2 + 2x_3^2 + x_4^2 - 5x_1 - 5x_2 - 21x_3 + 7x_4,$$

subject to

$$g_1(x) = x_1^2 + x_2^2 + x_3^2 + x_4^2 + x_1 - x_2 + x_3 - x_4 - 8 \leqslant 0,$$

$$g_2(x) = x_1^2 + 2x_2^2 + x_3^2 + 2x_4^2 - x_1 - x_4 - 10 \leqslant 0,$$

$$g_3(x) = 2x_1^2 + x_2^2 + x_3^2 + 2x_1 - x_2 - x_4 - 5 \leqslant 0.$$

The starting point is $(0,0,0,0)$, and the optimum point is $x^* = (0,1,2,-1)$, with $f(x^*) = -44$. At optimality the constraint values are given by

$$g_1(x^*) = 0,$$
$$g_2(x^*) = -1,$$
$$g_3(x^*) = 0.$$

The problem was first reported in Rosen and Suzuki (1965). The ϕ function was defined in terms of multiplier functions given in the first paragraph of Section 3. Problems were run with tolerances in the constant terms of $\pm 5\%$ and $\pm 1\%$. (For example, for $g_1(x)$, ϵ_1 was given by $5\% (8) = 0.40$, and $1\% (8) = 0.08$.) The Davidon-Fletcher-Powell procedure with golden section line searches was used for all unconstrained minimizations. The stopping criterion for all runs was the simultaneous satisfaction of a relative and absolute accuracy of *EPS* in the objective function and an absolute accuracy of *EPS* in the variables. The runs were done on the IBM 360/50 computer. For tolerance

24. NON-LINEAR TOLERANCE PROGRAMMING

function runs, Table 1 presents the number of iterations (IT), the optimal value of x, the value of $f(x)$, the value of EPS, and the number of seconds of CPU time consumed in execution. The values b_1, b_2, b_3 indicate the right-hand sides of the subproblem for which x is an optimal solution. That is,

$$b_j = \begin{cases} g_j(x), & j \in J_1, \\ -\epsilon_j, & j \in J_2. \end{cases}$$

In the Davidon–Fletcher–Powell procedure the update matrix was reinitialized to the identity after $n+1$ iterations. These results were compared with runs

TABLE 1
Rosen–Suzuki results: ϕ function

Tolerance	IT	x_1	x_2	x_3	x_4	$f(x)$	b_1	b_2	b_3	EPS	Time (sec)
±5%	10	0·023	1·0104	2·0224	−0·9704	−43·892	0·0592	−0·500	0·118	10^{-3}	3·05
±1%	5	0·0056	0·9195	2·0272	−0·9921	−44·050	0·0452	−0·1	0·039	10^{-3}	1·83
±5%	14	0·0023	1·0103	2·0225	−0·9703	−43·897	0·0591	−0·1	0·118	10^{-6}	4·03
±1%	15	0·0032	1·002	2·0068	−1·0028	+44·072	0·0480	−0·1	0·038	10^{-6}	4·60

of the same problem employing two different penalty functions. The two functions, with signs appropriate for minimization form, were:

(1) A quadratic loss function

$$f(x) + \alpha \sum_{j=1}^{3} (g_j(x) + |g_j(x)|)^2/4.$$

(2) An interior function

$$f(x) - r \sum_{j=1}^{3} \ln(-g_j(x)).$$

All unconstrained minimizations of the above functions were performed with the DFP/golden section combination. Results are tabulated in Tables 2 and 3. Given the unconstrained optimizing point x, the values of b_j appearing in the tables are given as follows:

(1) Runs with quadratic loss function:

$$b_j = \begin{cases} 0 & g_j(x) \leq 0, \\ g_j(x) & \text{otherwise.} \end{cases}$$

F. J. GOULD

TABLE 2
Rosen–Suzuki results: quadratic loss function

Parameter values	IT	x_1	x_2	x_3	x_4	$f(x)$	b_1	b_2	b_3	EFS	Time (sec)
$\alpha = 10^3$	15	0·0005	0·998	2·0005	−0·9997	−44·003	4×10^{-4}	0	10^{-3}	10^{-3}	4·34
$\alpha = 10^3$	17	0·0005	0·99998	1·99994	−1·00007	−44·001	5×10^{-4}	0	10^{-3}	10^{-6}	4·43
$\alpha = 10^4$	2	−0·001	0·741	2·099	−0·810	−43·38	0	0	2×10^{-6}	10^{-3}	3·64
$\alpha = 10^4$	30	0·000018	1·00001	2·00002	−0·99997	−44·0000	5×10^{-5}	0	10^{-4}	10^{-6}	10·07
$\alpha = 10^5$	2	−0·001	0·741	2·099	−0·809	−43·37	0	0	3×10^{-7}	10^{-3}	0·95
$\alpha = 10^5$	34	−0·00026	1·0001	2·0001	−0·9997	−44·0000	5×10^{-6}	0	10^{-5}	10^{-6}	11·5

TABLE 3
Rosen–Suzuki results: interior function

Parameter values	IT	x_1	x_2	x_3	x_4	$f(x)$	b_1	b_2	b_3	EPS	Time (sec)
$r = 1·000$	6	0·474	0·869	1·938	−0·737	−41·63	−1·08	−2·95	−0·518	—	
$r = 0·625$	14	−0·004	0·969	2·008	−0·975	−43·86	−0·067	−1·2	−0·030	—	
$r = 0·0039$	19									—	
⋮										⋮	
$r = 0·152\cdot 10^{-4}$	29	0·00029	0·9997	1·9998	−1·0002	−43·99997	-10^{-5}	−1·0009	-7×10^{-6}	—	6·43
$r = 1·0$	3	0·295	0·639	1·83	−0·927	−41·63	−0·84	−3·36	−0·16	10^{-6}	1·85
$r = 0·1$	10	0·362	0·544	1·82	−0·944	−41·43	−0·78	−3·5	−0·001	10^{-6}	4·23
$r = 10^{-3}$	8	0·432	0·448	1·83	−0·630	−39·87	−1·3	−5·0	-10^{-7}	10^{-6}	3·77
$r = 10^{-5}$	1	0·437	0·437	1·83	−0·612	−39·73	−1·4	−5·1	-10^{-6}	10^{-6}	1·00

24. NON-LINEAR TOLERANCE PROGRAMMING

(2) Runs with interior function:

$$b_j = g_j(x).$$

4.2. *Comments on Rosen–Suzuki Test Results*

In Table 2 the runs for $\alpha = 10^4$, $EPS = 10^{-3}$ and $\alpha = 10^5$, $EPS = 10^{-3}$ give misleading b_j values because the associated x is not a true penalty function minimizer. This was indicated in the computer printouts by the fact that the partial derivatives of the penalty function were large at the final x. The mechanism here is something one often observes with penalty function techniques. The parameter values $\alpha = 10^4$, $\alpha = 10^5$ are quite large and consequently heavy penalties are imposed at points even a small distance from the constraint set. In two iterations one moves to a point just outside the boundary of the constraint set. In this region, because of the large values of α, the curvature of the penalty function is changing rapidly and the contours are very tight and elongated. One is thus in a zig-zag setting and tends to take a very short step. The rather loose value of $EPS = 10^{-3}$ permits the code to give a false indication of convergence. Note, for these cases, an EPS value of 10^{-6} greatly increases accuracy in the x values and increases the number of iterations required for convergence. We remark that for the cases $EPS = 10^{-6}$ the partial derivatives were all quite small upon termination. For the more moderate value $\alpha = 10^3$ there is not great distinction between the two EPS values because the penalty function contours are not so bad. With this moderate penalty one is allowed to move about with greater ease outside the constraint set. Short steps do not tend to occur until one is close to the true minimizing x. Finally, concerning Table 2, note the considerable additional time required to obtain just a little more accuracy (i.e. compare $\alpha = 10^3$, $EPS = 10^{-6}$ with $\alpha = 10^4$, $EPS = 10^{-6}$).

In Table 1, the second case (tolerance $\pm 1\%$, $EPS = 10^{-3}$) is faster than the first case, probably again because the value of EPS is too loose for the tight tolerance $\pm 1\%$ with effects similar to those discussed above. Notice the considerable increase in computing time required in tightening EPS to 10^{-6} for the $\pm 1\%$ runs. All of the tolerance function runs terminated with small partial derivatives, though those in the second case were larger than in the other three cases (i.e. on the order of 0·1).

The results in the top half of Table 3 were obtained by using the SUMT code (without the extrapolation option) developed by Fiacco, McCormick, and Mylander at the Research Analysis Corporation. The penalty function is the interior function of Section 4.1 and the procedure is sequential. The rows in the bottom half of the table represent 'single shot' unconstrained optimizations. The SUMT code also uses the DFP/golden section combination with resetting on the $n + 1$st iteration. The convergence criterion in the SUMT code is that the norm of the penalty function gradient should be less than 0·5.

4.3. *The Shell Dual (TP2) Test Problem*

This problem is well known and has been amply described in other papers. (See, for example, Colville, 1968.) It is a problem in 15 non-negative variables with a cubic objective function and 5 quadratic constraints which define a non-convex constraint set. It is generally considered quite difficult to solve, and many codes have failed on this problem. Starting from the usual feasible starting point, I have investigated the performance of SUMT (with the DFP/ golden section module) and the ϕ function. The optimal objective value is

TABLE 4
TP2 test results

Penalty function	Parameter	Tolerance	ϵ	IT	$f(x)$	EPS
ϕ	—	±1%	0·01	227	32·34	10^{-9}
ϕ	—	±5%	0·1	122	35·00	10^{-6}
SUMT	$r = 1$			96	44·52	—
SUMT	$r = 0·625$			147	33·06	—
SUMT	$r = 0·0039$			185	32·39	—
SUMT	$r = 0·244 \cdot 10^{-3}$			209	32·35	—
SUMT	$r = 0·152 \cdot 10^{-4}$			232	32·348	—

32·348. For the ϕ function runs, non-negativity of the variables was introduced as added constraints
$$h_j(x) = -x_j \leq 0, \quad j = 1, \ldots, 15.$$
The components of the ϕ function corresponding to these constraints were
$$\lambda(h_j(x)) = \begin{cases} 0, & h_j(x) \leq -2\epsilon, \\ (h_j(x) + 2\epsilon)^2, & -2\epsilon < h_j(x) \leq -\epsilon, \\ \dfrac{-2\epsilon^3}{h_j(x)} - \epsilon^2, & -\epsilon < h_j(x) < 0, \\ \infty, & h_j(x) \geq 0. \end{cases}$$
Runs were performed with several values of ϵ and with tolerances of ±1% and ±5% on the given right-hand sides. The results are reported in Table 4. Running times could not be compared because the runs were made on different equipment.

4.4. A Non-Convex Constrained Non-Linear Least Squares Problem

The problem is

$$\min_{\theta_1,\theta_2} \sum_{i=1}^{44} [y_i - \theta_1 - (0{\cdot}49 - \theta_1)\exp(-\theta_2(x_i - 8))]^2$$

subject to

$$g_1(\theta) = \theta_1 \theta_2 - 0{\cdot}49\theta_2 + 0{\cdot}09 \leq 0,$$
$$g_2(\theta) = -\theta_1 + 0{\cdot}40 \qquad \leq 0.$$

TABLE 5
Test data for least squares problem

i	x_i	y_i	i	x_i	y_i
1	8	0·49	23	22	0·41
2	8	0·49	24	22	0·40
3	10	0·48	25	24	0·42
4	10	0·47	26	24	0·40
5	10	0·48	27	24	0·40
6	10	0·47	28	26	0·41
7	12	0·46	29	26	0·40
8	12	0·46	30	26	0·41
9	12	0·45	31	28	0·41
10	12	0·43	32	28	0·40
11	14	0·45	33	30	0·40
12	14	0·43	34	30	0·40
13	14	0·43	35	30	0·38
14	16	0·44	36	32	0·41
15	16	0·43	37	32	0·40
16	16	0·43	38	34	0·40
17	18	0·46	39	36	0·41
18	18	0·45	40	36	0·38
19	20	0·42	41	38	0·40
20	20	0·42	42	38	0·40
21	20	0·43	43	40	0·39
22	22	0·41	44	42	0·39

The objective function, $f(\theta)$, is not convex. The data x_i, y_i are presented in Table 5. The problem was run with SUMT, with the quadratic loss (Q.L.) function, and with the ϕ function employing tolerances of $\pm 10\%$, i.e. $\epsilon_1 = 0{\cdot}009$, $\epsilon_2 = 0{\cdot}040$. Results with these three penalty functions were compared from five different starting points. It is to be noted that the constraint region is a slim 'pencil' and that the objective function $f(\theta)$ is nearly flat over this region. The interior method, SUMT, performs poorly for this problem

TABLE 6
Test results for least squares problem

θ_0	Penalty function	Parameter	IT	$f(\theta)$	θ_1	θ_2	b_1	b_2	Time (sec)
(0·42, 5·0)	Q.L.	$\alpha = 100$	2	0·0288	0·423	1·343	0·0002	0·0	18·20
	Tol	$\pm 10\%$	3	0·02850	0·422	1·275	0·0042	−0·022	9·60
	SUMT	$r_0 = 1$	42	0·03064	0·4219	$3·23 \times 10^5$	-22×10^3	−0·021	183·55
	SUMT	$r_0 = 10^{-4}$	28	0·03064	0·4219	961·00	−65·38	−0·021	95·7
(0·42, 3·0)	Tol	$\pm 10\%$	6	0·028141	0·420	1·224	0·0045	−0·020	15·96
(1, 1)	Q.L.	$\alpha = 100$	9	0·0284	0·4199	1·279	0·0003	0·0	41·09
(2, 2)	Q.L.	$\alpha = 100$	8	0·0284	0·4199	1·279	0·0003	0·0	40·50
(0·41, 1·5)	Tol	$\pm 10\%$	6	0·028141	0·4202	1·224	0·0045	−0·020	16·24
	SUMT	$r_0 = 10^{-4}$	47	0·028445	0·4199	1·284	$-0·82 \times 10^{-7}$	−0·019	17·73

unless started quite close to the solution. The failure of SUMT from remote points was verified for several different initial values of r (denoted r_0 in Table 6). The exterior and tolerance methods both performed strikingly well from a variety of starting points (denoted θ_0 in Table 6). The solution is given by $\theta_1^* = 0.4199$, $\theta_2^* = 1.284$, $f(\theta^*) = 0.0284$. Only the first constraint, $g_1(\theta)$, is binding at optimality. The results of the runs are presented in Table 6. For all of these runs the value of EPS was 10^{-6}.

4.5. *Another Non-Convex Example*

This is the only problem which was run with the multiplier functions defined in Section 3.1 and the corresponding functions for non-negativity conditions as presented in 3.3. The problem is

$$\min f(x) = (x_1 - 10)^3 + (x_2 - 20)^3$$

subject to

$$g_1(x_1, x_2) = -x_1 + 13 \leqslant 0,$$
$$g_2(x_1, x_2) = -(x_1 - 5)^2 - (x_2 - 5)^2 + 100 \leqslant 0,$$
$$g_3(x_1, x_2) = (x_1 - 6)^2 + (x_2 - 5)^2 - 82.81 \leqslant 0,$$
$$g_4(x_1, x_2) = -x_2 \leqslant 0.$$

The objective function is not convex and the constraint region is a narrow winding (half-moon shaped) valley of the type described in Fletcher and McCann (1969). For interior runs and for the tolerance function the starting point is (14·35, 8·6). The optimum point is $x^* = (14.095, 0.842961)$, with $f(x^*) = -6961.81$. At optimality the constraint values are given by

$$g_1(x^*) = -1.095,$$
$$g_2(x^*) = 0.00,$$
$$g_3(x^*) = 0.00,$$
$$g_4(x^*) = -0.842961.$$

For the exterior run a starting point of (20·1, 5·84) was used. This point is exterior to the constraint region and the same distance away from the optimal point as is the interior starting point. The ϕ function was run with tolerances in the constant terms 50%, 5%, 0·5% and 0·05%, and with values of ϵ in the non-negativity multipliers of 0·3, 0·1, 0·1, and 0·05 respectively. The interior function used was

$$f(x) - r \sum_{j=1}^{4} 1/g_j(x).$$

TABLE 7
Test results for non-convex example

Penalty function	Parameter	Tolerance	ϵ	IT	x_1	x_2	b_1	b_2	b_3	b_4	$f(x)$	EPS	Time (sec)
ϕ	—	50%	0·3	8	10·15	0·037	2·84	48·77	0	−0·037	−7955	10^{-6}	0·3
ϕ	—	5%	0·1	20	13·41	0·058	0	4·78	0	−0·058	−7890	10^{-6}	0·5
ϕ	—	0·5%	0·1	42	13·67	0·075	0	0·456	0·379	−0·075	−7859	10^{-6}	0·8
ϕ	—	0·05%	0·05	53	14·09	0·837	0	0·003	0·001	−0·05	−6967	10^{-6}	1·0
EXT	10^5	—	—	31	14·08	0·830	0	0·005	0·006	0	−6975	10^{-6}	0·6
INT	10^{-2}	—	—	240	14·09	0·849	−1·097	−0·003	−0·002	−0·849	−6955	10^{-6}	1·59

All unconstrained minimizations were performed on the IBM 370/165 computer using the DFP/golden sections routine. Results are reported in Table 7.

5. Conclusions and Summary

It should first be pointed out that all of the results of our numerical experiments are dependent upon the unconstrained optimization technique which we used (DFP/golden sections) and the comparative performances of the penalty functions studied could conceivably differ with different unconstrained methods. Furthermore, we have looked at far too small a class of test problems to draw any general conclusions. The results we have seen indicate, for the tolerance function, that as the tolerance becomes smaller the approximation to the true solution to P_0 becomes more satisfactory, but the number of iterations required (and the computation time) also increases. It also appears that the interior function is particularly ill-suited for thin or narrow winding constraint regions. In such cases both the tolerance function and the exterior function performed well (see, for example, the least-squares result and the results from the non-convex problem reported in Section 4.5). As mentioned in Section 3.5, the tolerance function has a theoretical attribute of being able to guarantee, in some sense, a prespecified accuracy to the true solution to P_0 in a single unconstrained optimization. Beyond this, especially in comparison to the exterior function, it is not yet clear whether or not there is any real computational advantage.

Acknowledgement

I am grateful to one of the referees for his discussion of equality constraints.
    I would also like to thank Mr. R. Coppins for the final test problem (Section 4.5) as well as for the computational results on that problem.

References

Bellman, R. E., and Zahdeh, L. A. (1970). Decision making in a fuzzy environment. *Managmt Sci.* **17**, B-141–B-164.
Colville, A. R. (1968). A comparative study on nonlinear programming codes. *IBM New York Scientific Center Report No. 320-2949.*
Evans, J. P., and Gould, F. J. (1970). Stability in nonlinear programming. *Operat. Res.* **18**, 107–118.
Evans, J. P., Gould, F. J., and Tolle, J. W. (1971). Exact penalty functions and duality in nonlinear programming. Submitted to *Mathematical Programming.*
Fiacco, A. V., and McCormick, G. P. (1964). The sequential unconstrained minimization technique for nonlinear programming, a primal-dual method. *Managmt Sci.* **10**, 360–366.

Fletcher, R., and McCann, A. P. (1969). Acceleration techniques for nonlinear programming. *In*: 'Optimization' (R. Fletcher, ed.), p. 213. Academic Press, London.

Gould, F. J. (1969). Extensions of Lagrange multipliers in nonlinear programming. *SIAM J. appl. Math.* **17**, 1280–1297.

Gould, F. J. (1970). A class of inside-out algorithms for general programs. *Managmt Sci.* **16**, 350–356.

Gould, F. J. (1971). Proximate linear programming: an experimental study of a modified simplex method for solving linear programs with inexact data. 'Institute of Statistics Mimeo Series No. 789.' Dept. of Statistics, University of North Carolina, Chapel Hill, North Carolina, 27514, USA.

Pietrzykowski, T. (1969). An exact potential method for constrained maxima. *SIAM J. Numer. Anal.* **6**, 299–304.

Roode, J. D. (1968). 'Generalized Lagrangian Functions in Mathematical Programming.' Thesis, Leiden, The Netherlands.

Rosen, J. B., and Suzuki, S. (1965). Construction of nonlinear programming test problems. *Comm. ACM.* **8**, 113.

Rosenbrock, H. H. (1960). An automatic method for finding the greatest or least value of a function. *Comput. J.* **3**, 175–184.

Zangwill, W. I. (1967). Nonlinear programming via penalty functions. *Managmt Sci.* **13**, 344–358.

25. On the Convergence of the Logarithmic Barrier Function Method

R. MIFFLIN

Department of Administrative Sciences,
Yale University, New Haven, Connecticut, U.S.A.

Convergence of the logarithmic barrier function method developed by Lootsma (1967) and Fiacco and McCormick (1968) for solving non-linear programming problems is considered.

The problem of maximizing $f(x)$ over $S = \{x | g_i(x) \geq 0, i = 1, 2, \ldots, m\}$ where f and g_i for $i = 1, 2, \ldots, m$ are real-valued functions defined on R^n is replaced by a sequence of essentially unconstrained problems of approximately maximizing

$$P(x, r_k) = f(x) + r_k \sum_{i=1}^{m} \ln g_i(x)$$

over $\hat{S} = \{x | g_i(x) > 0, i = 1, 2, \ldots, m\}$ for a sequence of positive numbers $\{r_k\}$ converging to zero. For $k = 1, 2, \ldots$ a feasible point sequence $\{x^k\}$ in $\hat{S}$ generated by the algorithm is defined by $\|\nabla_x P(x^k, r_k)\| \leq \epsilon r_k$, where ϵ is a given non-negative number. A multiplier sequence $\{u^k\}$ is defined by $u^k = (u_1^k, u_2^k, \ldots, u_m^k)$ where

$$u_i^k = \left(\frac{r_k}{g_i(x^k)}\right) \quad \text{for } i = 1, 2, \ldots, m.$$

Since the sub problem convergence parameter ϵ is allowed to be positive here, x^k need not be an exact solution to the kth unconstrained maximization sub-problem as in Fiacco and McCormick (1968) and Lootsma (1967, 1968, 1969) and therefore may be found by finite step procedures.

In addition to assuming that $\hat{S}$ is non-empty the following assumptions are considered:

(1) S is convex and compact.
(2) f and $g = (g_1, g_2, \ldots, g_m)$ are concave and continuously differentiable on S.

(3) These exists a Kuhn-Tucker point (x^*, u^*) such that

(a) $L(x) = f(x) + \sum_{i=1}^{m} u_i^* g_i(x)$ is strongly concave (Poljak, 1966) on S.

(b) $\nabla L(x)$ satisfies a Lipschitz condition on S.

(c) $\nabla g_i(x^*)$ for $i \in A = \{i \mid g_i(x^*) = 0, \ 1 \leq i \leq m\}$ are linearly independent vectors.

(d) $u_i^* > 0$ for all $i \in A$.

Under assumptions (1) and (2) it is shown that

$$p^* \leq \varliminf_{k \to \infty} \left(\frac{f(x^*) - f(x^k)}{r_k} \right) \leq \varlimsup_{k \to \infty} \left(\frac{f(x^*) - f(x^k)}{r_k} \right) \leq m - q^*$$

where p^* is the number of positive elements in a Kuhn-Tucker multiplier vector u^*, and q^* is the number of positive constraint values for an optimal solution x^*. With the addition of assumption 3(a) a positive number θ depending on ϵ is found such that for $k = 1, 2, \ldots$

$$\|x^* - x^k\| \leq \theta (r_k)^{1/2}$$

and together with assumptions 3(b) and 3(c) a positive number ζ depending on ϵ is found such that for $k = 1, 2, \ldots$

$$|u_i^* - u_i^k| \leq \zeta (r_k)^{1/2} \qquad \text{for } i \in A.$$

When the nondegeneracy assumption 3(d) which implies $p^* = m - q^*$ is added then the above convergence rates are improved by finding positive numbers $\alpha, \beta, \gamma, \delta$ and η depending on ϵ such that for $k = 1, 2, \ldots$

$$\left(\frac{p^*}{1 + \alpha r_k} \right) r_k \leq f(x^*) - f(x^k) \leq (p^* + \alpha r_k) r_k,$$

$$p^* \eta r_k \leq \|x^* - x^k\| \leq \beta r_k,$$

$$|u_i^* - u_i^k| \leq \gamma r_k, \qquad \text{for } i \in A,$$

$$\left(\frac{1}{g_i(x^*) + \delta r_k} \right) r_k \leq u_i^k \leq \left(\frac{1 + \alpha r_k}{g_i(x^*)} \right) r_k, \qquad \text{for } i \notin A,$$

$$\left(\frac{1}{u_i^* + \gamma r_k} \right) r_k \leq g_i(x^k) \leq \left(\frac{1 + \alpha r_k}{u_i^*} \right) r_k, \qquad \text{for } i \in A,$$

and

$$|g_i(x^*) - g_i(x^k)| \leq \delta r_k, \qquad \text{for } i \notin A.$$

25. THE LOGARITHMIC BARRIER FUNCTION METHOD

Convergence which is linear in r_k as above has been established implicitly by Fiacco and McCormick (1968) and Lootsma (1968, 1969) using the implicit function theorem under the additional assumptions that $\epsilon = 0$ and the problem functions are twice continuously differentiable. Here the convergence results are established explicitly and without assuming exact solutions to the subproblems.

Details and proofs may be found in Mifflin (1972).

References

Fiacco, A. V., and McCormick, G. P. (1968). 'Nonlinear Programming, Sequential Unconstrained Minimization Techniques'. Wiley, New York.

Lootsma, F. A. (1967). Logarithmic programming: a method of solving nonlinear-programming problems. *Philips Res. Repts.* **22**, 329–344.

Lootsma, F. A. (1968). Extrapolation in logarithmic programming. *Philips Res. Repts.* **23**, 108–116.

Lootsma, F. A. (1969). Hessian matrices of penalty functions for solving constrained-optimization problems. *Philips Res. Repts.* **24**, 322–331.

Mifflin, R. (1972). 'Convergence Bounds for Nonlinear Programming Algorithms'. Administrative Sciences Report No. 57, Yale University.

Poljak, B. T. (1966). Existence theorems and convergence of minimizing sequences in extremum problems with restrictions. *Dokl. Akad. Nauk SSR* **166**, 287–290 = *Soviet Math. Dokl.* **7**, 72–75.

26. A Class of Methods for Non-linear Programming. III: Rates of Convergence

R. Fletcher

*Theoretical Physics Division, U.K.A.E.A. Research Group,
Atomic Energy Research Establishment, Harwell, England*

Summary

The first two papers in this series have developed a class of methods for non-linear programming in which an equality constrained optimization problem is replaced by the single unconstrained optimization of a function ϕ having the same number of variables. In this paper more detailed consideration is given to the case when second derivatives of the objective and constraint functions can be evaluated. It is shown that second derivatives of ϕ at the solution depend only upon second derivatives of the objective and constraint functions. This enables a sharper version of an earlier theorem to be proved. Of more importance, it also suggests a Newton-like iteration for the minimization of ϕ which does not require the calculation of any third derivatives. It is shown that this iteration converges at a superlinear (and usually quadratic) rate.

1. Introduction

It has been shown by Fletcher (1970) (Part I of this series) that the problem

$$\text{Minimize } F(\mathbf{x}), \quad \mathbf{x} \in E^n \tag{1a}$$

$$\text{Subject to } c_i(\mathbf{x}) = 0, \quad i = 1, 2, \ldots, m \leqslant n \tag{1b}$$

can be solved by minimizing the function

$$\phi(\mathbf{x}) = F - \mathbf{c}^T N^+ \mathbf{g} + q \mathbf{c}^T N^+ N^{+T} \mathbf{c} \tag{2}$$

if the preassigned parameter q is chosen so that it is larger than a certain threshold value. Details of a practical algorithm have been considered by Fletcher and Lill (1971) (Part II of this series) in the case when only first derivatives of F and c_i are given. In (2), $\mathbf{g} = \nabla F$ is the gradient of $F(\nabla = (\partial/\partial x_1, \partial/\partial x_2, \ldots, \partial/\partial x_n)^T)$ and $N = [\mathbf{n}_1, \mathbf{n}_2, \ldots, \mathbf{n}_m]$ is the matrix whose columns are the normal vectors $\mathbf{n}_i = \nabla c_i$ of the constraint functions. The matrix N^+ is defined to be the full rank generalized inverse matrix of N, that is

$$N^+ = (N^T N)^{-1} N^T. \tag{3}$$

A factor of $\frac{1}{2}$ which multiplied q in Parts I and II of this series has been dropped for convenience. Second derivatives of F, c_i and ϕ will be represented by the operator

$$\nabla^2 = \left[\frac{\partial^2}{\partial x_i \, \partial x_j} \right];$$

for instance $\nabla^2 F(\mathbf{x})$ will denote the symmetric (Hessian) matrix whose i, jth element is $\partial^2 F/(\partial x_i \, \partial x_j)$ evaluated at $\mathbf{x}$. The projection matrices

$$P = NN^+ \tag{4a}$$

and

$$\hat{P} = I - P \tag{4b}$$

will be used. Subscripts refer to vector or matrix elements, and a bracketed superscript (as in $\mathbf{x}^{(k)}$) will denote an iteration number. Use will be made of norms, and in particular $\|\cdot\|$ will represent the vector L_2 norm or the matrix norm induced by the vector norm, although the results apply in various norms with minor changes. The notation $O(\cdot)$ and $o(\cdot)$ will be used, as described by Hardy (1960).

It is convenient to denote a solution to (1) by $\boldsymbol{\xi}$, and it will be assumed that F, and the derivatives represented by ∇F and $\nabla^2 F$ exist and are continuous at $\boldsymbol{\xi}$, that is

$$F \in C_2(\boldsymbol{\xi}) \tag{5a}$$

and similarly that

$$c_i \in C_2(\boldsymbol{\xi}) \qquad \text{for } i = 1, 2, \ldots, m. \tag{5b}$$

It will also be assumed that the constraints satisfy the independence condition

$$\operatorname{rank}(N(\boldsymbol{\xi})) = m. \tag{6}$$

Equation (6) is a necessary and sufficient condition for the existence of the matrix $[N^T(\boldsymbol{\xi}) N(\boldsymbol{\xi})]^{-1}$ and hence of $N^+(\boldsymbol{\xi})$ as defined in (3). Given this assumption, then it is well known that the solution $\boldsymbol{\xi}$ must satisfy the necessary conditions

$$\mathbf{c}(\boldsymbol{\xi}) = \mathbf{0} \tag{7a}$$

and

$$\hat{P}(\boldsymbol{\xi}) \mathbf{g}(\boldsymbol{\xi}) = \mathbf{0}. \tag{7b}$$

Another way of stating (7b) is that there exists a constant vector $\boldsymbol{\lambda} \in E^m$ such that

$$\mathbf{g}(\boldsymbol{\xi}) - N(\boldsymbol{\xi})\boldsymbol{\lambda} = \mathbf{0}, \tag{8}$$

and clearly by (7b) and (4), the components of $\boldsymbol{\lambda}$ are the elements of the vector $N^+(\boldsymbol{\xi})\mathbf{g}(\boldsymbol{\xi})$.

The aim of this paper is to take a closer look at the situation when second derivatives of the objective and constraint functions are available. This was not done in Parts I and II, where terms involving $\nabla^2 c_i$ in particular were neglected whenever possible. In Section 2 of this paper the exact first and second derivatives of ϕ are examined, and it is seen that $\nabla^2 \phi(\boldsymbol{\xi})$ can be evaluated *using only first and second derivatives* of F and c_i. Now one of the disadvantages of this class of methods has been in the extra order of derivatives of F and c_i required to determine derivatives of ϕ; that is: ϕ requires first derivatives of F and c_i, $\nabla \phi$ requires second derivatives of F and c_i, and so on. However, the fact that $\nabla^2 \phi(\boldsymbol{\xi})$ only involves second derivatives of F and c_i is most significant. It suggests an iteration in which $\nabla^2 \phi(\mathbf{x})$ is approximated by a matrix $\Gamma(\mathbf{x})$, which comprises the terms of $\nabla^2 \phi(\mathbf{x})$ which would be non-zero at $\mathbf{x} = \boldsymbol{\xi}$. Calculation of Γ does not then require derivatives of F and c_i higher than second. In Section 3 it is shown that if the initial approximation $\mathbf{x}^{(1)}$ is sufficiently close to $\boldsymbol{\xi}$, then the iteration will converge, and at a superlinear (and usually quadratic) rate. Now, even if the exact second derivative matrix were calculated, using third derivatives of F and c_i, then a quadratic rate of convergence is the best that could be expected. Thus it is shown that quadratic convergence is usually obtained using derivatives which are an order less than might have been anticipated.

Although no more than second derivatives are needed to define the Newton-like iteration, it is important to examine the assumptions which need to be made to achieve these rates of convergence. The weakest assumption that can be made is that derivatives exist as in (5). In this case it is shown (in Section 3) that the iteration has superlinear convergence. Furthermore, if third derivatives exist, then it is straightforward to show that the rate of convergence is quadratic. However, at the expense of more analysis, it can be shown that quadratic convergence holds under the less restrictive assumption that the second derivatives of F and c_i satisfy a certain Lipschitz condition. In fact, this analysis is outlined in Section 3, because there is at least one important class of functions, namely cubic splines, which are included under the less restrictive assumption.

The final section considers some possibilities for incorporating the iteration into a general purpose algorithm. Another result derived in Section 3 enables it to be shown that the Newton-like algorithm is not incompatible with a technique used in unconstrained optimization for forcing the gradient of the objective function to zero.

2. Derivatives of $\phi(x)$

If equation (2) is differentiated directly, the resulting equation for $\nabla\phi$ will involve terms like $\partial(N^+)/\partial x_i$. This is inconvenient if the gradient is to be computed, so an explicit expression for $\nabla\phi$ in terms of the known derivatives of F and c_i will first be obtained. It is convenient to write (2) as

$$\phi = F - \mathbf{c}^T \boldsymbol{\pi} \tag{9}$$

(the dependence upon $\mathbf{x}$ being implicit) where $\boldsymbol{\pi}$ can be written using (3) as the solution of the equations

$$(N^T N)\boldsymbol{\pi} = N^T \mathbf{g} - q\mathbf{c}. \tag{10}$$

Note that this definition of $\boldsymbol{\pi}$ implies from (7a) and (8) that

$$\boldsymbol{\pi}(\boldsymbol{\xi}) = \boldsymbol{\lambda}, \tag{11}$$

and hence that

$$\mathbf{g}(\boldsymbol{\xi}) = N(\boldsymbol{\xi})\boldsymbol{\pi}(\boldsymbol{\xi}). \tag{12}$$

To obtain $\nabla\phi$, let Λ be the $m \times n$ matrix whose elements are

$$\Lambda_{ip} = \partial \pi_i / \partial x_p. \tag{13}$$

If (10) is operated upon by $\partial/\partial x_p$, and terms are collected, then Λ can be written as

$$N^T N\Lambda = M + N^T L - qN^T \tag{14}$$

where the symmetric $n \times n$ matrix L and the $m \times n$ matrix M are defined by

$$L_{jp} = \frac{\partial^2 F}{\partial x_j \partial x_p} - \sum_k \pi_k \frac{\partial^2 c_k}{\partial x_j \partial x_p} \tag{15}$$

and

$$M_{ip} = \sum_j \frac{\partial^2 c_i}{\partial x_p \partial x_j}(\mathbf{g} - N\boldsymbol{\pi})_j. \tag{16}$$

If equation (9) is now differentiated and (14) substituted, then

$$\nabla\phi = \mathbf{g} - N\boldsymbol{\pi} - LN^{+T}\mathbf{c} + qN^{+T}\mathbf{c} - M^T(N^T N)^{-1}\mathbf{c} \tag{17}$$

is obtained.

From this equation, an expression for the Hessian matrix $\nabla^2\phi(\boldsymbol{\xi})$ can be obtained. Assume for the moment that the functions F and c_i ($i = 1, 2, \ldots, m$) have third derivatives, then because $\mathbf{c}(\boldsymbol{\xi}) = \mathbf{0}$, differentiation of (17) yields

$$\nabla^2\phi(\boldsymbol{\xi}) = L - N\Lambda - (L - qI)N^{+T}N^T - M^T N^+ \tag{18}$$

$$= L - P(L - qI) - (L - qI)P - M^T N^+ - N^{+T}M \tag{19}$$

from (14) and (4a). Now, $M(\xi) = 0$ by virtue of (16) and (12), so the equation

$$\nabla^2 \phi(\xi) = L - P(L - qI) - (L - qI)P$$

is obtained. A final rearrangement using (4b) gives

$$\nabla^2 \phi(\xi) = \hat{P}L\hat{P} - PLP + 2qP \qquad (20)$$

where the terms on the right-hand sides of equations (18) to (20) are all evaluated at ξ.

In fact it is possible to prove this result even when the weaker assumptions (5) are made in place of the existence of third derivatives. In this case the difficulty will be to justify the use of equations like

$$[\nabla(\alpha(\mathbf{x}) c_i(\mathbf{x}))]\xi = \alpha(\xi) \nabla c_i(\xi) \qquad (21)$$

when $\alpha(\mathbf{x})$ need not be differentiable. Fortunately it is possible to show that this equation is true provided only that $\alpha(\mathbf{x})$ is continuous. By the differentiability of c_i, and $c_i(\xi) = 0$ there follows

$$c_i(\mathbf{x}) = \mathbf{h}^T \nabla c_i(\xi) + o(\|\mathbf{h}\|),$$

where $\mathbf{h} = \mathbf{x} - \xi$; and by the continuity of $\alpha(\mathbf{x})$

$$\alpha(\mathbf{x}) = \alpha(\xi) + o(1).$$

Combining these results and using $o(a)O(b) \Rightarrow o(ab)$ and $o(a) \Rightarrow O(a)$, yields

$$\alpha(\mathbf{x}) c_i(\mathbf{x}) = \mathbf{h}^T \nabla c_i(\xi) \alpha(\xi) + o(\|\mathbf{h}\|).$$

By definition of a derivative, and because $c_i(\xi) = 0$, the first derivative of $\alpha(\mathbf{x}) c_i(\mathbf{x})$ at ξ has been identified and is given by (21). All the terms in equation (17) which would otherwise have yielded third derivatives can be treated in this way. Thus (18) is valid if the terms which multiply $\mathbf{c}$ in (17) are continuous at ξ, which is so if assumptions (5) hold.

Equation (20) is the principal aim of this section and has some interesting and useful properties. It is seen that the Hessian in the general case is just the same as the Hessian for the quadratic/linear case developed in Part I, except that $\nabla^2 F(\xi)$ is replaced by $L(\xi)$. This is very satisfactory, because $L(\xi)$ is just the Hessian of the true Lagrangian function $\mathscr{L}(\mathbf{x},\boldsymbol{\lambda}) = F(\mathbf{x}) - \boldsymbol{\lambda}^T \mathbf{c}(\mathbf{x})$ by virtue of (11). It is also possible to prove much more readily than in Part I that there exists a suitable choice for q such that $\phi(\mathbf{x})$ has a local minimum at ξ, and to determine what is the smallest possible value of q.

Theorem 1

If in addition to assumptions (5) and (6), problem (1) satisfies the sufficient condition for a solution

$$\mathbf{v}^T L(\xi) \mathbf{v} > 0 \qquad (22)$$

for all $\mathbf{v}$ ($\neq \mathbf{0}$) such that $N^T \mathbf{v} = \mathbf{0}$, then the choice

$$q > \tfrac{1}{2} \| P(\xi) L(\xi) P(\xi) \| \tag{23}$$

implies that $\nabla^2 \phi(\xi)$ is strictly positive definite, and hence that ξ is a local minimum of $\phi(\mathbf{x})$.

Proof

Let $\mathbf{z}$ be an arbitrary non-zero vector, which can be written $\mathbf{z} = \mathbf{u} + \mathbf{v}$, where $\mathbf{u} = P\mathbf{z}$ and $\mathbf{v} = \hat{P}\mathbf{z}$. Then using (20) in which L, P and $\hat{P}$ are evaluated at ξ

$$\mathbf{z}^T \nabla^2 \phi(\xi) \mathbf{z} = \mathbf{v}^T L \mathbf{v} + \mathbf{u}^T (2qP - PLP) \mathbf{u}. \tag{24}$$

By virtue of (22) and (23), each term on the right-hand side of (24) is strictly positive unless either $\mathbf{v} = \mathbf{0}$ or $\mathbf{u} = \mathbf{0}$ respectively. But $\mathbf{u} = \mathbf{v} = \mathbf{0}$ implies $\mathbf{z} = \mathbf{0}$, which is a contradiction. Hence

$$\mathbf{z}^T \nabla^2 \phi(\xi) \mathbf{z} > 0 \tag{25}$$

for all non-zero $\mathbf{z}$, and hence $\nabla^2 \phi(\xi)$ is strictly positive definite. The necessary condition for a local minimum

$$\nabla \phi(\xi) = \mathbf{0} \tag{26}$$

is a consequence of equations (17), (12) and (7a). Therefore, ξ is a local minimum of $\phi(\mathbf{x})$. Q.E.D.

It will be noticed that if q satisfies

$$q > \tfrac{1}{2} \| L(\xi) \| \tag{27}$$

then Theorem 1 holds *a fortiori*. Furthermore, if it happens that

$$q < \tfrac{1}{2} \| P(\xi) L(\xi) P(\xi) \|,$$

then the matrix $2qP - PLP$ and therefore $\nabla^2 \phi(\xi)$ is no longer positive definite or semidefinite, so that ξ is not a local minimum of $\phi(\mathbf{x})$. Hence expression (23) gives the smallest possible bound on q.

3. An Iteration with Superlinear (and Usually Quadratic) Convergence

The fact that $\nabla^2 \phi(\xi)$ exists and can be calculated using at most second derivatives of F and c_i is of considerable importance. Assuming that a suitable constant q has been chosen, it implies that a Newton-like iteration

$$\mathbf{x}^{(k+1)} = \mathbf{x}^{(k)} - [\Gamma(\mathbf{x}^{(k)})]^{-1} \nabla \phi^{(k)} \tag{28}$$

can be set up in which the matrix $\Gamma(\mathbf{x})$ is an approximation to $\nabla^2 \phi(\mathbf{x})$. This approximation is chosen to have two important features; (a) that $\Gamma(\mathbf{x}) \to \nabla^2 \phi(\xi)$ as $\mathbf{x} \to \xi$, and (b) that $\Gamma(\mathbf{x})$ involves at most second derivatives of F

and c_i. Feature (a) is important, as will be seen, in that it ensures that the rate of convergence of the iteration will be superlinear (and usually quadratic); and in turn (b) is important because to obtain quadratic convergence directly using Newton's method would require calculation of third derivatives. The choice of $\Gamma(\mathbf{x})$ is not unique, although by analogy with (20) the most obvious choice is

$$\Gamma = \hat{P}L\hat{P} - PLP + 2qP \qquad (29)$$

where the matrices are evaluated at $\mathbf{x}$. Equally equations (10) and (15) show that in (29) the matrix L can be replaced by the matrix

$$\nabla^2 F - \sum_i [N^+ \mathbf{g}]_i \nabla^2 c_i.$$

Yet another possibility would be to retain the terms $M^T N^+ + N^{+T} M$ which appear in (19). All the results to be proved in this section apply to Γ defined in any of these ways.

These results will be proved in the following theorem. It will also be shown (briefly) that if the second derivatives of F and c_i satisfy a certain Lipschitz condition then the rate of convergence is quadratic. This will be done by reference to simple theorems about iterative methods given by Ortega and Rheinboldt (1970). Another result (namely part (ii) of the theorem) which has significance in setting up an algorithm will also be proved.

Theorem 2

If the matrix $\Gamma(\mathbf{x})$ is continuous at $\mathbf{x} = \boldsymbol{\xi}$, if $\Gamma(\boldsymbol{\xi}) = \nabla^2 \phi(\boldsymbol{\xi})$, (as for example when using (29)) and if assumptions (5), (6), (22) and (23) hold, then:

(i) $\boldsymbol{\xi}$ is a point of attraction of the iteration (28) (that is $\mathbf{x}^{(k)} \to \boldsymbol{\xi}$ if any iterate, $\mathbf{x}^{(1)}$ say, is sufficiently close to $\boldsymbol{\xi}$—see Ortega and Rheinboldt, 1970), and the rate of convergence is superlinear in the sense that

$$\lim_{k \to \infty} \frac{\|\mathbf{h}^{(k+1)}\|}{\|\mathbf{h}^{(k)}\|} = 0 \qquad (30)$$

where

$$\mathbf{h}^{(k)} = \mathbf{x}^{(k)} - \boldsymbol{\xi}. \qquad (31)$$

(ii) Given any σ, $0 < \sigma < 1$, then there exists k' such that for all $k > k'$, the change in $\phi(\mathbf{x})$ on each iteration satisfies

$$\phi^{(k)} - \phi^{(k+1)} > \tfrac{1}{2}\sigma \boldsymbol{\delta}^{(k)T} \Gamma^{(k)} \boldsymbol{\delta}^{(k)} \qquad (32)$$

where

$$\boldsymbol{\delta}^{(k)} = \mathbf{x}^{(k+1)} - \mathbf{x}^{(k)}. \qquad (33)$$

(iii) If in addition, the functions F and c_i satisfy Lipschitz conditions like

$$\left| \frac{\partial^2 F(\mathbf{x})}{\partial x_i \partial x_j} - \frac{\partial^2 F(\boldsymbol{\xi})}{\partial x_i \partial x_j} \right| \leqslant \alpha \|\mathbf{x} - \boldsymbol{\xi}\| \tag{34}$$

for all i, j and for sufficiently small $\|\mathbf{x} - \boldsymbol{\xi}\|$, and if the elements of the matrix $\Gamma(\mathbf{x})$ satisfy a similar Lipschitz condition at $\mathbf{x} = \boldsymbol{\xi}$ [which also follows when using (29) for example], then the rate of convergence is quadratic in the sense that

$$\frac{\|\mathbf{h}^{(k+1)}\|}{\|\mathbf{h}^{(k)}\|^2} \leqslant a \tag{35}$$

for sufficiently large k, where $a > 0$ is a constant.

Note that (22) and (23) imply (25), so that $\nabla^2 \phi(\mathbf{x})$, and therefore $\Gamma(\mathbf{x})$, are non-singular in a neighbourhood of $\boldsymbol{\xi}$, so that the iteration (28) is well defined.

Proof

Following Ortega and Rheinboldt (1970), the iteration (28) can be expressed in the form

$$\mathbf{x}^{(k+1)} = G(\mathbf{x}^{(k)}). \tag{36}$$

where $G(\mathbf{x}) = \mathbf{x} - \Gamma(\mathbf{x})^{-1} \nabla \phi(\mathbf{x})$. Equation (21) shows that $G(\mathbf{x})$ is differentiable at $\mathbf{x} = \boldsymbol{\xi}$, because $\Gamma(\mathbf{x})$ is continuous, and because $\nabla \phi(\boldsymbol{\xi}) = \mathbf{0}$. In fact this derivative is the matrix $G'(\boldsymbol{\xi}) = I - \Gamma(\boldsymbol{\xi})^{-1} \nabla^2 \phi(\boldsymbol{\xi}) = 0$ by the assumptions on Γ. Thus by Theorem 10.1.6 of Ortega and Rheinboldt (1970), $\boldsymbol{\xi}$ is a point of attraction of the iteration (28) and the rate of convergence is superlinear in the sense of (30). Thus part (i) of the theorem is proved.

To prove the second part of the theorem it is noted that the existence of $\nabla^2 \phi(\boldsymbol{\xi})$ gives the result

$$\phi(\mathbf{x}) - \phi(\boldsymbol{\xi}) - \mathbf{h}^T \nabla \phi(\boldsymbol{\xi}) - \tfrac{1}{2} \mathbf{h}^T \nabla^2 \phi(\boldsymbol{\xi}) \mathbf{h} = o(\|\mathbf{h}\|^2). \tag{37}$$

Now (33) can be written

$$\boldsymbol{\delta}^{(k)} = \mathbf{h}^{(k+1)} - \mathbf{h}^{(k)}$$

which together with (30) becomes

$$\mathbf{h}^{(k)} = -\boldsymbol{\delta}^{(k)} + o(\|\mathbf{h}^{(k)}\|). \tag{38}$$

Furthermore, the continuity of $\Gamma^{(k)}$ gives

$$\Gamma^{(k)} = \nabla^2 \phi(\boldsymbol{\xi}) + o(1). \tag{39}$$

Substituting (26), (38) and (39) into equation (37) for $\mathbf{x} = \mathbf{x}^{(k)}$ yields

$$\phi(\mathbf{x}^{(k)}) - \phi(\boldsymbol{\xi}) - \tfrac{1}{2}\boldsymbol{\delta}^{(k)\,T}\Gamma^{(k)}\boldsymbol{\delta}^{(k)} = o(\|\mathbf{h}^{(k)}\|^2). \tag{40}$$

Similarly, equation (37) with $\mathbf{x} = \mathbf{x}^{(k+1)}$, together with (30) gives

$$\phi(\mathbf{x}^{(k+1)}) - \phi(\boldsymbol{\xi}) = o(\|\mathbf{h}^{(k)}\|^2). \tag{41}$$

Combining (40) and (41) gives

$$\phi(\mathbf{x}^{(k)}) - \phi(\mathbf{x}^{(k+1)}) - \tfrac{1}{2}\boldsymbol{\delta}^{(k)T}\Gamma^{(k)}\boldsymbol{\delta}^{(k)} = o(\|\mathbf{h}^{(k)}\|^2)$$
$$= o(\|\boldsymbol{\delta}^{(k)}\|^2) \tag{42}$$

by (38). Now the continuity of $\Gamma^{(k)}$ and equation (25) show that $\Gamma^{(k)}$ is strictly positive definite for sufficiently large k, so the result (32) follows from the definition of $o(\cdot)$.

The proof of part (iii) of the theorem will only be summarized, and the first step is to establish the result

$$\|\nabla\phi(\mathbf{x}) - \nabla^2\phi(\boldsymbol{\xi})\mathbf{h}\| = O(\|\mathbf{h}\|^2). \tag{43}$$

This can be obtained from the Lipschitz conditions (34) by some suitable manipulation of the $O(\cdot)$ terms. In addition, the Lipschitz condition on $\Gamma(\mathbf{x})$ implies that

$$\Gamma(\mathbf{x}) = \nabla^2\phi(\boldsymbol{\xi}) + O(\|\mathbf{h}\|). \tag{44}$$

Now (25) and the continuity of $\Gamma(\mathbf{x})$ imply that $\Gamma(\mathbf{x})^{-1}$ is bounded for sufficiently small $\|\mathbf{x} - \boldsymbol{\xi}\|$. Hence, using (36), there follows

$$\|G(\mathbf{x}^{(k)}) - G(\boldsymbol{\xi})\| = \|\mathbf{x}^{(k)} - [\Gamma^{(k)}]^{-1}\nabla\phi^{(k)} - \boldsymbol{\xi}\|$$
$$\leq \|\Gamma^{(k)-1}\|\|\Gamma^{(k)}\mathbf{h}^{(k)} - \nabla\phi^{(k)}\|$$
$$\leq \|\Gamma^{(k)-1}\|\|\nabla^2\phi(\boldsymbol{\xi})\mathbf{h}^{(k)} - \nabla\phi^{(k)}\| + O(\|\mathbf{h}^{(k)}\|^2)$$

using (44). Therefore

$$\|G(\mathbf{x}^{(k)}) - G(\boldsymbol{\xi})\| = O(\|\mathbf{h}^{(k)}\|^2)$$

by (43). Theorem 10.1.6 of Ortega and Rheinboldt (1970) can again be invoked to show that (35) holds, which completes the proof. Q.E.D.

4. Algorithmic Considerations

Although the primary aim of this paper is theoretical, it is important to show that the above rates of convergence are likely to be retained in any practical algorithm into which iteration (28) is incorporated. Of course, (28) is not suitable on its own as a general algorithm, and the aim must be to construct an algorithm in which (28) is used for all sufficiently large k.

The sort of possibilities which might be kept in mind are for instance to introduce a Marquardt-like parameter μ, solving the system

$$(\Gamma^{(k)} + \mu^{(k)} I) \boldsymbol{\delta}^{(k)} = -\boldsymbol{\nabla}\phi^{(k)} \tag{45}$$

at each iteration, and ensuring that $\mu^{(k)} = 0$ for sufficiently large k. Alternatively, the correction $\boldsymbol{\delta}^{(k)}$ in (28) could be biased towards the steepest descent direction in the manner of Powell (1970), if it failed to satisfy a test like $\|\boldsymbol{\delta}^{(k)}\| \leqslant \Delta^{(k)}$. An important additional feature in conjunction with this sort of scheme is the inclusion of a test which forces $\boldsymbol{\nabla}\phi^{(k)}$ to zero. For instance, if the quadratic function $q^{(k)}(\mathbf{x})$ which approximates to $\phi(\mathbf{x})$ at $\mathbf{x}^{(k)}$ is defined by $q^{(k)}(\mathbf{x}^{(k)}) = \phi^{(k)}$, $\boldsymbol{\nabla}q^{(k)}(\mathbf{x}^{(k)}) = \boldsymbol{\nabla}\phi^{(k)}$ and $\nabla^2 q^{(k)} = \Gamma^{(k)}$, then a good test is to accept $\mathbf{x}^{(k+1)}$ only if the reduction in $\phi(\mathbf{x})$ is at least some fixed multiple σ $(0 < \sigma < 1)$ of the predicted reduction in $q^{(k)}(\mathbf{x})$; that is if

$$\phi^{(k)} - \phi^{(k+1)} \geqslant \sigma[q^{(k)}(\mathbf{x}^{(k)}) - q^{(k)}(\mathbf{x}^{(k+1)})]. \tag{46}$$

This test has been used successfully in a number of applications, see Powell (1970) for instance. If (46) fails, then $\mathbf{x}^{(k+1)}$ is recalculated, and in the above schemes this would be done either by increasing $\mu^{(k)}$ or decreasing $\Delta^{(k)}$ respectively. Now $q^{(k)}(\mathbf{x}^{(k)}) - q^{(k)}(\mathbf{x}^{(k+1)})$ in iteration (28) is just $\frac{1}{2}\boldsymbol{\delta}^{(k)T} \Gamma^{(k)} \boldsymbol{\delta}^{(k)}$, and so the significance of Theorem 2(ii) is that test (46) will always be satisfied for sufficiently large k, and hence that use of this test does not prevent the rates of convergence, referred to above, from being obtained.

Of course, it would be most desirable if the algorithm based upon say (45) and (46) could be analysed with a view to proving convergence to a solution of (1) under some mild assumptions. Unfortunately this is unlikely to be possible because there may be local minima of $\phi(\mathbf{x})$ which do not correspond to solutions of (1). Furthermore, there is the problem of ensuring that a sufficiently large value of q as required by (23) can be obtained. On the other hand, it is expected that the use of devices such as those described in this section, in conjunction with (28), will considerably increase the likelihood of convergence. It is therefore hoped to investigate the use of (28) in some such algorithm in a later paper.

Acknowledgements

I would like to thank M. J. D. Powell who read the manuscript and made a number of valuable suggestions for its improvement; and in particular for pointing out that the existence of $\nabla^2 \phi(\boldsymbol{\xi})$ could be proved without assuming that $\nabla^2 F$ and $\nabla^2 c_i$ satisfied Lipschitz conditions.

References

Fletcher, R. (1970). A class of methods for non-linear programming with termination and convergence properties. *In*: 'Integer and Non-linear Programming' (J. Abadie, ed.). North-Holland Publishing Co., Amsterdam.

Fletcher, R., and Lill, S. A. (1971). A class of methods for non-linear programming. II: Computational experience. *In*: 'Non-linear Programming' (B. Rosen, ed.). Academic Press, New York.

Hardy, G. H. (1960). 'A Course of Pure Mathematics'. 10th edition. Cambridge University Press, Cambridge, England.

Ortega, J. M., and Rheinboldt, W. C. (1970). 'Iterative Solution of Non-linear Equations in Several Variables'. Academic Press, New York.

Powell, M. J. D. (1970). A hybrid method for non-linear equations. *In*: 'Numerical Methods for Non-linear Algebraic Equations' (P. Rabinowitz, ed.). Gordon and Breach, London.

27. Generalization of an Exact Method for Solving Equality Constrained Problems to deal with Inequality Constraints

SHIRLEY A. LILL

*Department of Computational and Statistical Science,
University of Liverpool, Liverpool, England*

Summary

This work arises from the theoretical development of a class of methods for non-linear programming, introduced by Fletcher in 1969. It is an extension of previous work on equality constrained problems to more general problems with inequality constraints. Several computational strategies are reported together with results on a variety of test functions.

1. Introduction

This paper presents the results of applying one of a class of methods for non-linear programming to inequality constrained problems. The basis of the class of methods is that, for equality constrained problems, it is possible to construct penalty functions that have *unconstrained* minima at the solutions to the constrained problems. Thus only one unconstrained minimization of the penalty function is required compared with the several minimizations needed for conventional penalty functions (see Fiacco and McCormick, 1968). The theoretical details and justifications of the methods have already been presented (Fletcher, 1970a), and will be referred to as 'paper I'. A report on the computational experience of using different methods in the class on a variety of equality constrained problems has also been given (Fletcher and Lill, 1970), and will be referred to as 'paper II'. This enabled recommendations to be made as to the best penalty function to use and also the best way to organize the unconstrained minimization process. The present paper extends this recommended approach to inequality constrained problems, comparing possible strategies and giving their results on well known test problems.

The inequality constrained problem can be expressed as

$$\text{Minimize the objective function } F(\mathbf{x}), \quad \mathbf{x} \in E^n,$$
$$\text{Subject to the constraints } c_i(\mathbf{x}) \geq 0, \quad i = 1(1)m. \tag{1}$$

The first derivative of F with respect to $\mathbf{x}$ will be denoted by $\mathbf{g}$ (the gradient) and the hessian matrix of second derivatives by G. The first derivative of the ith constraint c_i with respect to $\mathbf{x}$ will be denoted by $\mathbf{n}_i$ (the normal vector).

The method for dealing with inequalities is to use an equality method with an exchange algorithm. A *basis* of *active* constraints that are being considered as equalities is held. Minimization then proceeds as for the equality case except that tests are regularly made to see whether any of the non-active (*passive*) constraints should be included, or whether any of the active constraints may be removed. An outline of the equality minimization process is given in Section 2. In Section 3 details of the changes in the penalty function and its derivatives at a change of basis are given, and in Section 4 decisions on when to make change of basis are discussed.

2. The Equality Minimization Process

Full details of the choices made in setting up the penalty function for the equality problem and carrying out an unconstrained minimization, and the reasons behind these choices are given in paper II, so a brief outline only will be given here. Assume that there is a basis of p ($\leq n$) active constraints, then, following the recommendation in paper II, the penalty function is

$$\phi = F - \mathbf{c}^T N^+ \mathbf{g} + \tfrac{1}{2} q \mathbf{c}^T N^+ N^{+T} \mathbf{c} \qquad (2)$$

where $\mathbf{c}$ is the $p \times 1$ vector of active constraint values, N^+ is a $p \times n$ matrix, the generalized inverse of a matrix N whose columns are the normals to the active constraints, and T indicates transpose. The columns of N are assumed to be normalized so that $\mathbf{n}_i^T \mathbf{n}_i = 1$, $i = 1(1)p$, and also they are assumed to be linearly independent so that $N^+ = (N^T N)^{-1} N^T$. In paper I it was shown that, under certain conditions, for an appropriate choice of q, ϕ has a stationary point at $\boldsymbol{\xi}$, the solution of the constrained problem, such that its hessian $\nabla^2 \phi(\boldsymbol{\xi})$ is positive definite. Thus a single unconstrained minimization of ϕ will locate $\boldsymbol{\xi}$. When the constraints are linear this condition on q is $q > \|G(\boldsymbol{\xi})\|$. To ensure that q is sufficiently large when there are non-linear constraints and since $\|G(\boldsymbol{\xi})\|$ is not generally known, q is initially calculated from the formula $q = s\|G(\mathbf{x})\| + r$, and is recalculated only when $\|G\|$ increases significantly or negative curvature is detected. For simplicity $s = 10$ is used in all the tests, with $r = 1$ initially and r increasing by a factor of 10 if negative curvature is detected.

A rank 1 variable metric method (for a list of references see Powell, 1970a) is used to minimize ϕ, which means that the gradient of ϕ, $\nabla \phi$, must be available. The formula used for $\nabla \phi$ is given by:

$$\nabla \phi = \mathbf{g} - NN^+ \mathbf{g} - GN^{+T} \mathbf{c} + qN^{+T} \mathbf{c} \qquad (3)$$

which is exact for a problem with only linear constraints. This formula requires the hessian G to be calculated, which may not be available analytically, or may be considered too lengthy to evaluate. Therefore two versions of the algorithm have been tested, one using an analytic formula for G throughout, the other using differences to set an initial G and updating this at each iteration by the formula due to Powell (1970b). When the function is quadratic and the constraints are linear the hessian of ϕ is

$$\nabla^2 \phi = G - NN^+ G - GNN^+ + qNN^+. \qquad (4)$$

As recommended in paper II the inverse of this matrix is used as an initial approximation to H, the variable metric.

The matrix N^+ is set up by the recurrence relation due to Fletcher (1969). One active constraint normal is added at a time, at a cost of $3np$ multiplications per constraint. To avoid complete recalculation of N^+ when there are non-linear constraints, and hence varying normals, N is ordered so that its first l columns represent the active linear constraints and the remaining columns the active non-linear constraints. Thus L^+, the constant inverse of the first l columns of N, need only be calculated once and N^+ can be built up from it whenever necessary by the addition of only $p - l$ normals.

3. The Mechanics of a Change of Basis

When a change of basis occurs the penalty function is altered. This means that ϕ and $\nabla\phi$ have to be recalculated and also that the variable metric H now applies to the wrong function. Complete recalculation of these quantities would be excessive in time, so recurrence formulae are used where possible to reduce the required number of operations.

The new generalized inverses of the constraint matrix, $N_{(p+1)}^+$ for an addition of a constraint and $N_{(p-1)}^+$ for a subtraction, can be readily calculated from $N_{(p)}^+$ the old generalized inverse, using recurrence relations due to Fletcher (1969). Therefore the recalculation of ϕ and $\nabla\phi$ is reduced to of order $n^2 + 6np$ multiplications. When the constraint is linear, L^+ must also be updated. In paper II it is shown that, using (4), the change in the hessian of ϕ following a change of basis can be approximated by the difference of two rank 1 matrices. Thus the change in the variable metric, which is exact in the quadratic programming case, is obtained by two applications of rank 1 formulae involving of order $4n^2 + 2np$ multiplications.

An advantage of using this particular form for the penalty function is that q depends only on $\|G\|$ and not on the constraints so it is unaffected by a change of basis.

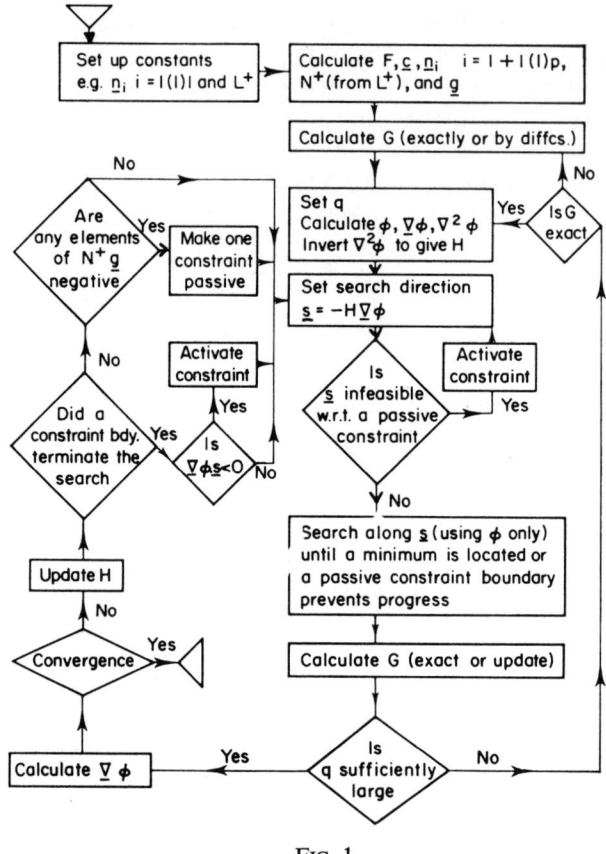

Fig. 1.

A flow diagram of the method is outlined in Fig. 1, and a more detailed breakdown of the calculation of ϕ and the operations involved in a change of basis is given in Fig. 2.

4. The Timing of a Change of Basis and Results

The problem of deciding when to make a change of basis is critical to the success of an algorithm. Cases can arise where constraints are repeatedly added to and then removed from the basis. This phenomenon, known as zig-zagging, slows down convergence, and may even cause convergence to a non-stationary point.

The basic criterion for deciding when to *add* a constraint into the basis is straightforward: starting from a feasible point, the search continues until a passive constraint is encountered, then, if the function value is decreasing

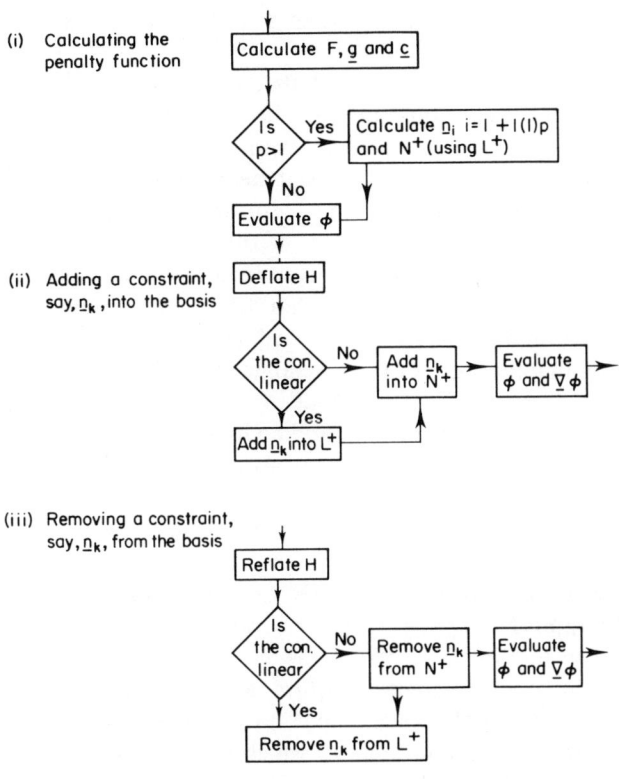

FIG. 2.

along the direction of search, out of the feasible region, the constraint is made active. By adopting this approach the search is assured to stay within the feasible region with respect to the passive constraints. However, it may be infeasible with respect to the active constraints, but only by small amounts since ϕ is being minimized. When constraints are removed from the basis this property can break down since it is possible for an active constraint that is violated (i.e. $c_i < 0$) to be made passive. To account for this possibility a test is made on the direction of search: if the search is moving away from the feasible region with respect to any of the violated, passive constraints then the offending constraints are remade active. Practical experience shows that this test is only invoked occasionally. The reason for this could be that constraints are only made passive if it is likely that the objective function can be reduced by moving into the feasible region away from the constraint boundary. Therefore, subsequent directions of search are likely to point into the feasible region.

The basic criterion for deciding when to *drop* a constraint is based on the Kuhn-Tucker condition that the Lagrange multipliers at the solution must all be positive. It was shown in paper I that the elements of the vector $N^+\mathbf{g}$ provide continuous approximations to the Lagrange multipliers so that the penalty functions can be considered as Lagrangian functions. This property of the penalty functions is very useful since it means that the Lagrange multipliers do not have to be specially calculated to test for possible passive constraints.

However, it is by no means clear how often this test should be applied. On the one hand the signs could be examined at the end of every iteration. On the other hand, finite convergence of a 'basis and exchange' algorithm for quadratic programming has been proved (Fletcher, 1970b) which simply involves examining the signs of the Lagrange multipliers at the solution of the current basis. Also, evidence by McCormick (1969) indicates that finite convergence of a projection-type method on linearly constrained problems can be obtained by only including new information, and hence allowing movement away from an active constraint, after a minimum has been located. These approaches are combined here so that constraints are allowed to become passive only after a successful linear minimization.

The exchange rules given above have been used to extend the program for equality constraints from paper II to a program that will deal with inequalities. Since this was to be primarily a test on the feasibility of the 'basis and exchange' approach on this type of penalty function, as few alterations as possible were made to the program. This meant that no improvements were made to the linear search to reduce the number of function evaluations in the degenerate case of a constraint being encountered along the line. Also, when more than one Lagrange multiplier was negative, constraints were removed on a very simple 'last in first out' basis. The problems used to test the program are the three-variable 'parcel' problems due to Rosenbrock (1960), referred to as PP1 for the problem with 1 active constraint, and PP3 for the problem with 3 active constraints, and the first three problems of the Colville (1968) report, referred to as TP1, TP2, and TP3. Problem TP1 has a non-linear objective function of 5 independent variables and 15 linear constraints, 4 of which are active at the solution. Problem TP2 is the dual of TP1, and has a non-linear objective function of 15 variables with 15 linear and 5 non-linear constraints. At the solution, 6 of the linear and all the non-linear constraints are active. The objective function of TP3 is a quadratic in 5 independent variables and there are 10 linear and 5 non-linear constraints; 3 of the linear and 2 of the non-linear constraints are active at the solution. To test, in particular, the efficiency of the exchange procedure, a quadratic programming problem, for which approximation (4) is exact but which requires several changes of basis, is also included. The problem, referred to as Quad./lin., has 3 variables and 11 constraints, 3 of which are active at the solution.

Results on these test problems showed a significant improvement over conventional penalty function methods where several unconstrained minimizations are performed (for example see, Fletcher and McCann, 1969). The use of the 'basis and exchange' approach was successful for all but TP2, the largest of the problems attempted. The pattern observed was that a certain number of iterations were required to locate the correct basis and this was followed by rapid convergence to the solution, similar to that described in paper II. On four of the tests, the only constraints added into the basis were those active at the solution and there were only occasional linear searches before the correct basis was obtained. On Quad/lin., several extra constraints

TABLE 1
Version with G exact

	Basis		Solution		
Problem	a	b	c	d	e
PP1	1	1	6	13	$\sim 10^{-7}$
PP3	3	3	6	10	$\sim 10^{-9}$
TP1	6	8	10	17	$\sim 10^{-6}$
TP2	66	148	99	239	$\sim 10^{-6}$
TP3	5	6	9	16	$\sim 10^{-6}$
Quad/lin.	7	8	7	8	$\sim 10^{-8}$

[a] Number of iterations to locate final basis.
[b] Number of *penalty* function evaluations to locate final basis.
[c] Number of iterations to locate solution.
[d] Number of *penalty* function evaluations to locate solution.
[e] Absolute accuracy in x_i.

were activated and later made passive, whereas with TP2 zig-zagging occurred and the method failed to obtain the correct basis after 150 iterations.

These results indicated two possible areas for improvement. Firstly, a linear search that requires only one function evaluation when a constraint prevents progress should considerably reduce the number of function evaluations needed to locate the final basis. Secondly, when more than one constraint has a negative Lagrange multiplier, a more realistic choice of dropping that constraint which corresponds to the largest negative multiplier, should reduce the number of changes of basis.

The results of a program incorporating both these suggestions are given in Table 1. A 50% reduction in the number of function evaluations needed to locate the correct basis, due to the improved linear search, was observed on all the problems. There was a further reduction for the Quad/lin. problem because the choice of redundant constraint lead directly to the correct basis and no constraint was reactivated once it had been made passive.

The results in Table 2 are for the same version of the program as Table 1, but with the approximation for G. Apart from TP2 there is little difference

TABLE 2
Version with G approximate

Problem	Basis		Solution		
	a	b	c	d	e
PP1	1	1	5	10	~0
PP3	3	3	3	3	~10^{-9}
TP1	6	9	10	18	~10^{-7}
TP2			No convergence		
TP3	6	8	10	19	~10^{-6}
Quad/lin.	7	9	7	9	~10^{-8}

[a] Number of iterations to locate final basis.
[b] Number of *penalty* function evaluations to locate final basis.
[c] Number of iterations to locate solution.
[d] Number of *penalty* function evaluations to locate solution.
[e] Absolute accuracy in x_i.

TABLE 3
Pattern of zig-zagging constraints for TP2 with exact matrix G

1	2	3	4	5	6*	7*	8	9*	10	11	12*	13*	14	15*	16*	17*	18*	19*	20*
A					A	A	A		A	A	A	A	A	A	A	A	A	A	A
R					R				R	R	R								R
					A					A									
					R					R									
					A					A									
					R					R									
					A					A									
					R					R									
					A					A									
					R					R			RESTART LINE						
					A					A			R						A
					R														

The numbers at the head of the table represent the 20 constraints of TP2, those marked * are active at the solution.
A—indicates the addition of a constraint.
R—indicates the removal of a constraint.

between the figures in the two tables. This behaviour is in accordance with the findings in paper II, where it was concluded that the approximation to G is of less importance than the other approximations made in solving the problem. In fact, when the calculated G is used, PP1 and PP3 are solved faster than when it is exact. However, this is a special case, because the updating formula for G on this problem is such that, once a constraint is satisfied, subsequent searches are confined exactly to the hyperplane of that constraint.

TABLE 4
Pattern of zig-zagging constraints for TP2 with the approximate matrix G

```
 1  2  3  4  5  6* 7* 8  9* 10 11 12* 13* 14 15* 16* 17* 18* 19* 20*

 A  A        A  A  A  A  A  A  A      A   A   A   A   A   A   A   A
 R  R     R  R                 R  R               R                 R
 A  A        A                             A      A                 A
 R  R        R                                    R                 R
 A  A        A                                    A                 A
 R           R                                    R                 R
 A           A                                    A                 A
             R                                    R                 R
             A                                    A                 A
             R                                    R                 R
             A                                    A                 A
             R                                    R                 R
             A                                    A                 A
             R                                    R                 R
             A                                    A                 A
             R                                                      R
             A                                                      A
             R                                                      R
             A                                                      A
```

The numbers at the head of the table represent the 20 constraints of TP2, those marked * are active at the solution.
A—indicates the addition of a constraint.
R—indicates the removal of a constraint.

The only test problem that presents any real difficulties is TP2. When the exact G is used there is a considerable amount of zig-zagging before the final basis is located, and for the approximation to G, the final basis is not located after 150 iterations. The zig-zagging patterns are described pictorially in Tables 3 and 4. In both cases the majority of constraints active at the solution are added into the basis and remain there, but a small number are repeatedly added and then removed. The space of the search $(n - p)$ is small, so that any ad-hoc anti-zig-zag device based on it, such as requiring a constraint to remain in the basis if it has been made passive during the last $n - p$ iterations, is ineffective. In fact any devices which prevent constraints from being made passive, and hence prevent an increase in the space of the search, are restrictive.

The significant difference in the patterns in Tables 3 and 4 is that in Table 3 the program is restarted and within a few iterations the correct basis is established. Now, one feature of the equality method is that it may be periodically restarted and the conditions for this are also incorporated in the inequality program. The first condition for restarting is when the parameter q of the penalty function is too small, so that the curvature of ϕ in the space orthogonal to the constraints may not be positive. The second condition is when the linear search breaks down but the current point does not have a zero gradient. In this case the variable metric, which has been updated using a linear approximation to the constraints at the starting point, is no longer appropriate. The

TABLE 5
The final version on problem TP2

Problem	Basis			Solution	
	a	b	c	d	e
Exact G	52	98	61	131	$\sim 10^{-5}$
Approx. G	42	86	67	161	$\sim 10^{-5}$

[a] Number of iterations to locate final basis.
[b] Number of *penalty* function evaluations to locate final basis.
[c] Number of iterations to locate solution.
[d] Number of *penalty* function evaluations to locate solution.
[e] Absolute accuracy in x_i.

program for TP2 using an exact G restarts for the latter reason and so calculates a new variable metric valid for the current basis at the current point. The outcome of this restart is that the final basis is rapidly located, indicating that the old metric is indeed no longer valid. However the continued zig-zagging in the program with the approximation to G prevents the conditions for a restart from ever being satisfied.

Furthermore, the directions of search are repeatedly found to be infeasible with respect to the violated, passive constraints, another indication that the metric is invalid. Therefore a condition for restarting the program in the inequality case is required and the obvious choice is the occurrence of infeasible directions of search. The results of a program using this test which show a marked improvement over the previous results are given in Table 5. These results, together with those for the remaining test functions from Tables 1 and 2, are very encouraging and suggest that this is a viable method for non-linear programming.

References

Colville, A. R. (1968). 'A Comparative Study on Non-linear Programming Codes'. IBM, New York Scientific Center Report No. 320–2949.

Fiacco, A. V., and McCormick, G. P. (1968). 'Nonlinear Programming: Sequential Unconstrained Minimization Techniques'. John Wiley and Sons, Inc., New York.

Fletcher, R. (1969). A technique for orthogonalization. *J. Inst. Maths. Applics.*, **5**, p. 162.

Fletcher, R. (1970a). A class of methods for non-linear programming with termination and convergence properties. *In*: 'Integer and Non-linear Programming' (J. Abadie, ed.). North-Holland Publishing Co., Amsterdam. (Referred to as 'paper I'.)

Fletcher, R. (1970b). 'A General Quadratic Programming Algorithm'. Harwell report TP 401.

Fletcher, R., and Lill, S. A. (1970). A class of methods for non-linear programming: II. Computational experience. *In*: 'Nonlinear Programming' (J. B. Rosen, O. L. Mangasarian, and K. Ritter, eds.). Academic Press, New York and London. (Referred to as 'paper II'.)

Fletcher, R., and McCann, A. P. (1969). Acceleration techniques for non-linear programming. *In*: 'Optimization' (R. Fletcher, ed.). Academic Press, London and New York.

McCormick, G. P. (1969). Anti-zig-zagging by bending. *Managmt Sci.* **15**, No. 5, p. 315.

Powell, M. J. D. (1970a). Rank one methods for unconstrained optimization. *In*: 'Integer and Non-linear Programming' (J. Abadie, ed.). North-Holland Publishing Co., Amsterdam.

Powell, M. J. D. (1970b). A new algorithm for unconstrained optimization. *In*: 'Nonlinear Programming' (J. B. Rosen, O. L. Mangasarian, and K. Ritter, eds.). Academic Press, London and New York.

Rosenbrock, H. H. (1960). An automatic method for finding the greatest or least value of a function. *Comput. J.*, **3**, p. 175.

28. A Hybrid Algorithm for Non-linear Programming

M. R. OSBORNE AND D. M. RYAN

Australian National University, Canberra, Australia

1. Introduction

1.1 Inequality Constrained Problems

We consider first the use of barrier functions to solve the inequality constrained problem (ICP)

$$\min_{x \in R_1} f(x)$$

where $R_1 = \{x \mid x \in E^n, g_i(x) \geq 0, i = \overline{1,l}\}$ and $f(x)$ and $g_i(x) \in C^{(2)}, i = \overline{1,l}$. We assume R_1°, the interior of R_1, is non-empty and characterize barrier functions $\Phi(x)$ by the properties:

(i) $\Phi(x)$ is a scalar valued function on R_1°, and
(ii) for any sequence of points $\{x^k\} \in R_1^\circ$ such that $x^k \to w$, where $g_i(w) = 0$ for some i, then

$$\lim_{k \to \infty} \Phi(x^k) = \infty.$$

Fiacco and McCormick (1968, p. 47), show that by minimizing the barrier objective function

$$T(x, r) = f(x) + r\Phi(x), \qquad r > 0, \tag{1}$$

over R_1° for a decreasing null sequence $\{r^k\}$ of parameter values, the sequence of minima $\{x^k\}$ exist and converge under quite general conditions to a local solution x^* of the ICP. Further information can be gained by requiring that

$$\Phi(x) = \sum_{i=1}^{l} \phi_i(g_i(x)) \tag{2}$$

where the $\phi_i(\cdot)$ satisfy

(i) $\phi_i(\cdot) \in C^{(2)}$ for $x \in R_1^\circ$,
(ii) if $\lim_{k \to \infty} g_i(x^k) = 0$ then $\lim_{k \to \infty} \phi_i(g_i(x^k)) = \infty$,

(iii) $\dfrac{d\phi_i(g_i(x))}{dg_i} < 0$ and $\dfrac{d^2 \phi_i(g_i(x))}{dg_i^2} > 0$, $x \in R_1^\circ$, and

(iv) $\dfrac{d^2 \phi_i}{dg_i^2}$ is a monotonic decreasing function of $g_i(x)$.

Examples

(a) $\Phi(x) = \sum_{i=1}^{l} \dfrac{1}{g_i(x)}$; the inverse barrier function discussed by Carrol (1961) and Fiacco and McCormick (1968).

(b) $\Phi(x) = -\sum_{i=1}^{m} \log(g_i(x))$; the log barrier function discussed by Frisch (1955), Parisot (1961) and Lootsma (1967).

Let x^* be a local minimum of the ICP and let x^k minimize (1) on R_1° with $\Phi(x)$ defined by (2). Define

$$u_i^k = -r^k \frac{d\phi_i(g_i(x^k))}{dg_i}, \quad i = \overline{1, l}. \tag{3}$$

If the sequence $\{u^k\}$ is bounded, then the limit points give the multipliers of the Kuhn-Tucker conditions. If, however, the sequence has no finite limit points, then the Kuhn-Tucker conditions do not hold at x^* (Osborne and Ryan, 1970). This is a more general result than that given by Fiacco and McCormick (1968).

These results are of considerable importance, since they show that the barrier function methods are extremely robust and that they provide considerable structural information about the ICP. In particular, the methods do not require the solution of the ICP to be a Kuhn-Tucker point. This information is actually provided by the calculation.

The advantages of barrier function methods are offset, however, by two important disadvantages.

(i) The rate of convergence is dependent on $\{r^k\}$. For the inverse barrier function we have

$$\|x^k - x^*\| = O([r^k]^{1/2})$$

at best, and for the log barrier function we have
$$\|\mathbf{x}^k - \mathbf{x}^*\| = O(r^k)$$
at best. This implies that for a fixed r^k sequence, the rate of convergence depends on the barrier function chosen. The 'obvious' approach of choosing a fast converging r^k sequence leads to costly minimizations of $T(\mathbf{x}, r^k)$ for each k. Osborne and Ryan (1970) have considered the possibility of improving the rate of convergence by a careful choice of barrier function. They have proposed the family of barrier functions

$$\left.\begin{array}{l}\phi_i^{(1)} = \log(G_i - \log(g_i(\mathbf{x}))) \\ \phi_i^{(j+1)} = \log(\sigma_i^{(j)} + \phi_i^{(j)}), \quad j = 1, 2, \ldots\end{array}\right\} \quad (4)$$

where G_i is sufficiently large and $\sigma_i^{(j)} \geq 1$. It is readily seen that $\phi_i^{(j)}$, $j = 1, 2, \ldots$ are barrier functions for $0 < g_i(\mathbf{x}) < e^{G_i - 1}$. Further, it is possible to demonstrate an improved convergence for each j and Osborne and Ryan were able to show that if $\mathbf{x}^{(j)}$ minimized

$$T^{(j)}(\mathbf{x}, r) = f(\mathbf{x}) + r \sum_{i=1}^{l} \phi_i^{(j)}(g_i(\mathbf{x}))$$

on R_1°, then the limit points of the sequence $\{\mathbf{x}^{(j)}\}$ for fixed r are local minima of the ICP. This result implies that for any prescribed accuracy, there is a *barrier function* which estimates the solution of the ICP to within this accuracy in a single unconstrained minimization. However, the minimizations required are costly. This corresponds to the more obvious result that such accuray can be obtained for a *particular* barrier function with r^k chosen sufficiently small and positive.

(ii) The second disadvantage of barrier function methods involves the increasing difficulty of minimizing $T(\mathbf{x}, r^k)$ as r^k becomes small. This is reflected in numerical examples in which the number of unconstrained minimization iterations required to find $\mathbf{x}^k$ does not decrease markedly even though $\|\mathbf{x}^k - \mathbf{x}^{k-1}\|$ is decreasing as k increases. It has been suggested (Murray, 1969; Lootsma, 1969) that this behaviour can be at least partly explained by examining the condition number of the Hessian of $T(\mathbf{x}, r)$ at its minimum. It is possible to show that the condition number of $\nabla^2 T(\mathbf{x}(r), r)$ is given by

$$\frac{\gamma(r)}{\|\mathbf{x}(r) - \mathbf{x}^*\|}$$

where $\gamma(r) > 0$ is bounded. This result implies that each barrier function will generate similar conditioning problems as the active constraints are approached.

1.2. Equality Constrained Problems

We now consider the equality constrained problem (ECP)

$$\min_{x \in R_e} f(x)$$

where $R_e = \{x \mid h_i(x) = 0, \; i = \overline{1,m}\}$. In this case, a treatment very similar to that discussed for barrier functions can be developed by considering *penalty* objective functions of the form

$$T(x,r) = f(x) + \frac{1}{r} \sum_{i=1}^{m} h_i(x)^2 \tag{5}$$

where $r \to 0$ from above. In particular it can be shown that the penalty function method retains all the robustness and also the disadvantages previously described for barrier functions. The general constrained problem (GCP) can be solved by using a mixed objective function combining barrier terms for inequalities and penalty terms for equalities. This approach is comprehensively treated by Fiacco and McCormick (1968).

However, Powell (1969) has proposed a generalized penalty function method which has the following attractive properties:

(i) the penalty function parameter $1/r$ does not have to become arbitrarily large, and
(ii) in certain circumstances the rate of convergence can approach that of a second-order process.

Powell's objective function has the form

$$P(x, \theta, \sigma) = f(x) + (h(x) + \theta)^T S (h(x) + \theta) \tag{6}$$

where $S = \mathrm{diag}(\sigma_i)$. It should be noted that (6) reduces to (5) when $\theta = 0$ and $\sigma = (1/r)e$. Let x^k minimize $P(x, \theta^k, \sigma^k)$. Powell (1969) has shown that provided $\min_{i=\overline{1,m}} \sigma_i^k$ is sufficiently large, the relations

$$\left.\begin{array}{l}\sigma^{k+1} = \sigma^k \\ \theta^{k+1} = \theta^k + h(x^k)\end{array}\right\} \tag{7}$$

provide effective convergence in the sense that $\{x^k\}$ converges rapidly to x^*, the ECP solution. If sufficiently rapid convergence is not achieved, the relations

$$\left.\begin{array}{l}\sigma^{k+1} = \kappa \sigma^k \\ \theta^{k+1} = \frac{1}{\kappa} \theta^k\end{array}\right\} \tag{8}$$

are applied with some constant $\kappa > 1$ to improve convergence. In practice (8) is seldom required and convergence is obtained by repeated application of (7). This suggests that the Hessian of $P(\mathbf{x}, \boldsymbol{\theta}, \boldsymbol{\sigma})$ may be better conditioned and this observation is reflected in numerical results which show that the minimizations of Powell's method quickly become trivial. This behaviour will be evident in the results to be discussed in Sections 3 and 4.

In a manner similar to barrier functions, Powell's method also provides structural information as a byproduct of the sequence of minimizations. In this case it can be seen that

$$v_i^k = 2\sigma_i^k(h_i(\mathbf{x}^k) + \theta_i^k) \tag{9}$$

will estimate the generalized Lagrange multiplier of the Kuhn-Tucker conditions associated with the constraint $h_i(\mathbf{x}) = 0$. Powell's method does not possess the robustness of the barrier function methods since it has only been justified when $\nabla h_i(\mathbf{x})$, $i = \overline{1, m}$ are linearly independent. This, however, is a standard assumption required by many theoretical results and methods for the ECP.

2. A Hybrid Algorithm

Motivation for the development of hybrid methods for the GCP is provided by the disadvantages of the standard barrier and penalty function approaches and the excellent convergence properties of Powell's method. We adopt the view that the robustness of the barrier function approach makes it valuable at least in the initial phase of problem solution when it is often the case that comparatively little structural information concerning the solution is available (e.g. which constraints are active or inactive). To improve its ultimate convergence properties, some modification of the standard barrier function method must be made.

One approach, discussed extensively by Fiacco and McCormick (1968) and Lootsma (1968), is to determine the form of dependence of $\mathbf{x}^k$ on r^k and to use this as a basis for extrapolation to predict $\mathbf{x}^{k+1}$ and also $\mathbf{x}^*$ if required. Standard polynomial extrapolation has proved to be successful for the inverse and log barrier functions but the dependence for the log family (4) is much more complex.

An alternative approach, and that adopted here, is to identify constraints active at the solution and then considering them as equalities, to apply an efficient penalty function method such as Powell's method described in Section 1.2. Constraints identified as inactive at the solution can be ignored, since they will not effect the solution. The hybrid objective function to be considered has the form

$$M(\mathbf{x}, \boldsymbol{\theta}, \boldsymbol{\sigma}, r^k) = f(\mathbf{x}) + r^k \sum_{i \in P_k} \phi_i(g_i(\mathbf{x})) + \sum_{i \in Q_k} \sigma_i(h_i(\mathbf{x}) + \theta_i)^2 \tag{10}$$

where at the kth stage the unclassified inequality constraints belong to the index set P_k and the equality constraints, including those inequalities identified as active, belong to the index set Q_k. The functions $\phi_i(\cdot)$, $i \in P_k$, are barrier functions satisfying the conditions (2) and r^k is a barrier function parameter.

The kth stage of the hybrid algorithm begins with the application of Powell's method to the ECP

$$\min_{\mathbf{x} \in R_e} \left\{ f(\mathbf{x}) + r^k \sum_{i \in P_k} \phi_i(g_i(\mathbf{x})) \right\} \tag{11}$$

where $R_e = \{\mathbf{x} | h_i(\mathbf{x}) = 0, \ i \in Q_k\}$. This amounts to repeated minimization of (10) for values of $\boldsymbol{\theta}$ and $\boldsymbol{\sigma}$ generated by (7) or (8) to produce values θ_i^k, σ_i^k,

TABLE 1
Classification of active and inactive constraints

Conditions		Classification
$g_i(\mathbf{x}^k) \to 0$ $u_i^k \to u_i^* \geq 0$	$r^k \to 0$	Constraint $g_i(\mathbf{x})$ is active at $\mathbf{x}^*$
$g_i(\mathbf{x}^k) \to \kappa_i > 0$ $u_i^k \to 0$	$r^k \to 0$	Constraint $g_i(\mathbf{x})$ is inactive at $\mathbf{x}^*$
$g_i(\mathbf{x}^k) \to 0$ $u_i^k \to \infty$	$r^k \to 0$	Kuhn-Tucker conditions do not hold at $\mathbf{x}^*$

$i \in Q_k$, such that $M(\mathbf{x}^k, \boldsymbol{\theta}^k, \boldsymbol{\sigma}^k, r^k)$ is minimized and $h_i(\mathbf{x}^k) = 0$, $i \in Q_k$. If all inequality constraints have been classified as either active or inactive, this step completes the solution. If, however, P_k is not empty an attempt is made to classify the constraints of P_k. Any constraints classified as active are transferred to Q_k and initial values of θ^k and σ^k are estimated for the new equality constraints. Constraints classified as inactive are simply removed from the summation of (10). As in the barrier function methods, the r^k value is now decreased to give r^{k+1} and the $(k+1)$th stage is begun.

In attempting to classify inequality constraints, use is made of the structural information provided by the barrier function method in the form of constraint values, $g_i(\mathbf{x}^k)$, and the generalized Lagrange multipliers u_i^k. The basic scheme is set out in Table 1. The testing for convergence of constraint values and generalized Lagrange multiplier estimates is computed using the values

$(\mathbf{x}^k, \boldsymbol{\theta}^k, \boldsymbol{\sigma}^k)$ at successive values of r^k. The following tests, based on Table 1 have, without exception, given correct classification on standard test problems.

(i) Constraints are considered *active* if the conditions

$$g_i(\mathbf{x}^k)/g_i(\mathbf{x}^{k-1}) < \gamma r^k/r^{k-1}, \qquad \gamma > 1,$$

and (12)

$$|u_i^k - u_i^{k-1}|/(1 + u_i^k) < \epsilon, \qquad 0 < \epsilon \ll 1,$$

are both satisfied.

(ii) Constraints are considered inactive if the conditions

$$u_i^k/u_i^{k-1} < \gamma r^k/r^{k-1}, \qquad \gamma > 1,$$

and (13)

$$|\phi_i(g_i(\mathbf{x}^k)) - \phi_i(g_i(\mathbf{x}^{k-1}))|/[1 + |\phi_i(g_i(\mathbf{x}^k))|] < \epsilon, \qquad 0 < \epsilon \ll 1,$$

are both satisfied.

The choice of γ and ϵ in (12) and (13) will depend on the severity of the test required. In practice, values of $\gamma = 1 \cdot 1$ and $\epsilon = 0 \cdot 01$ have been satisfactory when r^k is generated by $r^k = 10^{1-k}$, $k = 1, 2, \ldots$. If a constraint does not satisfy (12) or (13) it remains unclassified and a further classification attempt is made at r^{k+1}.

It should be noted that if the Kuhn-Tucker conditions do not hold, inactive constraints will be recognized but for active constraints, the GLM test of (12) will be difficult to satisfy. In general these constraints remain unclassified. Also, if the condition of strict complementarity does not hold (i.e. $u_i^* = 0$ for some active constraint) the inactive constraint will be recognized but for active constraints with $u_i^* = 0$, the GLM test of (12) will again be difficult to satisfy if ϵ is too small.

When a constraint is identified as active, we make a transfer of the form $h_\nu(\mathbf{x}) \equiv g_p(\mathbf{x})$. The transfer is effected using the barrier function estimates of the GLM, $\mathbf{u}^k$, associated with the active inequality constraints to provide an initial estimate of the θ and σ parameters of Powell's method. We choose θ_ν^k and σ_ν^k such that

$$2\sigma_\nu^k(h_\nu(\mathbf{x}^k) + \theta_\nu^k) = r^k \frac{\mathrm{d}\phi_p(g_p(\mathbf{x}^k))}{\mathrm{d}g_p}.$$

This choice ensures that $\mathbf{x}^k$ is a stationary point of (10). The first application of (7) for correcting on θ_ν^k implies that θ_ν^k should be chosen according to

$$\theta_\nu^k = \frac{r^k}{2\sigma_\nu^k} \frac{\mathrm{d}\phi_p(g_p(\mathbf{x}^k))}{\mathrm{d}g_p}$$

for some σ_ν^k. The aim is to choose σ_ν^k to ensure a good rate of convergence in Powell's method and our strategy is to pick σ_ν^k so that θ_ν^k is about ten times $h_\nu(\mathbf{x}^k)$. Under these circumstances (assuming the GLM estimate is accurate to several figures) σ_ν^k rarely exceeds 100 and the rate of convergence of Powell's method is very rapid indeed. It is not uncommon for $h_\nu(\mathbf{x}^k)$ to be reduced by a factor of 10^{-4} in the first minimization of Powell's method.

It is necessary to consider the possibility of a constraint being misclassified. The error is clearly most important if an inactive constraint is classified as active (recognition of the other case is trivial). This error can be detected by monitoring the sign of the generalized Lagrange multiplier estimate provided by Powell's method. The estimate will change sign if the corresponding constraint has been misclassified. The constraint can then be transferred back to the index set P_k for reclassification. It can be shown using (7) that when the sign change occurs, the current point will be feasible with respect to the constraint and the barrier function is therefore well defined. Note that more than a simple test on zero is required to prevent an active constraint with a very small or zero generalized Lagrange multiplier being repeatedly classified and rejected.

In making the constraint transfer, the Hessian of $M(\mathbf{x}^k, \boldsymbol{\theta}^k, \boldsymbol{\sigma}^k, r^k)$ can be modified to take account of the change in M. This is similar to the modifications proposed by Powell (1969) in the context of his method. It is not difficult to show that the change is dominated by the rank-one correction

$$\delta H^k = \left[2\sigma_\nu^k - r^k \frac{d^2 \phi_p}{dg_p^2} \right] \nabla g_p(\mathbf{x}^k) \nabla g_p(\mathbf{x}^k)^T.$$

The application of this correction is particularly important if an estimate of the inverse Hessian is available (for example provided by the DFP method for unconstrained minimization). The importance of using all available Hessian information is demonstrated numerically by the following example.

Example

Consider Colville's (1968) Problem 3. This is a five variable problem in which the objective function is subject to ten linear inequality constraints, six of which are active, and six non-linear inequality constraints, two of which are active. In Table 2 we compare the performance of the hybrid algorithm when

  (i) the inverse Hessian is reset to the unit matrix after each minimization, and
  (ii) the latest estimate of the Hessian is used at all times and is updated by rank-one modifications whenever possible.

The columns headed 'Reset' give the number of iterations 'it' and function evaluations 'ev' required to minimize $M(\mathbf{x},\boldsymbol{\theta},\boldsymbol{\sigma},r)$ when the inverse Hessian was reset as described in strategy (i). The columns headed 'Corrected' give the corresponding number of iterations and function evaluations when the inverse Hessian was corrected as described in strategy (ii). The column headed k and κ indicate the κth Powell iteration with the solution of (11) at r^k. An indication is also given of the constraint classification at each stage with the columns headed 'un', 'in' and 'ac' giving the number of constraints at the beginning of each unconstrained minimization which are unclassified, inactive or active, respectively.

TABLE 2
Colville Problem 3. Comparison of resetting or correcting inverse Hessian estimates

Iteration		Classification			Reset		Corrected	
k	κ	un	in	ac	it	ev	it	ev
1	1	16	0	0	14	34	14	34
2	1	16	0	0	28	72	12	26
3	1	7	7	2	18	41	11	25
	2				7	14	3	5
	3				1	1		
	4				2	3		
4	1	0	11	5	7	13	4	7
	2				6	11	2	3
	3				2	3	1	1
	4				4	7	2	3
	5						2	3
Totals					89	199	51	107

Remarks

(i) Convergence tests used in the computations are discussed in Section 3. The same tests were used for both the 'corrected' and 'reset' results and the accuracy attained by both methods was identical.

(ii) Although not without limitations, the use of numbers of iterations and function evaluations can be considered as fairly environment independent measures of performance. In general, neither the computation required for constraint selection nor that required for the consequent updating of the Hessian is significant in the overall calculation.

(iii) The apparent discrepancies in Table 2 at iterations $k = 3$, $\kappa = 3$ and $\kappa = 4$ and $k = 4$, $\kappa = 5$ are due to the convergence tests being satisfied at slightly different points.

(iv) The barrier function used was $\phi(g_i(\mathbf{x})) = \log(10 - \log(g_i(\mathbf{x})))$.

The results of Table 2 (which are typical of results from other problems) demonstrate clearly the advantage of retaining and updating the Hessian whenever possible. For this reason, the strategy is incorporated in the hybrid method.

3. Numerical Results

In this section we compare the performance of three barrier function algorithms on four standard test problems. The algorithms considered are

(i) the hybrid algorithm described in Section 2,
(ii) the basic barrier function algorithm, and
(iii) an extrapolated barrier function method.

The first two methods use the $\log(10 - \log(\cdot))$ barrier function while in the third the $\log(\cdot)$ barrier function is the natural choice. The extrapolated barrier function method was implemented in the manner described by Fletcher and McCann (1969). This uses linear extrapolation on the results of the first two minimizations and quadratic extrapolation based on the last three minimizations subsequently.

In implementing these algorithms, the DFP code already referred to was used for the unconstrained minimizations. The following convergence tests were used.

(i) Minimization of $M(\mathbf{x}, \mathbf{\theta}, \mathbf{\sigma}, r)$. The current point is accepted as a minimum if either

(a) $\|\nabla M\| \leq n^2 10^{-13}$, \hfill (14)

(b) $\max_{i=\overline{1,n}} \dfrac{|x_i^{\alpha+1} - x_i^{\alpha}|}{1 + |x_i^{\alpha+1}|} \leq 10^{-8}$.

The second condition is required for small r^k values.

(ii) Powell's method. The current point is accepted as a solution of the ECP (11) if

$$\max_{i \in Q_k} |h_i(\mathbf{x})| \leq 10^{-8}.$$

In addition to the Colville (1968) Problem 3 already described we also consider Colville Problems 1 and 2 and the Hexagon problem of Murray (1969).

28. A HYBRID ALGORITHM FOR NON-LINEAR PROGRAMMING

Colville Problem 1 involves the minimization of a cubic function of 5 variables subject to non-negativity conditions on the variables and 10 linear inequality constraints four of which are active. Colville Problem 2, the dual of Colville

TABLE 3
Colville Problem 1

k	κ	Hybrid method					k	B.F.		E.B.F	
		un	in	ac	it	ev		it	ev	it	ev
1	1	15	0	0	25	66	1	25	66	16	40
2	1	15	0	0	13	27	2	13	27	12	29
3	1	4	10	1	10	22	3	11	32	14	30
	2				3	5	4	10	22	12	26
4	1	1	11	3	5	19	5	7	15	9	21
	2				2	12	6	19	41	7	13
5	1	0	11	4	4	10	7	6	13	3	6
	2				2	3	8	4	9	2	4
	3				2	3	9			2	4
							10			2	4
Totals					66	167		95	225	79	177

TABLE 4
Colville Problem 2

k	κ	Hybrid method					k	B.F.		E.B.F.	
		un	in	ac	it	ev		it	ev	it	ev
1	1	20	0	0	34	74	1	34	74	31	61
2	1	20	0	0	26	55	2	26	55	22	48
3	1	10	3	7	20	41	3	21	47	21	44
	2				7	13	4	19	42	16	33
	3				2	3	5	18	39	13	29
4	1	2	7	11	7	13	6	21	48	7	15
	2				4	7	7	18	39	2	3
	3				2	3	8	24	53	2	4
5	1	1	8	11	4	7	9	21	43	2	4
	2				3	5	10			2	4
6	1	0	9	11	3	5					
	2				2	3					
Totals					114	229		202	440	118	245

Problem 1, has 15 variables, 15 linear inequality constraints 6 of which are active and 5 non-linear inequalities all of which are active. The Hexagon problem has 9 variables, 13 linear inequalities none of which are active and 15 non-linear inequalities 6 of which are active.

TABLE 5
Colville Problem 3

k	κ	Hybrid method					k	B.F.		E.B.F.	
		un	in	ac	it	ev		it	ev	it	ev
1	1	16	0	0	14	34	1	14	34	13	27
2	1	16	0	0	12	26	2	12	26	12	25
3	1	7	7	2	11	25	3	11	23	16	31
	2				3	5	4	8	17	6	12
4	1	0	11	5	4	7	5	13	31	2	3
	2				2	3	6	9	21	3	7
	3				1	1	7	4	8	2	4
	4				2	3	8	3	8	2	4
	5				2	3	9	2	5	2	4
							10			2	6
Totals					51	107		76	173	60	123

The results are presented in Tables 3–4 in the same format as Table 2, except that the 'Reset' column is replaced by columns headed 'B.F.' and 'E.B.F.' giving the comparative results of the barrier function and extrapolated barrier function methods, respectively.

The results are typical of results obtained from other test problems. It should be noted that the same accuracy was attained by each method. The test (14) on

TABLE 6
Hexagon problem

k	κ	Hybrid method					k	B.F.		E.B.F.	
		un	in	ac	it	ev		it	ev	it	ev
1	1	28	0	0	10	20	1	10	20	12	27
2	1	28	0	0	23	50	2	23	50	24	51
3	1	16	21	0	15	33	3	15	36	16[a]	35[a]
4	1	6	20	2	13	29	4	14	30	13[a]	28[a]
	2				4	7	5	11	27	7	14
5	1	0	22	6	4	7	6	9	20	2	3
	2				3	5	7	11	27	2	4
							8	7	16	2	7
							9	16	36	2	4
							10			2	3
Totals					72	151		116	262	82	176

[a] Extrapolated point infeasible.

the change in **x** ensures an accuracy of either eight significant figures or eight decimal places (whichever is least) is obtained in the barrier function method—this is a consequence of the known dependence of $\mathbf{x}^k$ on r^k—and the calculations bear this out well. The slower convergence of the log barrier function as $r^k \to 0$ is evident in the extra one or two minimizations required to attain the accuracy. Although the evidence presented here is too limited to be conclusive, it shows the hybrid algorithm consistently superior to the extrapolation method by between 10 and 20%. Both methods show a significant improvement over the basic barrier function algorithm. It should also be noted that the minimizations of Powell's method in the hybrid algorithm are extremely simple and small in number reflecting the fast rate of convergence.

4. Linear Constraints

A particular feature of the GCP which can influence the performance of barrier function methods is found in the occurrence of linear constraints. Fiacco and McCormick (1968) have pointed out the advantages of treating linear constraints explicitly by projection methods instead of including them as non-linear constraints in barrier function methods. We have recently adapted the DFP code to incorporate the gradient projection approach. The use of projected gradient methods was first proposed by Rosen (1960) who adapted the method of steepest descent. More recently Goldfarb and Lapidus (1968) used the DFP algorithm, while Murtagh and Sargent (1969) used a rank-one method and a rather different organization of the computation. In our approach (Ryan, 1971) we have basically used the DFP algorithm while incorporating some of the definite advantages of the Murtagh and Sargent algorithm. These advantages virtually amount to preserving the Hessian information whenever possible—in certain circumstances information can be lost in the Goldfarb-Lapidus approach.

In this section we report preliminary experience with an algorithm in which the non-linear constraints are treated by the hybrid method, and the linear constraints are treated by the adapted DFP algorithm. Lootsma has suggested that we ought also consider the adapted DFP algorithm in conjunction with extrapolation procedures. This idea might well prove fruitful, in particular as the basis of active linear constraints is obtained at an early stage of the computation in each of our examples. However, we have not considered this suggestion further as yet.

In implementing the algorithm, the convergence test (14) has been replaced by

$$\|PH\nabla M\| \leq n^2 \, 10^{-13}$$

and

$$\gamma \leq 10^{-8}$$

and

$$\max_{i=\overline{1,n}} \frac{|x_i^{\alpha+1} - x_i^{\alpha+1}|}{1 + |x_i^{\alpha+1}|} \leq 10^{-8}$$

where P is the projection operator, H the inverse Hessian estimate and γ is a function of the smallest Lagrange multiplier (see Goldfarb and Lapidus, 1968). This convergence test produces results of similar accuracy to those already presented. Direct comparison is therefore valid.

TABLE 7
Colville Problem 1 (projection)

k	κ	Non-linear			Linear		it	ev
		un	in	ac	in	ac		
1	1	0	0	0	11	4	7	15

TABLE 8
Colville Problem 2 (projection)

k	κ	Non-linear			linear		it	ev
		un	in	ac	in	ac		
1	1	5	0	0	8	7	44	120
2	1	5	0	0	8	7	13	27
3	1	2	0	3	9	6	17	39
	2				9	6	4	7
4	1	0	0	5	9	6	4	7
	2				9	6	3	5
Totals							85	205

The results are presented in Tables 7–10. Similar column headings to those of Tables 4–6 are used and a further column headed 'Linear' is included to give the number of constraints which are inactive 'in' or active 'ac' *after* each minimization.

Because Colville problem 1 involves only linear constraints the results can be compared with results provided by the Goldfarb-Lapidus and Murtagh-Sargent methods and published by Murtagh and Sargent (1969).

 (i) Goldfarb-Lapidus: 10 iterations; 13 function evaluations.
 (ii) Murtagh-Sargent: 7 iterations; 7 function evaluations.

The results of Table 10 are very similar to those of Table 6. This is to be expected, since no linear constraints are active. However, the results of Tables 8 and 9 demonstrate the potential advantages of this approach and underline the importance of using the special properties of constraint linearity when linear constraints are active.

TABLE 9
Colville Problem 3 (projection)

k	κ	Non-linear			Linear		it	ev
		un	in	ac	in	ac		
1	1	6	0	0	7	3	11	22
2	1	6	0	0	7	3	7	16
3	1	2	4	0	7	3	6	13
4	1	0	4	2	7	3	3	6
	2				7	3	2	3
	3				7	3	1	1
	4				7	3	2	4
Totals							32	65

TABLE 10
Hexagon Problem (projection)

k	κ	Non-linear			Linear		it	ev
		un	in	ac	in	ac		
1	1	15	0	0	13	0	10	23
2	1	15	0	0	13	0	27	61
3	1	12	3	0	13	0	13	31
4	1	2	9	4	13	0	9	19
	2				13	0	3	5
5	1	0	9	6	13	0	5	9
	2				13	0	2	3
Totals							69	151

References

Carrol, C. W. (1961). The created response surface technique for optimizing non-linear restrained systems. *Operat. Res.* **9**(2), 169–184.

Colville, A. R. (1968). 'A Comparative Study on Nonlinear Programming Codes'. IBM New York Scientific Center Tech. Rep. No. 320-2949.

Fiacco, A. V., and McCormick, G. P. (1968). 'Nonlinear Programming: Sequential Unconstrained Minimization Techniques', Wiley, New York.

Fletcher, R., and McCann, A. P. (1969). Acceleration techniques for nonlinear programming. *In*: 'Optimization' (R. Fletcher, ed.). Academic Press, London.

Frisch, K. R. (1955). 'The Logarithmic Potential Method of Convex Programming'. Memorandum May 13, 1955, University Institute of Economics, Oslo.

Goldfarb, D., and Lapidus, L. (1968). Conjugate gradient method for nonlinear programming problems with linear constraints. *I. & E. C. Fundamentals* **7**(1), 142–151.

Lootsma, F. A. (1967). Logarithmic programming: a method of solving nonlinear programming problems. *Philips Res. Rept.* **22**(3), 329–344.

Lootsma, F. A. (1968). Extrapolation in logarithmic programming. *Philips Res. Rept.* **23**, 108–116.

Lootsma, F. A. (1969). Hessian matrices of penalty functions for solving constrained-optimization problems. *Philips Res. Rept.* **24**, 322–331.

Murray, W. (1969). Behaviour of Hessian matrices of barrier and penalty functions arising in optimization. *Nat. Phys. Lab. Rept.* Ma77.

Murtagh, B. A., and Sargent, R. W. H. (1969). A constrained minimization method with quadratic convergence. *In*: 'Optimization' (R. Fletcher, ed.). Academic Press, London.

Osborne, M. R., and Ryan, D. M. (1970). On penalty function methods for nonlinear programming problems. *J. Math. Anal. Appls* **31**, 559–578.

Parisot, G. R. (1961). Rèsolution numérique approchée du problème de programmation linéaire par application de la programmation logarithmique. *Revue fr. Rech. opér.* **20**, 227–259.

Powell, M. J. D. (1969). A method for nonlinear constraints in minimization problems. *In*: 'Optimization' (R. Fletcher, ed.). Academic Press, London.

Rosen, J. B. (1960). The gradient projection method for nonlinear programming, part 1: linear constraints. *J. Soc. ind. appl. Math.* **8**(1), 181–217.

Ryan, D. M. (1971). 'Transformation Methods in Nonlinear Programming.' Ph.D. Thesis, Australian National University.

29. Constrained Minimization Using Recursive Equality Quadratic Programming

M. C. BIGGS

*Numerical Optimization Centre, The Hatfield Polytechnic,
Hatfield, Hertfordshire, England*

Summary

This paper describes an exterior point algorithm for constrained minimization. It is related to penalty function methods, the best known of which is SUMT, described both in its interior point and its exterior point version by Fiacco and McCormick. SUMT proceeds by minimizing a sequence of penalty functions, since it can be shown that, under appropriate conditions, the trajectory of penalty function minima passes through the solution to the constrained problem. An approach devised by Murray avoids this repeated unconstrained minimization by using a simple QP subproblem to approximate the minima of a sequence of penalty functions. The algorithm derived in this paper has the same motivation as Murray's but uses an alternative QP subproblem. Numerical experiments confirm that the QP methods can converge in fewer function evaluations and less computer time than SUMT.

1. Introduction

The solution of constrained minimization problems via a sequence of quadratic programming problems has been suggested by Murray (1969a) and Fletcher (1970a). We describe a further implementation of this proposal and follow Murray's ideas in seeking a sequence of infeasible points approximating the trajectory of penalty function minima, which, under suitable conditions, passes through the solution of the constrained problem. The present method differs from that of Murray in several important ways. It employs an alternative subproblem (whose theoretical justification is given in Sections 2 and 3) which is in fact an equality constrained QP subproblem instead of the inequality constrained one proposed by Murray. There is a computational advantage here in that the solution to an inequality constrained problem cannot, in general, be obtained as a single algebraic expression whereas the solution to an equality constrained QP subproblem can be written down explicitly. This point is considered in Section 4. The formulation of a QP subproblem will involve some or all of the constraints of the main problem, and we call these constraints the active set or basis. The active basis used by Murray's method consists only of those constraints that are currently violated; our method, on the other hand, may have non-violated constraints in its active set. The use of a larger constraint basis has proved valuable in practice for dealing with non-linear constraints. A procedure for selecting active constraints is described, together with other computational details, in Section 5. Section 6 consists of a brief discussion of the relationship, in theory and practice, between our algorithm and that of Murray. Finally, some encouraging computational experience is reported in Section 7.

2. Properties of Penalty Functions

The problem considered is

$$\text{minimize } F(\mathbf{x}), \quad \mathbf{x} = (x_1, \ldots, x_n)^T$$
$$\text{subject to } b_i(\mathbf{x}) \geq 0, \quad i = 1, 2, \ldots, m, \tag{1}$$

where $F(\mathbf{x})$ and all $b_i(\mathbf{x})$ are assumed to be twice continuously differentiable. It is also assumed that there is a unique solution $\mathbf{x}^*$ to (1) where the gradients of the binding constraints are linearly independent (in particular this implies that not more than n constraints are effective at the solution) and the Lagrange multipliers associated with the active constraints are strictly positive. At the solution all constraints must be satisfied and the projection of the gradient of $F(\mathbf{x})$ upon the binding constraint manifold must be zero.

The penalty function $P(\mathbf{x}, r)$ is defined as

$$P(\mathbf{x}, r) = F(\mathbf{x}) + \frac{1}{r} \mathbf{v}^T \mathbf{v}$$

where $\mathbf{v}$ is the vector of constraints violated at $\mathbf{x}$.

It can be shown (Pietrzykowski, 1962) that if $P(\mathbf{x}, r_k)$ is strictly convex for each $r_k > 0$ and has a minimum $\mathbf{x}_k^*$, the sequence $\mathbf{x}_k^*$ ($k = 1, 2, \ldots$) converges to $\mathbf{x}^*$ if and only if r_k ($k = 1, 2, \ldots$) is a sequence of positive numbers decreasing to zero. The SUMT exterior point method of solving (1) (Fiacco and McCormick, 1968) exploits this property by minimizing a sequence of $P(\mathbf{x}, r_k)$ for decreasing r_k and using the solutions $\mathbf{x}_k^*$ to make extrapolations along the trajectory of minima towards $\mathbf{x}^*$. This process however involves the minimization of functions which become more difficult as $r_k \to 0$ (Murray, 1967). Some properties of penalty functions will now be considered which enable us to avoid these minimizations.

29. CONSTRAINED MINIMIZATION

In what follows we shall denote $F(\mathbf{x}_k)$ by F_k, $\mathbf{b}(\mathbf{x}_k)$ by $\mathbf{b}_k$, etc. The set of violated constraints at $\mathbf{x}_k$ will be denoted by I_k, and $\mathbf{v}_k$ and $\mathbf{c}_k$ are the vectors of violated and non-violated constraints.

Penalty functions for inequality constrained problems have the awkward property that their curvature is discontinuous at each constraint. This means that Taylor series predictions about the penalty function have to be used with some caution. At the point $\mathbf{x}_k$ we have

$$P(\mathbf{x}_k, r_k) = F_k + \frac{1}{r_k} \mathbf{v}_k^T \mathbf{v}_k$$

and

$$\nabla P(\mathbf{x}_k, r_k) = \mathbf{f}_k + \frac{2}{r_k} W_k^T \mathbf{v}_k$$

where $\mathbf{f}$ is the gradient of F and W is the Jacobian of $\mathbf{v}$. If the *same* constraints are violated at a neighbouring point $\mathbf{x}_k + \mathbf{p}$ the truncated Taylor series estimate of the gradient of the penalty function is

$$\nabla P(\mathbf{x}_k + \mathbf{p}, r_k) = \mathbf{f}_k + B_k \mathbf{p} + \frac{2}{r_k}(W_k^T \mathbf{v}_k + W_k^T W_k \mathbf{p}) \tag{2}$$

where B is the Hessian matrix of F.

If, however, some of the constraints violated at $\mathbf{x}_k$ are no longer violated at $\mathbf{x}_k + \mathbf{p}$, then the equivalent expression for $\nabla P(\mathbf{x}_k + \mathbf{p}, r_k)$ is

$$\nabla P(\mathbf{x}_k + \mathbf{p}, r_k) = \mathbf{f}_k + B_k \mathbf{p} + \frac{2}{r_k}(W_k^T \mathbf{v}_k + W_k^T \boldsymbol{\gamma}_k) \tag{3}$$

where the elements of $\boldsymbol{\gamma}_k$ are

$$\left. \begin{array}{c} \gamma_{k,i} = \min(-v_{k,i}, \mathbf{p}^T \mathbf{w}_i), \\ \mathbf{w}_i \text{ being the } i\text{th row of } W_k. \end{array} \right\} i \in I_k$$

Finally, if at $\mathbf{x}_k + \mathbf{p}$ constraints are violated which were not violated at $\mathbf{x}_k$ then, using $\mathbf{c}_k$ the vector of constraints not violated at $\mathbf{x}_k$ and D the Jacobian of $\mathbf{c}$, the more general expression for $\nabla P(\mathbf{x}_k + \mathbf{p}, r_k)$ is

$$\nabla P(\mathbf{x}_k + \mathbf{p}, r_k) = \mathbf{f}_k + B_k \mathbf{p} + \frac{2}{r_k}(W_k^T \mathbf{v}_k + W_k^T \boldsymbol{\gamma}_k + D_k^T \boldsymbol{\delta}_k) \tag{4}$$

where $\boldsymbol{\gamma}_k$ is defined as before and the elements of $\boldsymbol{\delta}_k$ are

$$\left. \begin{array}{c} \delta_{k,i} = \min(0, c_{k,i} + \mathbf{p}^T \mathbf{d}_i) \text{ for each } c_{k,i}, \\ \mathbf{d}_i \text{ being the } i\text{th row of } D_k. \end{array} \right\}$$

Assume first that the violated set I_k is the same as the constraint set active at the solution of (1). Then the minimum of $P(x, r_k)$ lies outside all the constraints in I_k and equation (2) may be used. The more general situation, when the violated set I_k is not the same as the solution set, will be considered in Section 3.

If $\mathbf{x}_k + \mathbf{p}$ is the minimum of $P(\mathbf{x}, r_k)$, then

$$0 = \nabla P(\mathbf{x}_k + \mathbf{p}, r_k) = \mathbf{f}_k + B_k \mathbf{p} + \frac{2}{r_k}(W_k^T \mathbf{v}_k + W_k^T W_k \mathbf{p})$$

i.e.

$$\left(B_k + \frac{2}{r_k} W_k^T W_k\right) \mathbf{p} = -\mathbf{f}_k - \frac{2}{r_k} W_k^T \mathbf{v}_k. \tag{5}$$

This expression yields the Newton correction for minimizing $P(\mathbf{x}, r_k)$. Because of the ill-conditioning of the matrix

$$B_k + \frac{2}{r_k} W_k^T W_k$$

(Murray, 1967) this choice of $\mathbf{p}$ can lead to computational difficulties. We therefore look at some modifications of equation (5), particularly with a view to isolating the effect of the penalty term in determining the position of the minimum of $P(\mathbf{x}, r_k)$. Premultiplying (5) by $W_k B_k^{-1}$ and rearranging gives

$$\left(\frac{r_k}{2} I + W_k B_k^{-1} W_k^T\right) W_k \mathbf{p} = -\frac{r_k}{2} W_k B_k^{-1} \mathbf{f}_k - W_k B_k^{-1} W_k^T \mathbf{v}_k.$$

If $$M_k = \frac{r_k}{2} I + W_k B_k^{-1} W_k^T, \quad \text{then}$$

$$W_k \mathbf{p} = -\frac{r_k}{2} M_k^{-1} W_k B_k^{-1} \mathbf{f}_k - M_k^{-1} W_k B_k^{-1} W_k^T \mathbf{v}_k. \tag{6}$$

Adding

$$\frac{r_k}{2} M_k^{-1} \mathbf{v}_k$$

to the first term on the right-hand side of (6) and subtracting

$$\frac{r_k}{2} M_k^{-1} \mathbf{v}_k$$

from the second term gives an alternative expression

$$W_k \mathbf{p} = -\frac{r_k}{2} M_k^{-1}(W_k B_k^{-1} \mathbf{f}_k - \mathbf{v}_k) - \mathbf{v}_k. \tag{7}$$

Formulas (6) and (7) do not determine $\mathbf{p}$ completely unless there are as many violated constraints as there are variables. But if $\mathbf{x}_k + \mathbf{p}$ minimizes $P(\mathbf{x}, r_k)$

then, within our approximations, **p** must satisfy (6) and (7). It is therefore possible to regard these equations as constraints which limit the choice of **p**.

Before this point is discussed further the above analysis will be repeated under the simplifying assumption (which is reasonable when r is small) that the curvature of F can be neglected in comparison with the curvature of the penalty term $(1/r)\mathbf{v}^T\mathbf{v}$. The truncated Taylor series estimate now gives

$$0 = \nabla P(\mathbf{x}_k + \mathbf{p}, r_k) = \mathbf{f}_k + \frac{2}{r_k}(W_k^T \mathbf{v}_k + W_k^T W_k \mathbf{p}).$$

Hence **p** must satisfy

$$W_k^T W_k \mathbf{p} = -\frac{r_k}{2}\mathbf{f}_k - W_k^T \mathbf{v}_k.$$

Premultiplying by either W_k or $W_k B_k^{-1}$ gives, respectively

$$W_k \mathbf{p} = -\frac{r_k}{2}(W_k W_k^T)^{-1} W_k \mathbf{f}_k - \mathbf{v}_k \qquad (8)$$

or

$$W_k \mathbf{p} = -\frac{r_k}{2}(W_k B_k^{-1} W_k^T)^{-1} W_k B_k^{-1} \mathbf{f}_k - \mathbf{v}_k. \qquad (9)$$

Equations (6)–(9) give estimates of the change in the violated constraint vector $\mathbf{v}_k$ caused by a move to the minimum of $P(\mathbf{x}, r_k)$. They give no information about the component of **p** that lies in the null space of W_k, however, and in this space the behaviour of $F(\mathbf{x})$ will be considered. The vector **p** can be determined as the particular solution of one of (6)–(9) which minimizes $F(\mathbf{x})$ in the null space of the violated constraints. More simply, the step to the minimum of $P(\mathbf{x}, r_k)$ can be approximated by choosing **p** to minimize some quadratic approximation to $F(\mathbf{x})$ subject to one of the conditions (6)–(9). This strategy converts the problem of choosing a direction of search into an equality constrained quadratic programming problem.

In subsequent sections we shall only consider the use of (8) and (9) since these appear to be more convenient computationally than (6) and (7).

3. Lagrange Multipliers and 'Active' Constraints

Let $\boldsymbol{\lambda}^*$ denote the vector of Lagrange multipliers associated with the binding constraints at $\mathbf{x}^*$. It is well known that the vectors $(W_k W_k^T)^{-1} W_k \mathbf{f}_k$ and $(W_k B_k^{-1} W_k^T)^{-1} W_k B_k^{-1} \mathbf{f}_k$ can be regarded as approximations to the Lagrange multipliers associated with the violated constraints at $\mathbf{x}_k$. In the discussion which follows let $\hat{\boldsymbol{\lambda}}_k$ denote either of these approximations. (It should be noted that

$\hat{\boldsymbol{\lambda}}_k$ can only be calculated if the rows of W_k, i.e. the constraint normals, are independent. We have stipulated that problem (1) has a solution where this condition is satisfied, and hence there is some region about the solution where the relationships will be meaningful.) Using $\hat{\boldsymbol{\lambda}}_k$ the expressions (8) and (9) can both be written in the form

$$W_k \mathbf{p} = -\frac{r_k}{2} \hat{\boldsymbol{\lambda}}_k - \mathbf{v}_k. \qquad (10)$$

Fiacco and McCormick (1968) show that if at the minimum $\mathbf{x}_k^*$ of $P(\mathbf{x}, r_k)$ all elements of $\hat{\boldsymbol{\lambda}}(\mathbf{x}_k^*)$ are positive then the relationship

$$\mathbf{v}(\mathbf{x}_k^*) = -\frac{r_k}{2} \hat{\boldsymbol{\lambda}}(\mathbf{x}_k^*) \qquad (11)$$

holds between the approximate multipliers, the vector of violated constraint values and the current penalty parameter. Equation (10) is a first order condition that the constraints at $\mathbf{x}_k + \mathbf{p}$ should satisfy such a condition.

In Section 2 the temporary assumption was made that I_k was the same as the binding constraint set. In general this will not be so and hence $\hat{\boldsymbol{\lambda}}_k$ may be a very poor approximation to $\boldsymbol{\lambda}^*$. Indeed the two vectors may have different numbers of elements. Moreover, there is no reason why the approximation should not have some negative elements even though problem (1) is such that $\boldsymbol{\lambda}^*$ has all positive elements. The negativity of an approximate Lagrange multiplier gives an indication—but not a conclusive one—that the corresponding constraint is not binding at the solution $\mathbf{x}^*$. It will be seen that if any elements of $\hat{\boldsymbol{\lambda}}_k$ are negative then (10) is a first-order condition that at $\mathbf{x}_k + \mathbf{p}$ the corresponding constraints will no longer be violated. Now it was observed in Section 2 that equation (2), from which (10) was derived is not strictly appropriate to the situation where the displacement $\mathbf{p}$ causes any constraints to be shed. Hence if there are negative elements in $\hat{\boldsymbol{\lambda}}_k$, (10) is no longer a necessary condition that $\mathbf{x}_k + \mathbf{p}$ minimizes $P(\mathbf{x}, r_k)$. We argue that it is still useful, however, (particularly while the binding constraints have still to be identified) because it provides a simple and efficient means of dropping from the violated set any constraints which appear to be non-binding.

So far only the violated set at $\mathbf{x}_k$ has been considered. By using equation (4) of Section 2, however, it is possible to extend the analysis to take account of some non-violated constraints. It would clearly be useful if binding constraints could be identified and used to control the move $\mathbf{p}$ even when not in the set I_k. In the algorithm of Section 5 we shall describe one possible way of choosing 'active' constraints which are not necessarily violated and we consider here how to make use of such a basis. Assume that all the violated constraints at $\mathbf{x}_k$ are in fact binding and that $\bar{\mathbf{c}}_k$ is the vector of non-violated constraints that are also being assumed to be binding. Then from (4)

$$0 = \nabla P(\mathbf{x}_k + \mathbf{p}, r_k) = \mathbf{f}_k + \frac{2}{r_k}(W_k^T \mathbf{v}_k + W_k^T W_k \mathbf{p} + \bar{D}_k^T(\bar{\mathbf{c}}_k + \bar{D}_k \mathbf{p}))$$

where $\bar{D}$ is the Jacobian of $\bar{\mathbf{c}}$.
Hence if we define

$$\mathbf{g}_k = \begin{pmatrix} \mathbf{v}_k \\ \bar{\mathbf{c}}_k \end{pmatrix}, \quad A_k = \begin{pmatrix} W_k \\ \bar{D}_k \end{pmatrix}$$

as a vector of 'active' constraints and its Jacobian then we can obtain in the same way as (8) and (9)

$$A_k \mathbf{p} = -\frac{r_k}{2}(A_k A_k^T)^{-1} A_k \mathbf{f}_k - \mathbf{g}_k, \tag{12}$$

$$A_k \mathbf{p} = -\frac{r_k}{2}(A_k B_k^{-1} A_k^T)^{-1} A_k B_k^{-1} \mathbf{f}_k - \mathbf{g}_k. \tag{13}$$

Let the vector $\hat{\boldsymbol{\lambda}}_k$ now stand for either of the Lagrange multiplier approximations $(A_k A_k^T)^{-1} A_k \mathbf{f}_k$ or $(A_k B_k^{-1} A_k^T) A_k B_k^{-1} \mathbf{f}_k$ associated with the active constraints. If $\hat{\boldsymbol{\lambda}}_k > \mathbf{0}$ then, within our assumptions, (12) or (13) must be satisfied if $\mathbf{x}_k + \mathbf{p}$ minimizes $P(\mathbf{x}, r_k)$. If $\hat{\boldsymbol{\lambda}}_k \not> \mathbf{0}$ then (12) and (13) are first-order conditions that at $\mathbf{x}_k + \mathbf{p}$ any apparently redundant constraints will have been shed.

In the next section we discuss the use of (12) and (13) as constraints in a quadratic programming problem for determining $\mathbf{p}$.

4. A QP Subproblem

In previous sections it has been suggested that a method of choosing $\mathbf{p}$ would be to minimize a quadratic approximation to F subject to linear equality constraints on $\mathbf{p}$ to represent the effect of the penalty term

$$\frac{1}{r} \mathbf{v}^T \mathbf{v}$$

in $P(\mathbf{x}, r)$ the QP subproblem proposed is therefore

$$\text{minimize } \{\tfrac{1}{2}\mathbf{p}^T B_k \mathbf{p} + \mathbf{f}_k^T \mathbf{p}\}$$

$$\text{subject to } A_k \mathbf{p} = -\frac{r_k}{2} \hat{\boldsymbol{\lambda}}_k - \mathbf{g}_k. \tag{14}$$

When B_k is positive definite in the subspace defined by the constraint in (14) the solution $\mathbf{p}_k$ to this QP subproblem can be written down immediately (Fletcher, 1971, for example).

$$\mathbf{p}_k = B_k^{-1}(A_k^T \mathbf{u}_k - \mathbf{f}_k), \tag{15}$$

$$\mathbf{u}_k = (A_k B_k^{-1} A_k^T)^{-1}\left(A_k B_k^{-1} \mathbf{f}_k - \frac{r_k}{2} \hat{\boldsymbol{\lambda}}_k - \mathbf{g}_k \right). \tag{16}$$

Since the operator $(A_k B_k^{-1} A_k^T)^{-1}$ occurs in both (16) and (13) the number of matrix operations per iteration may be reduced if $\hat{\boldsymbol{\lambda}}_k$ is, from now on, defined by

$$\hat{\boldsymbol{\lambda}}_k = (A_k B_k^{-1} A_k^T)^{-1} A_k B_k^{-1} \mathbf{f}_k. \tag{17}$$

A relationship has been shown to exist, under certain assumptions, between the problem of minimizing a penalty function and the problem of solving an equality constrained QP subproblem. However the fact remains that the solution $\mathbf{p}_k$ given by (15) is such that $\mathbf{x}_k + \mathbf{p}_k$ only *approximates* the minimum of $P(\mathbf{x}, r_k)$. In this section and in Section 5 we consider safeguards to ensure that, in some sense, the points $\mathbf{x}_k$ ($k = 1, 2, \ldots$) follow the trajectory of true penalty function minima.

The rate at which the feasible region is approached will obviously depend upon the rate at which r_k is reduced. The QP formulation permits the choice of any r_k, but if r_k is taken too large the constraint in (14) may force $\mathbf{x}_k + \mathbf{p}_k$ to be more infeasible than $\mathbf{x}_k$. The condition for (14) to force a reduction in *all* violated constraint values is

$$r_k < \min\left\{ -\frac{2v_{k,i}}{\hat{\lambda}_{k,i}} \right\}, \tag{18}$$

the minimum being taken over all i for which $v_{k,i} < 0$ and $\hat{\lambda}_{k,i} > 0$. This condition can make r_k very small if by chance a single constraint is only just violated; and this may not be desirable when the search is far from the solution. It might in fact be good strategy to move away from some constraints. If we stipulate only that $\mathbf{p}_k$ should tend to reduce the total penalty then we require $\mathbf{p}_k^T W_k^T \mathbf{v}_k < 0$. This implies

$$-\frac{r_k}{2} \mathbf{v}_k^T \hat{\boldsymbol{\lambda}}_k - \mathbf{v}_k^T \mathbf{v}_k < 0.$$

If $\mathbf{v}_k^T \hat{\boldsymbol{\lambda}}_k < 0$, this gives the restriction

$$r_k < -\frac{2 \mathbf{v}_k^T \mathbf{v}_k}{\mathbf{v}_k^T \hat{\boldsymbol{\lambda}}_k}. \tag{19}$$

If $\mathbf{v}_k^T \hat{\boldsymbol{\lambda}}_k > 0$ then any positive value of r_k will be suitable for use in (14).

Choosing r_k in this way does not mean that the constraint penalty will decrease steadily from iteration to iteration. The move $\mathbf{p}_k$ may cross a constraint that was not previously violated and hence the value of $\mathbf{v}_{k+1}^T \mathbf{v}_{k+1}$ may exceed $\mathbf{v}_k^T \mathbf{v}_k$. The constraints of the QP subproblem do however impose a restriction upon the amount by which a non-violated 'active' constraint may be violated.

5. Description of an Algorithm

We consider in detail a computational algorithm implementing the ideas outlined in previous sections. The algorithm uses the first derivatives of both the function and the constraints. The Hessian matrix of $F(\mathbf{x})$ is not calculated directly, however, but an approximation to its inverse is used and updated as the search proceeds. In what follows B_k^{-1} will be understood to represent the inverse Hessian approximation at iteration k.

At the initial point $\mathbf{x}_0$ the function and constraints and their first derivatives are calculated. An initial inverse Hessian estimate (usually $B_0^{-1} = I$) is also stored. The algorithm then proceeds as follows for $k = 0, 1, 2, \ldots$.

Step 1: Choice of 'Active' Constraints

All violated constraints at $\mathbf{x}_k$ are included in the active basis. For $k > 0$ the constraints active at the previous point $\mathbf{x}_{k-1}$ are also considered. Any of these for which the corresponding element of the Lagrange multiplier approximation $\hat{\boldsymbol{\lambda}}_{k-1}$ is positive is also placed in the active set. In this way some non-violated constraints may be treated as active. This has the advantage in all problems of preventing constraints 'disappearing' from the basis due to rounding error. Moreover, when the feasible region is non-convex, a constraint may leave the violated set even though the QP subproblem (14) indicates that it should continue to be violated. A basis of active, rather than violated, constraints is valuable in ensuring that effective constraints are not disregarded too hastily, and this helps prevent 'zig-zagging'.

Step 2: Choice of Penalty Parameter r_k

An initial estimate r_0 must be supplied at the start of the algorithm. This is usually over-ridden by one of the following automatic procedures.

If there are no violated constraints then $r_k = r_{k-1}$. If some constraints are violated then, providing the rows of W_k are independent, we obtain r_k by using expression (19). For each iteration a factor α_k ($\alpha_k < 1$) is defined and the penalty parameter is given by

$$r_k = -\alpha_k \frac{2\mathbf{v}_k^T \mathbf{v}_k}{\mathbf{v}_k^T \hat{\boldsymbol{\lambda}}_k}. \tag{20}$$

If $\mathbf{v}_k^T \hat{\boldsymbol{\lambda}}_k$ is positive then (20) yields a negative, and hence impermissible, value of r_k. This happens only rarely when, for instance, I_k contains a large number of non-binding constraints. Under these circumstances, and also when the rows of W_k are not independent, r_k is chosen to minimize

$$\|\nabla P(\mathbf{x}_k, r)\| \quad \text{or} \quad \left\|\frac{\nabla P(\mathbf{x}_k, r)}{r}\right\|$$

with respect to r (Murray, 1969b). If this also fails then the choice is $r_k = \alpha_k r_{k-1}$.

In exceptional circumstances it may be necessary to reduce r_k below the value obtained by any of these methods. Should this be the case it should be sufficient to reset $r_k = 0.5\, r_k$ once or twice until $\mathbf{p}_k^T \nabla P(\mathbf{x}_k, r_k) < 0$.

Step 3: Choice of Vector $\mathbf{p}_k$

Suppose that j is the number of constraints in the active set. Then we have to distinguish three cases.

(a) $j = 0$. There are no active constraints and $\mathbf{p}_k$ is defined as the quasi-Newton correction by

$$\mathbf{p}_k = -B_k^{-1} \mathbf{f}_k. \tag{21}$$

(b) $0 < j \leqslant n$ and the rows of A_k are independent. In this, the most usual case, $\mathbf{p}_k$ is obtained as the solution of (14). Using (15) and (16)

$$\mathbf{p}_k = B_k^{-1} A_k^T \hat{\boldsymbol{\lambda}}_k - B_k^{-1} A_k^T (A_k B_k^{-1} A_k^T)^{-1} \left(\frac{r_k}{2} \hat{\boldsymbol{\lambda}}_k + \mathbf{v}_k\right) - B_k^{-1} \mathbf{f}_k. \tag{22}$$

(c) $j > n$ or the rows of A_k are not independent. Under these circumstances (14) cannot be used because $\hat{\boldsymbol{\lambda}}_k$ cannot be determined. Rather than attempt to isolate an independent subset of active constraints we have chosen to calculate $\mathbf{p}_k$ by considering the penalty function $P(\mathbf{x}, r_k)$ directly and employing equation (5).

$$\mathbf{p}_k = -\left(B_k + \frac{2}{r_k} W_k^T W_k\right)^{-1} \left(\mathbf{f}_k + \frac{2}{r_k} W_k^T \mathbf{v}_k\right). \tag{23}$$

Our assumption that the constraints effective at the solution are independent implies that situation (c) is not likely to occur near the solution where small values of r_k would be appropriate. Hence we do not expect any problems due to ill-conditioning of the Hessian matrix of the penalty function.

Step 4: Choice of Move Along $\mathbf{p}_k$

If our assumptions are justified then the move $\mathbf{p}_k$ defined by (21), (22) or (23) will yield a good approximation to the minimum of $P(\mathbf{x}, r_k)$. For various reasons, however, such as inaccuracies in the inverse Hessian estimate B_k^{-1} or non-linearity in the constraints, the point $\mathbf{x}_k + \mathbf{p}_k$ may not be an improvement, in any sense, over the previous point $\mathbf{x}_k$. The algorithm therefore generates the new point $\mathbf{x}_{k+1}$ from $\mathbf{x}_{k+1} = \mathbf{x}_k + s\mathbf{p}_k$ where s is a scaling factor. Naturally we hope that $s = 1$ can be used for most iterations. The value of s is determined by a line search along $\mathbf{p}_k$ using the values of the penalty function $P(\mathbf{x}, r_k)$. It should be stressed that the algorithm does not seek the least value of $P(\mathbf{x}, r_k)$ along the search vector $\mathbf{p}_k$, but simply ensures that a satisfactory reduction is made in the value of the penalty function.

A point $\mathbf{x}_k + s\mathbf{p}_k$ is regarded as satisfactory if for some specified $\epsilon_1 > 0$

$$P(\mathbf{x}_k, r_k) - P(\mathbf{x}_k + s\mathbf{p}_k, r_k) > \epsilon_1 s \mathbf{p}_k^T \nabla P(\mathbf{x}_k, r_k). \tag{24}$$

The method used for the search is the 'dominant degree' technique (Biggs, 1970, 1971a) which was adopted because, under some circumstances, it can be used to improve the matrix updating procedure described below. It reduces the work involved in the search if an initial estimate of s is obtained by considering those inactive constraints c_i for which $\mathbf{p}_k^T \nabla c_i < 0$ and setting

$$s = \min\left\{1, \min\left\{\frac{-c_i - r_k}{\mathbf{p}_k^T \nabla c_i}\right\}\right\}.$$

This choice is intended to ensure that the amount by which any new constraint can be violated is comparable with the magnitude of r_k.

Since the linear search can, in a sense, be regarded as a measure of the validity of our approximations, it is used to govern the choice of the factor α_k which appears in equation (20) defining r_k. If the value of s suggested by the line search of the kth iteration is such that $0.95 \leq s \leq 1.05$, then $\alpha_{k+1} = \beta \alpha_k$ (where $\beta < 1$ is a preset parameter). Otherwise, $\alpha_{k+1} = \alpha_k$.

Step 5: Updating the Inverse Hessian Approximation

When a new point $\mathbf{x}_{k+1}$ has been found and the function F_{k+1} and gradient $\mathbf{f}_{k+1}$ calculated then the approximate inverse Hessian is modified. We use a rank-two updating proposed by Biggs (1971a). Write

$$\mathbf{y}_k = \mathbf{f}_{k+1} - \mathbf{f}_k, \qquad \mathbf{z}_k = B_k^{-1} \mathbf{y}_k.$$

If $\mathbf{y}_k^T \mathbf{p}_k \leq 0$ then F is locally non-convex and in the present algorithm no updating takes place. Providing $\mathbf{y}_k^T \mathbf{p}_k > 0$ we make a further test, due to Fletcher (1970b), and if $\mathbf{y}_k^T \mathbf{p}_k \geq \mathbf{y}_k^T \mathbf{z}_k$ we write

$$B_{k+1}^{-1} = B_k^{-1} - \frac{\mathbf{p}_k \mathbf{z}_k^T}{\mathbf{p}_k^T \mathbf{y}_k} - \frac{\mathbf{z}_k \mathbf{p}_k^T}{\mathbf{p}_k^T \mathbf{y}_k} + \left(\xi s + \frac{\mathbf{y}_k^T \mathbf{z}_k}{\mathbf{p}_k^T \mathbf{y}_k}\right) \frac{\mathbf{p}_k \mathbf{p}_k^T}{\mathbf{p}_k^T \mathbf{y}_k}. \tag{25}$$

If $\mathbf{y}_k^T \mathbf{p}_k < \mathbf{y}_k^T \mathbf{z}_k$, then we write

$$B_{k+1}^{-1} = B_k^{-1} - \frac{\mathbf{z}_k \mathbf{z}_k^T}{\mathbf{y}_k^T \mathbf{z}_k} + \xi s \frac{\mathbf{p}_k \mathbf{p}_k^T}{\mathbf{p}_k^T \mathbf{y}_k}. \tag{26}$$

In both these formulae ξ is a scalar determined during the course of the 'dominant degree' line search and is intended to improve the accuracy of the Hessian approximations by allowing for non-quadraticity. Sometimes, ξ cannot be calculated, and then it is set equal to 1, giving, of course, the Broyden-Fletcher-Shanno and the Davidon-Fletcher-Powell updating formulae respectively. In fact in the present form of the algorithm, ξ can only be determined when the search is in the feasible region or in the null space of linear constraints.

Step 6: Convergence Tests

To allow for the possibility that there is an unconstrained solution the algorithm stops if a feasible point is found where $\|\mathbf{f}_k\|$ is less than some specified tolerance ϵ_2. More generally the solution will be constrained and the algorithm terminates when the following conditions are satisfied.

(i) All approximate Lagrange multipliers in the vector $\hat{\boldsymbol{\lambda}}_k$ must be positive.

(ii) The norm of the active constraint vector, $\|\mathbf{g}_k\|$, must be less than some specified tolerance ϵ_2.

(iii) *Either* the correction step length $\|\mathbf{p}_k\|$ is less than ϵ_2; *or* the norm of the projection of the gradient of F upon the active constraint manifold, $\|\mathbf{f}_k - A_k^T \hat{\boldsymbol{\lambda}}_k\|$, is less than ϵ_2.

The computational procedure summarized in the steps 1 to 6 above has been programmed and tested. Some results are given in Section 7.

6. Relationship with Murray's Method

The idea of using QP subproblems to approximate a sequence of exterior penalty function minima has been explored extensively by Murray (1969b) who suggests several possible algorithms and describes some numerical experiments. Further computational experience, using some of Murray's proposals, is given by Biggs (1971b). Clearly the overall strategy of the present method has much in common with Murray's original scheme, and the purpose of this section is to summarize the main points where the methods are different.

The analysis of Sections 2 and 3 leads to the choice of equation (13) as the constraint to be used in the QP subproblem. Murray (1969a, 1969b), by making the additional assumption that $\mathbf{x}_k$ is near the minimum of the penalty function $P(\mathbf{x}_k, r_{k-1})$, obtains as the corresponding constraint

$$W_k \mathbf{p} \geqslant -\left(1 - \frac{r_k}{r_{k-1}}\right) \mathbf{v}_k. \tag{27}$$

Expression (27) involves only the violated constraints, and this in itself is a significant difference from (13) which may involve some non-violated constraints. Moreover (27) implies a reduction in all violated constraint values while (13) makes it possible for the search to move away from some of the violated constraints. Finally (27) is an inequality because Murray considers the magnitude of the neglected terms in Taylor series expressions like equation (2). The equality (13) resulted from the assumption that the Taylor series predictions could be regarded as exact. The solution to the equality constrained subproblem can be written down algebraically [equation (22)]. This makes it easy for any number of problem constraints to be dropped at a single iteration. The strategy recommended by Murray for dealing with an inequality QP subproblem amounts to solving the corresponding equality constrained problem and then, by a modification of this solution, dropping one or two of the constraints that have negative multipliers. While the amount of work involved in the solution of the two QP subproblems is unlikely to be significantly different, it does seem that use of (13) may be more effective during the search for the correct basis of constraints because it allows redundant constraints to be shed more rapidly.

It should also be noted that Murray recommends that the QP subproblem should be based on a quadratic approximation to the Lagrangian function $F(\mathbf{x}) - \hat{\boldsymbol{\lambda}}^T \mathbf{v}$. We have preferred not to follow this because we were not sure of the value of updating an approximation to the inverse Hessian of $F(\mathbf{x}) - \hat{\boldsymbol{\lambda}}^T \mathbf{v}$ when the constraints appearing in $\mathbf{v}$ may change from iteration to iteration.

7. Computational Experience

A FORTRAN subroutine (REQP) implementing the algorithm of Section 5 has been written and applied to a number of well-known test problems. The examples used are as follows. *Problems* 1 *and* 2 are, respectively, cases I and IV of Rosenbrock's post office parcel problem as given by Box (1966). *Problem* 3 is a non-linearly constrained version of the post office parcel problem used by Davies (1968). *Problems 4 and 5* are both taken from Fiacco and McCormick's (1964) paper: problem 4 (the cattle-feed problem) is a worked example and problem 5 is the first function quoted in the appendix. No starting point is given for problem 5 in the source paper, and we have used $\mathbf{x}_0 = (1,1,2)$. *Problems 6, 7 and 8* are the first three problems in the set used by Colville (1968). Where the source paper gives more than one starting point we have always chosen the feasible one. *Problem 9* is the problem of maximizing the area of a hexagon subject to constraints on its diameter. The formulation is given by Murray (1969b), and the starting point is the one appearing in his Table VII.

An implementation of Murray's method has also been applied to these problems. Furthermore, in order to give a comparison between the QP methods and the more usual penalty function approach, the problems were attempted using an interior-point and an exterior-point version of SUMT. The program embodying Murray's method is fully described elsewhere (Biggs, 1971b). It follows similar computational steps to those described in Section 5, with, of course, the important differences mentioned in Section 6. In particular, the convergence tests are the same for both programs. The interior-point SUMT algorithm uses the barrier function

$$B(\mathbf{x},r) = F(\mathbf{x}) + r \sum_{i=1}^{m} \frac{1}{b_i(\mathbf{x})}.$$

The exterior-point version of SUMT uses the penalty function

$$P(\mathbf{x},r) = F(\mathbf{x}) + \frac{1}{r} \sum_{i=1}^{m} [\min(0, b_i(\mathbf{x}))]^2.$$

The unconstrained minimizations of $B(\mathbf{x},r)$ (or $P(\mathbf{x},r)$) are performed using an efficient variable metric method (Biggs, 1971a). Extrapolations are made along the trajectory of minima to predict the solutions and, if necessary, the starting point for the next unconstrained minimization. The convergence test is that recommended by Fiacco and McCormick (1968) based upon the dual function of $F(\mathbf{x})$. If G_k denotes the dual of $F(\mathbf{x})$ at $\mathbf{x}_k^*$, the minimum of $B(\mathbf{x},r_k)$ [or $P(\mathbf{x},r_k)$], then a point $\mathbf{x}$ which is either feasible or acceptably close to the feasible region is the solution if, for a specified $\epsilon_2 > 0$,

$$F(\mathbf{x}) - G_k < \epsilon_2 G_k.$$

The calculations were all done in single precision arithmetic on a DEC PDP-10 time sharing computer. The tolerance ϵ_2 appearing in the convergence tests for all the methods considered was usually taken as $5*10^{-5}$. Exceptions to this are noted in the tables below. With $\epsilon_2 = 5*10^{-5}$ SUMT gave solutions correct to about 5 figures in all variables. The QP methods, for the same value of ϵ_2, were slightly more accurate but the results were comparable. To illustrate this we show, in Table 1, the optimum value of $F(\mathbf{x})$ found by the methods for each problem. The correct value of $F(\mathbf{x}^*)$ appears in the final column of the table.

A comparison of the function evaluations required by the various algorithms for convergence is given in Table 2. In this comparison we have included some figures quoted from Davies (1968) showing the performance of his projection method. Davies gives function and constraint evaluations separately, a distinction which we have not made in our own experiments.

Several points emerge from these results. The QP methods clearly fulfil expectations in requiring far fewer function evaluations than either version of

TABLE 1
Solutions obtained by the methods

Problem	Optimum function value obtained				
	Interior SUMT	Exterior SUMT	Murray's method	REQP	Correct $F(x^*)$
1	$-3\cdot4560*10^3$	$-3\cdot4560*10^3$	$-3\cdot4560*10^3$	$-3\cdot4560*10^3$	$-3\cdot4560*10^3$
2	$-3\cdot2995*10^3$	$-3\cdot3000*10^3$	$-3\cdot3000*10^3$	$-3\cdot3000*10^3$	$-3\cdot3000*10^3$
3	$-2\cdot2626*10^3$	$-2\cdot2627*10^3$	$-2\cdot2627*10$	$-2\cdot2627*10$	$-2\cdot2627*10$
4	$2\cdot9889*10$	$2\cdot9888*10$	$2\cdot9889*10$	$2\cdot9889*10$	$2\cdot9889*10$
5	$-4\cdot5855$	$-4\cdot5860$	$-4\cdot5858$	$-4\cdot5858$	$-4\cdot5858$
6	$-3\cdot2348*10$	$-3\cdot2349*10$	$-3\cdot2349*10$	$-3\cdot2349*10$	$-3\cdot2349*10$
7	$3\cdot2350*10$	$3\cdot2346*10$	$3\cdot2344*10$	$3\cdot2349*10$	$3\cdot2349*10$
8	$-3\cdot0664*10^4$	$-3\cdot0667*10^4$	$-3\cdot0666*10^4$	$-3\cdot0666*10^4$	$-3\cdot0666*10^4$
9	[a]	$-6\cdot7499$	$-6\cdot7498$	$-6\cdot7498$	$-6\cdot7498$

[a] Problem 9 was not attempted using the interior point version of SUMT as the starting point chosen is infeasible.

SUMT. It is worth noting, however, that when the constraints are linear (problems 1, 2 and 6) the best performances are those reported for the projection method. Murray's method and REQP appear to be comparable in most cases, although REQP usually has an advantage. The most striking result is that for problem 7, which is clearly the most difficult of the examples. This is a problem with non-convex constraints and the success of REQP appears to be due, in part, to its method of selecting an active constraint basis. It was found that premature convergence could occur if ϵ_2 were set as $5*10^{-5}$ for problem 9, and hence the tighter convergence criterion was imposed. This is a problem with convex constraints and this means that the expected decrease in constraint penalty resulting from use of QP subproblem (14) will not be achieved in practice. Murray (1969b) reports that it is worth setting $B_0^{-1} = 0\cdot01 I$

TABLE 2
A comparison of function evaluations required for convergence

	Number of function evaluations required for convergence				
Problem	Interior SUMT	Exterior SUMT	Murray's method	REQP	Function/constraint evaluations by Davies's projection method
1	64	83	12	11	4/4[a]
2	73	87	7	7	4/4[a]
3	74	86	31	25	11/34[a]
4	152	97	26	16	N.R.
5	147	387	24	14	N.R.
6	110	233	14	8	8/8[a]
7	511	900	128	47	227/421[a]
8	84	191	14	10	10/22[a]
9	[b]	431	22[c]	19[c]	N.R.

[a] Quoted from Davies (1968).
[b] Problem 9 was not attempted using the interior point version of SUMT as the starting point chosen is infeasible.
[c] The convergence parameter ϵ_2 was set equal to $5*10^{-7}$.
N.R. No result given by Davies (1968).

TABLE 3
A comparison of computer times

	Computer time (seconds) required for convergence			
Problem	Interior SUMT	Exterior SUMT	Murray's method	REQP
1	2·0	5·0	1·5	1·5
2	4·8	4·4	1·0	1·0
3	4·9	6·6	2·2	1·7
4	14·5	6·5	2·0	1·8
5	5·8	26·9	2·5	1·9
6	14·1	15·6	3·1	1·9
7	97·0	126·0	79·0	40·0
8	10·2	11·0	2·5	2·2
9	[a]	33·7	8·0[b]	7·0[b]

[a] Problem 9 was not attempted using the interior point version of SUMT as the starting point chosen is infeasible.
[b] The convergence parameter ϵ_2 was set equal to $5*10^{-7}$.

for this example, to restrict the moves made during the first few iterations. This strategy yielded the performance figures quoted in Table 2. When B_0^{-1} was taken as I (the initial setting used in all the other problems) the QP methods required about twice as many function evaluations for convergence.

Since the work done per function evaluation by each of the methods is different it is also interesting to compare the times taken to solve the problems. Table 3 gives this information. Clearly, the economy of the QP methods as regards function evaluations outweighs the fact that they perform more 'housekeeping' per iteration than the SUMT algorithms.

8. Conclusions

An algorithm for constrained minimization has been developed which proceeds in a similar way to that proposed by Murray (1969a), using a sequence of quadratic programming problems to approximate a sequence of exterior penalty function minima. The QP subproblem used by the method of this paper, however, is subject to equality constraints and is different from the one used in Murray's approach. Computer programs have been written implementing both the algorithm described here (REQP) and Murray's technique. Their performance on a number of problems has been compared with that of SUMT in both its interior-point and exterior-point versions. It has been found that the QP methods consistently converge in fewer function evaluations and less computer time than SUMT. Moreover, the algorithm of this paper is generally somewhat faster and more economical than the program embodying Murray's technique.

There is some evidence that, when the constraints are linear, projection techniques may be even more successful than the QP methods: but for the general non-linear programming problem algorithms based on QP subproblems—particularly the equality constrained QP subproblem proposed here—appear to be among the most rapid and efficient available.

Acknowledgements

This work was partly supported by the British Aircraft Corporation, Stevenage, England. It also performs part of the author's research for a higher degree of the University of London.

References

Biggs, M. C. (1970). 'A New Method of Linear Minimization'. Hatfield Polytechnic, Numerical Optimization Centre. Technical Report No. 8.

Biggs. M. C. (1971a). Minimization algorithms making use of non-quadratic properties of the objective function. *J. Inst. Maths. Appls.* December 1971.

Biggs, M. C. (1971b). 'Computational Experience with Murray's Method for Constrained Minimization'. Hatfield Polytechnic, Numerical Optimization Centre. Technical Report No. 23.

Colville, A. R. (1968). 'A Comparative Study on Non-linear Programming Codes'. IBM report 320–2949.

Davies, D. (1968). 'The Use of Davidon's Method in Non-linear Programming'. Imperial Chemical Industries. Report MSDH/68/110.

Box, M. J. (1966). A comparison of several current optimization methods, and the use of transformations in constrained problems. *Comput. J.* **9**, 67–77.

Fiacco, A. V., and McCormick, G. P. (1964). Computational algorithm for the sequential unconstrained minimisation technique for non-linear programming. *Managmt Sci.* **10**, 601–617.

Fiacco, A. V., and McCormick, G. P. (1968). 'Nonlinear Programming, Sequential Unconstrained Minimisation Techniques'. Wiley, New York.

Fletcher, R. (1969). 'A Class of Methods for Nonlinear Programming with Convergence and Termination Properties'. U.K. Atomic Energy Authority Report TP 368.

Fletcher, R. (1970a). 'An Efficient Globally Convergent Algorithm for Unconstrained and Linearly Constrained Optimisation Problems'. The 7th Mathematical Programming Symposium, The Hague.

Fletcher, R. (1970b). A new approach to variable metric algorithms. *Comput. J.* **31**, 317–322.

Fletcher, R. (1971). A general quadratic programming algorithm. *J. Inst. Maths. Appls.* **7**, 76–91.

Murray, W. (1967). 'Ill-Conditioning in Barrier and Penalty Functions Arising in Constrained Non-linear Programming'. The 6th Mathematical Programming Symposium, Princeton University.

Murray, W. (1969a). An algorithm for constrained minimisation. *In*: 'Optimization' (R. Fletcher, ed.). Academic Press, London.

Murray, W. (1969b). 'Constrained Optimisation'. National Physical Laboratory Report MA79.

Pietrzykowski, T. (1962). 'On a Method of Approximate Final Conditional Maxima'. Inst. Maszyn Matematcyoznych PAN, Algorythmy VI, 1962.

Author Index

Numbers in *italic* type are those pages where references are listed at the end of chapters.

A

Abadie, J., 266, *276*, 285, *295*, *343*
Ablow, C. M., 317, *343*
Ahmad, M. S., 200, *201*
Allran, R. R., 343, *343*
Arrow, K. J., 314, *343*

B

Balas, E., 265, 266, 276, *276*, 277
Bard, Y., 47, *66*, 84, 89, *96*, 149, 159, 160, *168*
Barnes, V., 195, 198, *201*
Bartels, R. H., 253, *253*
Barrow, D. E., 200, *201*
Beale, E. M. L., 77, *96*, 218, *219*, 248, *254*, 268, 277
Beightler, C. S., *347*
Bell, M. D., 266, *278*
Bellman, R. E., 350, *365*
Bellmore, M., *343*
Beltrami, E. J., 317, *343*
Benders, J. F., 335, 336, 337, 341, 342, *343*
Ben Israel, A., *220*
Beuhler, R. J., 89, *97*
Biggs, M. C., 48, *66*, 159, 162, 163, 167, 168, *168*, 421, 423, 424, *427*
Boot, J. J. G., 266, *277*
Borowski, N., 46, *67*
Box, M. J., 79, 89, *96*, 99, *113*, 115, *135*, 162, *169*, 316, *343*, 423, *428*
Bracken, J., *343*
Branin, F. H., 231, 232, 233, *237*
Brigham, G., 317, *343*
Brooks, R., 335, *343*
Brown, K. M., 28, *33*, 65, *66*
Broyden, C. G., 17, *17*, 21, 22, 24, 25, 26, 30, *33*, 37, *38*, 47, 53, *66*, 89, *96*, 149, 150, 152, 154, 157, 159, 161, 168, *169*, 235, *237*
Buchholdt, H. A., 200, *201*
Bui Trong Lieu, 318, 326, *343*
Bulirsch, R., 320, *343*
Butler, T., 317, *343*
Butz, A. R., 342, *343*
Buys, J. D., 339, 341, *345*

C

Cabot, A. V., 213, *219*
Camp, G. D., 317, *344*
Carpentier, J., *344*
Carroll, C. W., 316, *344*, *409*
Charnes, A., 301, 308, *312*
Collatz, L., 19, *33*, 309, *312*
Colville, A. R., 162, *169*, *344*, 360, *365*, 388, *393*, 402, 404, *409*, 424, *428*
Conn, A. R., 323, *344*
Cooper, L., 213, *220*
Cooper, W. W., 301, 308, *312*
Cottle, R. W., 213, *219*, 252, *254*, 301, 304, *312*
Courant, R., 317, *344*
Courtillot, M., 266, *277*
Cragg, E. E., 86, *96*
Curtin, J. F., 89, *96*

D

Dantzig, G. B., 248, 252, *254*, 301, 304, *312*
Davidon, W. C., 1, *17*, *33*, 46, 48, 60, *66*, 100, *113*, 149, *169*
Davies, D., 89, *96*, 115, *135*, 316, 342, *343*, *344*, 424, 425, 426, *428*
Dennis, J. B., 293, *295*
Dennis, J. E., 20, 22, 23, 24, 28, *33*, *34*, 37, *38*, *38*, 65, *66*

Dinkelbach, W., 266, *277*
Dixon, L. C. W., 2, 16, 17, *17*, 54, *66*, 149, 151, 155, 168, *169*
Dorn, W. S., 320, *344*
Dragan, I., 266, *277*
Driebeek, N. J., 268, *277*
Duffin, R. J., 214, *219*

E

Eaves, B. C., 304, *312*
Eckhardt, U., 303, 304, *312*
Eisemann, K., 255, *263*
Engvall, J. L., 78, 83, *97*
Evans, J. P., 318, 343, *344*, 349, 354, *365*
Everett, H., 334, 335, *344*

F

Falk, J. E., 211, 214, 215, 218, *219*, *220*, 337, 338, 341, *344*
Faure, P., 326, 329, *344*
Fiacco, A. V., 227, *230*, 314, 315, 316, 317, 318, 319, 320, 321, 322, 324, 325, 326, 327, 328, *344*, *346*, *365*, 367, 368, *369*, 383, *393*, 395, 396, 397, 399, 407, *409*, 412, 416, 424, *428*
Fielding, K., 167, *169*
Fletcher, R., 1, 3, 4, *17*, 25, *34*, 37, *38*, 39, 40, *42*, 47, 48, 50, 56, 59, *66*, 89, *97*, 143, 144, *147*, 149, 150, 152, 154, 157, 159, 160, 168, *169*, 189, *189*, *220*, 239, 241, 243, 248, 253, *254*, 285, 288, 291, 294, *295*, *296*, 297, *300*, 316, 322, 323, 338, 341, *345*, 363, *366*, 371, *380*, 383, 385, 388, 389, *393*, 404, *410*, 411, 418, 422, *428*
Francis, R. L., 213, *219*
Fricks, R. E., *220*
Frisch, K. R., *410*
Frisch, R., 316, *345*
Fromowitz, S., 314, 315, *346*

G

Gaspar, T., 273, *277*
Gentry, J. W., 48, *66*
Geoffrion, A., 265, 267, *277*, 335, *343*
George, M. D., 211, *220*
Gill, P. E., 65, *66*, 292, *296*
Ginsburgh, V., 265, *277*

Glover, F., 265, 276, *277*
Goldfarb, D., 25, *34*, 37, *38*, 150, *169*, 239, 242, 244, *254*, 288, 290, *296*, 407, 408, *410*
Goldstein, A. A., 55, *66*, 89, *97*, 157, 158, *169*, 182, *189*
Golub, G., 173, 174, *189*
Golub, G. H., 253, *253*
Gonçalves, A. S., 255, 260, 261, *263*
Gould, F. J., *220*, 318, 334, 343, *344*, *345*, 349, 350, 351, 352, 354, *365*, *366*
Gran, R., *220*
Graves, G. W., 218, *220*
Graves, R. L., *345*
Greenberg, H. J., *343*
Greenstadt, J., 25, *34*, 50, 54, *66*, 159, *169*, 183, *189*
Griffith, R. E., 295, *296*
Guigou, J., 285, *295*

H

Haarhoff, P. C., 339, 341, *345*
Hadley, G., *345*
Hall, C. A., 306, *312*
Hammer, P. L., 266, *277*
Hansen, P., 266, 267, 273, 276, *277*, *278*
Hardy, G. H., *381*
Hartley, H. O., 211, *220*
Hesse, R., 213, *220*
Hestenes, M. R., 39, *43*, 339, 341, *345*
Hext, G. R., 115, *135*
Hillier, F. S., 266, *278*
Himmelblau, D. M., 46, *67*, 89, *97*
Himsworth, F. R., 115, *135*
Hofmann, W., 19, *34*
Holst, W. R., 75, *97*
Hooke, R., 89, *97*, 167
Howe, S., 334, *345*
Hu, T. C., 213, *220*
Huang, H., 35, *38*, 53, *67*, 89, *97*, 149, 150, 151, 154, 159, 160, 168, *169*
Huard, P., 318, 320, 326, 327, 329, 330, *343*, *345*
Hurwicz, L., 314, *343*
Hutchinson, D., 46, *67*, 100, *113*

J

Jarvis, J. J., *343*
Jeeves, T. A., 89, *97*, 167

AUTHOR INDEX

Jennings, L. S., 172, 173, *189*
John, F., 314, *345*
Johnson, I. L., 50, 59, 60, *67*
Johnson, S. E. J., 343, *343*
Jones, A., 89, *97*, 198, *201*, *345*

K

Kahan, W., 173, *189*
Kate, A., *220*
Kelley, E. J., 266, *278*
Kelley, H. J., 50, 53, 59, 60, 65, *67*, 297, *300*
Kempthorne, O., 89, *97*
Kettler, P. C., 159, *170*
Kjellström, G., 137, *147*
Kleibohm, K., 209, *220*, 308, 311, *312*
Knopp, K., 23, *34*
Kortenek, K. O., 301, 308, *312*
Kowalik, J., 99, *113*, 115, *135*, 180, 184, *189*, 316, 326, 331, 332, *345*, *346*
Kreuser, J., 299, *300*
Krolak, P. D., *220*
Kuhn, H. W., 314, *346*
Künzi, H. P., 266, *278*

L

La Motte, L. R., 211, *220*
Lapidus, L., 407, 408, *410*
Laughhunn, D. J., 265, *278*
Laurent, P. J., 320, *346*
Lawler, E., 266, *278*
Lemke, C. E., 293, *296*, 301, 303, 304, 311, *312*
Leon, A., 89, *97*
Levenberg, K., 195, 196, *201*
Levy, A. V., 86, 89, *97*, 151, 154, 159, 160, *169*
Lill, S. A., *113*, 159, 297, *300*, 371, *381*, 383, *393*
Little, E. A., 200, *201*
Lootsma, F. A., 315, 319, 320, 321, 322, 323, 325, 326, 328, 331, *346*, 367, 368, *369*, 396, 397, 399, *410*

M

Mangasarian, O. L., 214, *220*, 305, *312*, 314, 315, 320, *346*, *393*
Mao, J. C. T., 266, 273, *278*

Marquardt, D. W., 197, *201*
Martin, A. V., 317, *343*
Matyas, J., 137, *147*, 211, *220*
Matz, A. W., 198, *201*
McCann, A. P., 316, 323, *345*, 363, *366*, 389, *393*, 404, *410*
McCormick, G. P., 47, *67*, *220*, 227, *230*, 285, 291, *296*, 314, 315, 316, 317, 318, 319, 320, 321, 322, 324, 325, 326, 327, 328, *343*, *344*, *346*, 352, *365*, 367, 368, *369*, 383, 388, *393*, 395, 396, 398, 399, 407, *409*, 412, 416, 424, *428*
McMillan, B. R., 200, *201*
Mead, R., 89, *97*, 100, *113*, 115, 118, *135*
Meinardus, G., 311, *312*
Meyer, R., *220*
Mifflin, R., 369, *369*
Morrison, D. D., 332, *346*
Motkus, I. B., 211, *220*
Mueller, R., 213, *220*
Muller, M. E., 138, *147*
Murray, W., 65, *66*, 286, 292, *296*, 316, 323, 324, 342, *346*, 397, 404, *410*, 411, 412, 414, 420, 422, 423, 424, 425, 427, *428*
Murtagh, B. A., 48, 55, 60, *67*, 89, *97*, 155, 159, 160, *169*, 197, 288, 290, *296*, 407, 408, *410*
Myers, G. E., 48, 50, 53, 59, 60, 65, *67*
Mylander, W. C., 213, *219*, *346*

N

Nelder, J. A., 89, *97*, 100, *113*, 115, 118, *135*

O

Oettli, W., 266, *278*
Ortega, J. M., 4, 13, *17*, 377, 378, 379, *381*
Osborne, M. R., 99, *113*, 115, *135*, 171, 172, 173, 180, 184, *189*, 322, 332, *346*, 396, 397, *410*

P

Panne van de, C., 248, 249, 251, *254*, 257, 261, *263*
Parisot, G. R., 316, 322, *346*, 396, *410*
Parkinson, J. M., 46, *67*, 100, *113*, 116, *135*
Pascual, L. D., *220*

Pearson, J. D., 3, *17*, 25, *34*, 35, 38, *38*, 47, *67*, 89, *97*, 159, *169*, 340, *346*
Peetersen van, A., 265, *277*
Peterson, C., 256, *278*
Peterson, E. L., 214, *219*
Pierson, B. L., 89, *97*
Pietrzykowski, T., 317, 323, *347*, 349, *366*, 412, *428*
Pincus, M., 215, *221*
Polak, E., 14, *17*
Poljak, B. T., 368, *369*
Pollard, G. P., 53, *67*
Pomentale, T., 316, *347*
Porsching, T. A., 306, *312*
Powell, M. J. D., 2, 3, 4, 12, *17*, 24, 25, 27, 28, 29, 30, *34*, 37, *38*, 47, 48, 56, *66*, *67*, 83, 85, 89, 96, *97*, 100, *113*, 143, 144, *147*, 149, 157, 159, 160, 162, *169*, 288, *295*, 322, 340, 341, *345*, *347*, 380, *381*, 384, 385, *393*, 398, 402, *410*
Price, J. F., 55, *66*, 89, *97*, 157, 158, *169*, 182, *189*

R

Rabinowitz, P., 304, *312*
Rajtora, S. G., 89, *97*
Ralston, A., 19, *34*
Ravindran, A., 304, *312*
Reeves, C. M., 39, 40, *42*, 50, *66*, 89, *97*, 285, *295*
Rheinboldt, W. C., 4, 13, *17*, 377, 378, 379, *381*
Ritter, K., 213, *221*, 223, *230*, *393*
Roode, J. D., 317, 338, 341, *347*, 349, *366*
Rosen, J. B., 242, 243, *254*, 285, 290, *296*, 297, *300*, 356, *366*, 407, *410*
Rosenbrock, H. H., 89, *97*, 162, 167, *169*, 326, *347*, 351, *366*, 388, *393*
Rubin, A. A., 266, *277*
Rudeanu, S., 266, *277*, *278*
Rutishauser, H., 173, *189*
Ryan, D. M., 322, 332, *346*, 396, 397, 407, *410*

S

Sargent, R. W. H., 48, 53, 55, 60, *67*, 89, *97*, 155, 159, 160, *169*, 288, 290, *296*, 407, 408, *410*
Saunders, M. A., 253, *253*

Scedrin, B. M., *221*
Schmidt, J. W., 81, *97*
Schnabel, B. K., 115, *135*
Schrack, G. F., 46, *66*
Schrager, R. I., 295, *296*
Schumer, M. A., 137, *147*
Shah, B. V., 89, *97*
Shanno, D. F., 25, *34*, 37, *38*, 150, 152, 159, 160, 163, *170*, 201, *201*
Small, R., 268, *277*
Smith, F. B., 201, *201*
Soland, R. M., 215, 218, *220*, *221*
Sorenson, H. W., 42, *43*
Souami, B., 304, *312*
Spendley, W., 99, *113*, 115, *135*
Speyer, J. L., 297, *300*
Steiglitz, K., 137, *147*
Stewart, G. W., 65, *67*, 89, *97*, 100, *113*, 159
Stewart, R. A., 295, *296*
Stiefel, E., 39, *43*
Stoer, J., 320, *343*
Stong, R. E., 316, *347*
Suzuki, S., 356, *366*
Swann, W. H., 89, *96*, 115, *135*, 316, *343*
Szegö, G. P., *221*
Szkopiak, Z. C., 200, *201*

T

Thiel, H., 249, *254*, 266, *277*
Tolle, J. W., 349, *365*
Tomlin, J. A., 218, *219*, *221*
Tomlin, J., 268, *277*, *278*
Tremolières, R., 318, 326, 327, 329, *347*
Tucker, A. W., 314, *346*
Tui, H., 213, *221*

U

Ueing, U., 229, *230*
Uzawa, H., 314, *343*

V

Veinott, A. F., 308, 311, *312*
Vetters, K., 81, *97*

W

Wallingford, B. A., 266, 273, *278*
Watters, L. J., 266, *278*

Wetterling, W., 309, *312*
Whinston, A. B., 218, *220*, 248, 249, 251, *254*, 257, 261, 262, *263*
White, B. F., 75, *97*, 205, *207*
Wilde, D. J., *347*
Wilkinson, J. H., 174, *189*, 235, *237*
Witzgall, C., 266, *278*
Wolfe, P., 2, 9, *17*, 152, 158, *170*, 248, *254*, 285, 288, 289, 295, *296*, 320, 335, 338, *345*, *347*
Wood, C. F., 87, *97*

Woolsey, E., 276, *277*
Wortman, J. D., 89, *97*

Z

Zahdeh, L. A., 350, *365*
Zangwill, W., 266, *278*, 317, 322, 338, *347*, 349, *366*
Zangwill, W. I., 74, 80, *97*
Zidov, N. P., *221*
Zoutendijk, G., 247, 252, *254*, 289, 290, 293, *296*, 308, 311, *312*, 338, 342, *347*
Zwart, P. B., *97*, 214, *221*

Subject Index

A

Active set, 288
Additive penalties, 265, 268–271, 276
Algorithm, additive, 265
Algorithm, BFS, 152, 153
Algorithm, Broyden's, 21, 22, 23, 28
Algorithm, conjugate descent, 14
   application, 14–16
Algorithm, conjugate gradient, 14, 50, 53
   properties, 50, 54
   sensitivity, 65
Algorithm, cutting-plane, 266, 295, 297, 311
Algorithm, DFP, 46–50
   batch-processing version, 59, 60
   Broyden generalization, 53
   comparison with Broyden formula, 59, 64
   comparison with Fletcher-Reeves algorithm, 50, 53
   effectiveness, 47, 50, 64
   for linear constraints, 407
   properties, 47
   sensitivity, 65
Algorithm, Fletcher-Reeves, 45, 50
   advantages, 50
   comparison with DFP algorithm, 50, 53
   effectiveness, 50, 53
Algorithm, Gauss-Newton, 151, 171, 180, 195, 197, 199
Algorithm, gradient projection, 239, 297
   tests, 299
Algorithm, heuristic, 209, 271
Algorithm, hybrid, 280, 395, 399–404
   for non-linear constraints, 407
   tests, 404–407, 408
Algorithm, implicit enumeration, 265, 266, 267
   tests, 267, 268, 271, 272, 273
Algorithm, Newton-like, 373

Algorithm, non-gradient
   evaluation, 101–105, 112
   failures, 112
   for high-dimensionality functions, 99
   performance of new, 109–112, 113
   reduced primary storage, 108, 109
   storage considerations, 105–108
   test functions, 100, 101
   test method, 100
Algorithm, primal-dual, 255
   efficiency, 261, 262
   in tableau form, 260
   restricted problems, 257
Algorithm, pseudo-complementary, 301, 302
   applications, 303–311
   complementary pivot method, 308, 311
   continuous case, 307, 308
   finite case, 303, 304
Algorithm, rank-one, 35, 64
Algorithm, rank-two, 37
Algorithm, unconstrained
   evaluation, 89–96
   evaluation criteria, 70–73
   failures, 91
   ranking, 94, 95, 96
   testing, 73–88
Algorithm, variable metric
   behaviour, 155–157
   Broyden family, 151–154, 168
   description, 2, 3
   Dixon's theorem, 16, 17
   efficiency, 168
   for convex functions, 11–13
   for uniformly convex function, 1
   for various conjectures, 4–11
   Huang family, 35, 36, 150, 151, 167, 168
   in convergence problems, 16, 17
   properties, 3, 4, 150–154
   step length, 149, 154, 155

SUBJECT INDEX

Algorithm, variable metric—*cont.*
  step length choice, 157–160, 161
  success, 1
  testing, 161–167
  use, 1–17
Angle test, 55
Anti-zigzagging, 293, 308, 391
Anti-zigzag rule, 247, 248
Average convergence rate, 137, 141–143, 147

B

Barrier function, 47, 324, 325, 395
Barrier function methods
  disadvantages, 396, 397
  extrapolated, 404
  performance with linear constraints, 407
  robustness, 399
  tests, 404–407
Boolean methods, 266
Bounded deterioration, 20, 38
Bounded second derivatives, 1, 12
Branch and bound method, 209, 215–219
Branching rule, 217

C

Cauchy's inequality, 37
Complementarity conditions, 256
Column-generating procedure, 335, 336, 337
Computation time, 72, 426
Conjugate directions, 39
  concept, 40
  construction, 40, 41
  construction from gradient vectors, 41, 42
Conjugate gradients, 39
Constraint basis, 240
Constraint qualification, 225
Contraction, 116, 119
Controlling parameters, 313, 326
Convergence criteria, 103, 112, 133, 134
Convex envelope, 209
Convex subfunction, 210
Convexity, uniform, 1
Cubic fitting, 48, 54
Cubic interpolation, 159

D

Damping, 195, 196
Damping term, 294
Decomposition methods, 295
Descent test, 56, 59
  failure, 59, 60
Detection, 138
Dimensionality, 99, 103, 104, 105, 106, 107, 109, 112, 113, 119, 137, 145, 193
Distance function, 326
Dominant degree, 167, 421, 422
Double rank method, 24–27
Dual variables, 281
Dynamic programming, 203, 205
  formulation, 205, 206

E

Equivalent function evaluation, 96
Essential singularities, 233
Execution time, 72, 91
Expansion, 116
  unlimited, 134
Exponential penalty function, 335
Exterior-point method, 313, 317, 341
  non-parametric, 330, 331, 333, 341
Exterior point penalty function, 213, 341
Extraneous singularities, 233
Extrapolation, 158, 320, 321, 328, 342

F

Feasible directions method, 252
Function minimization
  equality constraints, 282–288
  inequality constraints, 288–292
  linear constraints, 279, 280, 281
Function minimization, unconstrained
  using function and gradient evaluations, 45, 46

G

Generalized inverse, 172, 280
Geometric programming, 213
Global maximum, 231
Global minimum, 69, 231, 243
Global solution, 209, 223
  branch and bound approach, 215–219

SUBJECT INDEX 437

geometric programming, 213, 214
grid approach, 210
  in non-convex optimization problem, 223–229
  Lagrangian approach, 211, 212
  non-convex quadratic programming, 213
  penalty function approach, 212, 213
  random method, 211
  successive feasibility methods, 213
Golden section, 48, 89
Gram-Schmidt formulae, 39, 40
Grid method, 209

H

Hessian approximation, 29, 30 56, 285
Hessian approximation, inverse, 419
  dual, 30
  updating, 421, 422
Hypercubes method, 294

I

Ill-conditioning, 297, 323, 414
Implicit enumeration, 265, 266, 267
  efficiency, 276
Integer variable, 266
Interior-point method, 313, 316, 341
  non-parametric, 326–330, 333, 341
Interior point penalty function, 212, 341

J

Jacobian approximation, 22
Jacobian uniqueness conditions, 315, 323

K

Kantorovich theorem, 21, 297
Kuhn-Tucker conditions, 226, 280, 281, 289, 298, 388, 396, 399, 401
Kuhn-Tucker multipliers, 281
Kuhn-Tucker point, 289, 314, 315, 320, 323, 328, 333, 334, 368, 396

L

Lagrange multiplier variable, 239, 240
Lagrangian function, 211, 287, 297, 299, 333, 334, 335, 337, 338, 342, 375, 423

Lagrangian function, augmented, 338–340
Lagrangian method, 209, 313, 314, 333, 334, 337, 338, 340, 341, 342
Lagrangian-multiplier technique, 334
Lagrangian multipliers, 211, 281, 283, 284, 286, 287, 288, 290, 314, 331, 334, 335, 388, 389, 399, 415
Least squares, 69
Least squares, damped, 173
Least squares, linear, 172
  in non-linear problems, 175
  problems, 172–175
Least squares, non-linear, 27
  problems, 175–183
  special case, 184–189
Levenberg's parameter, 196
  modified, 198, 199, 200, 201
Lexicographical enumeration, 266
  efficiency, 276
Line search, 35, 48, 149, 157, 160, 161, 287, 294
Linear program, 243
Linear programming, 255, 279
Local minimum, 70
Logarithmic barrier function, 321, 327, 367, 368, 396, 399
Logarithmic programming, 316
Loss functions, 317, 341

M

MAP, 295
Method of centres, 326
  variants, 329
Minimizing trajectory, 320
  series expansion, 320
Mixed penalty function, 317, 324, 325
Monte Carlo technique, 137
Morrison parameters, 332
Moving truncations, 326
Multiple extrema, 231
  quasi-Newton methods, 235, 236
Multiplier function, 351, 352, 353, 354
Murray's method, 411, 412, 422, 243, 427

N

Newton-like iteration, 371, 376
  superlinear convergence, 376

SUBJECT INDEX

Newton's method, 19, 21, 25, 28, 232, 286, 287, 299, 377
   extensions, 239, 242
   in quadratic programming problems, 239–248
Newton-Raphson method, 191, 193, 194, 195, 196, 197
Newton vector, 233, 234
Noise, 115
Non-convex optimization, 223
   convex subproblems, 227–229
   local solutions, 225–227

O

One-variable minimization, 45
   accuracy, 64
   and convergence, 54–64
   avoidance, 54–64
   in DFP algorithm, 48
   replacement by stability text, 45
   termination, 48
Optimal control, 53
Orientation, 115, 118
Orthogonal factorization, 174
Orthogonal transformation, 172

P

Parabolic interpolation, 155
Penalty function, 212, 213, 214
   change of basis, 385, 386–392
   for equality constrained problems, 383, 384, 385
   mixed, 324
   properties, 412–415
Penalty function, exact, 349, 350, 351, 352
Penalty function method, 209, 297
   generalized, 398
Polynomial program, 266
Potential function, 323
Powell's MINFA, 29, 30
Principal Hessian matrix, 323, 324
Principal pivoting method, 252
Product form of the inverse, 255
Projection matrices, 280
Projection operator, 241

Q

Quadratic convergence, 103
Quadratic fitting, 48, 54
Quadratic programming, 239, 255, 279
Quadratic programming problems, 239–247
   Beale's method, 248, 249, 251, 252
   Dantzig's simplex method, 248, 249, 250, 251, 252, 253
   Wolfe's method, 248, 249
   with bounded variables, 255, 246, 257, 262
Quadratic termination, 35
Quasi-Newton equation, 19
Quasi-Newton formula, 35, 53
Quasi-Newton method, 36, 45, 54, 64, 192, 235, 285, 288
   properties, 54

R

Random search, 137
   convergence behaviour, 147
   effect of dimensionality, 145
   effect of Gaussian noise perturbations, 146
   effect of scaling, 145
   effect of uniform noise, 146
   Kjellstrom's, 139, 140, 141
   Matyas', 138, 139
   Schumer and Steiglitz's, 139
   test functions, 143, 144
Random walk, 139
Recursive equality quadratic programming, 411
   algorithm, 419–422
   tests, 423–427
Reduced costs, 281
Reduced gradient method, 282
Reflection, 116, 118
Rejection, 138
Restart, 39, 40
Robustness, 70, 100, 104, 112, 113

S

Search direction, 72
Secant approximation, 19
   results, 28–33

SUBJECT INDEX

Secant equation, 19, 21
Secant method, 19, 20, 21
   definitions, 20, 21
Separable programming, 209
Shrinkage, 117
Simplex method, 99, 100, 109
   basic operations, 116, 117
   efficiency, 115, 134
   in quadratic programming problems, 239, 249–252
   initial parameters, automatic setting, 125–128
   initial parameters, effect, 120–125
   modifications, 118, 128–133, 134
   Nelder and Mead version, 115, 116, 117, 118, 119, 120, 130, 131, 133, 134
   notation, 116
   PHS version, 134
   test functions, 119, 120
Simplex tableau, 249
Single rank method, 21–24
Special ordered sets, 218
Stability, 45, 152
Standard time, 74, 299
Steepest descent, 46, 55, 65, 119, 157, 193
Step length
   choice, 157–160
   estimation, 191, 192
Step reduction, 45
Storage, primary
   backing store effect, 100
   limitation, 99, 104
   reduced, 108, 109
   relation of dimensionality, 105, 106
Storage, secondary
   additional time requirements, 108
   use, 107, 108
SUMT, 161, 316, 319, 320, 326, 327, 411, 412, 424, 425, 427

T

Tableau, 255
Tangent parameters, 332
Termination criteria, 70, 90
Tolerance function, 355, 365
   robustness, 356
Tolerance programming, 349
   tests, 356–365
Translation, 130

U

Unidimensional search, 70
Unimodal function, 213
Update, 19, 25, 27, 31, 32, 37, 55, 64, 149, 161, 248
   Broyden-Fletcher-Goldfarb-Shanno, 37, 152
   Davidon-Fletcher-Powell, 24, 28, 29, 151
   Powell, 28
   symmetrization, 37
   Updating formulae, 17
   Broyden family, 151, 152, 154
   Huang family, 150, 151
Upper triangular form, 172

V

Variable reduction method, 285

Z

Zero-one programming, 265
   additive penalties, 268, 269, 270
   problems, 266, 273–276
   solutions, 265, 266
   uses, 265
Zigzagging, 289, 290, 291, 293, 294, 386, 390, 391, 419

Acknowledgements to the Referees

I wish to thank the following experts for graciously giving of their time to help in the reviewing of the papers presented at the Conference on Numerical Methods for Non-linear Optimization:

J. Abadie, E. Balas, Y. Bard, E. M. L. Beale, V. Belevitch, J. F. Benders, J. C. G. Boot, J. Bootsma, C. G. Broyden, L. Collatz, J. W. Daniel, W. C. Davidon, J. E. Dennis, L. C. W. Dixon, A. J. Douglas, J. H. R. M. Elst, A. V. Fiacco, R. Fletcher, R. Goffin, D. Goldfarb, G. H. Golub, J. de Groot, H. Halkin, M. R. Hestenes, D. M. Himmelblau, P. Huard, J. S. Kowalik, J. Kriens, F. Kruseman Aretz, H. Kwakernaak, L. S. Lasdon, G. de Leve, G. P. McCormick, A. Miele, W. Murray, W. C. Mylander, G. L. Nemhauser, W. Oettli, M. R. Osborne, C. van de Panne, J. D. Pearson, E. L. Peterson, T. Pietrzykowski, E. Polak, J. Ponstein, M. J. D. Powell, P. Rabinowitz, J. D. Roode, R. W. H. Sargent, D. F. Shanno, R. M. van Slyke, W. H. Swann, G. W. Veltkamp, G. A. Watson, W. Wetterling, A. Whinston, G. Zoutendijk.

Eindhoven, August 1972   *F. A. Lootsma*